Recent Advances in the Diversity and Taxonomy of Subterranean Arthropods

Recent Advances in the Diversity and Taxonomy of Subterranean Arthropods

Guest Editors

Srećko B. Ćurčić
Gordan S. Karaman

MDPI Basel • Beijing • Wuhan • Barcelona • Belgrade • Novi Sad • Cluj • Manchester

Guest Editors
Srećko B. Ćurčić
Faculty of Biology
University of Belgrade
Belgrade
Serbia

Gordan S. Karaman
Montenegrin Academy of
Sciences and Arts
Podgorica
Montenegro

Editorial Office
MDPI AG
Grosspeteranlage 5
4052 Basel, Switzerland

This is a reprint of the Special Issue, published open access by the journal *Diversity* (ISSN 1424-2818), freely accessible at: https://www.mdpi.com/journal/diversity/special_issues/Subterranean_Arthropods.

For citation purposes, cite each article independently as indicated on the article page online and as indicated below:

Lastname, A.A.; Lastname, B.B. Article Title. *Journal Name* **Year**, *Volume Number*, Page Range.

ISBN 978-3-7258-5029-7 (Hbk)
ISBN 978-3-7258-5030-3 (PDF)
https://doi.org/10.3390/books978-3-7258-5030-3

Cover image courtesy of Nikola Vesović

Contents

About the Editors vii

Preface ix

Srećko Ćurčić and Gordan Karaman
Recent Advances in the Diversity and Taxonomy of Subterranean Arthropods
Reprinted from: *Diversity* **2025**, *17*, 532, https://doi.org/10.3390/d17080532 1

David Eme, Kristen M. Westfall, Brynja Matthíasardóttir, Bjarni Kristófer Kristjánsson and Snæbjörn Pálsson
Contrasting Phylogeographic Patterns of Mitochondrial and Genome-Wide Variation in the Groundwater Amphipod *Crangonyx islandicus* That Survived the Ice Age in Iceland
Reprinted from: *Diversity* **2023**, *15*, 88, https://doi.org/10.3390/d15010088 7

Gergely Balázs, Špela Borko, Dorottya Angyal, Valerija Zakšek, Anna Biró, Cene Fišer and Gábor Herczeg
Not the Last Piece of the Puzzle: *Niphargus* Phylogeny in Hungary
Reprinted from: *Diversity* **2023**, *15*, 223, https://doi.org/10.3390/d15020223 31

Dieter Weber and Alexander M. Weigand
Groundwater Amphipods of the Hyporheic Interstitial: A Case Study from Luxembourg and The Greater Region
Reprinted from: *Diversity* **2023**, *15*, 411, https://doi.org/10.3390/d15030411 43

Sarahi Jaime, Adrián Cervantes-Martínez, Martha A. Gutiérrez-Aguirre, Gerardo Hernández-Flores, Roger A. González-Herrera, Gabriel Sánchez-Rivera, et al.
Toward an Integrative Overview of Stygobiotic Crustaceans for Aquifer Delimitation in the Yucatan Peninsula, Mexico
Reprinted from: *Diversity* **2025**, *17*, 77, https://doi.org/10.3390/d17020077 55

Robert S. Vargovitsh
Deep Troglomorphy: New Arrhopalitidae (Collembola: Symphypleona) of Different Life Forms from the Snezhnaya Cave System in the Caucasus
Reprinted from: *Diversity* **2022**, *14*, 678, https://doi.org/10.3390/d14080678 70

Nikola Vesović, Christo Deltshev, Plamen Mitov, Dragan Antić, Dalibor Z. Stojanović, Dejan V. Stojanović, et al.
The Diversity of Subterranean Terrestrial Arthropods in Resava Cave (Eastern Serbia)
Reprinted from: *Diversity* **2024**, *16*, 234, https://doi.org/10.3390/d16040234 98

Leonardo Latella and Stefano Brighenti
Exploring Ice Cave Biodiversity in Northeastern Italy
Reprinted from: *Diversity* **2024**, *16*, 364, https://doi.org/10.3390/d16070364 123

Javier Blasco-Aróstegui and Lorenzo Prendini
Redescription of *Euscorpius studentium* Based on Adult Specimens; Updated Classification of Cavernicolous Euscorpiidae; and Review of Cavernicolous Scorpions in the Balkans
Reprinted from: *Diversity* **2024**, *16*, 737, https://doi.org/10.3390/d16120737 138

Serban M. Sarbu, Traian Brad, Raluca I. Băncilă and Andrei Ştefan
Exploring Biodiversity and Food Webs in Sulfur Cave in the Vromoner Canyon on the Greek–Albanian Border
Reprinted from: *Diversity* **2024**, *16*, 477, https://doi.org/10.3390/d16080477 164

Sofija Vranić, Ljubodrag Vujisić, Nikola Vesović, Marina Todosijević, Miloš Pavićević, Dejan Radović and Srećko Ćurčić
On the Diversity of Semiochemicals of the Pygidial Gland Secretions of Subterranean Ground Beetles (Coleoptera: Carabidae)
Reprinted from: *Diversity* **2023**, *15*, 136, https://doi.org/10.3390/d15020136 . . . . **176**

Amos Frumkin and Ariel D. Chipman
A Subsurface Stepping Stone Hypothesis for the Conquest of Land by Arthropods
Reprinted from: *Diversity* **2024**, *16*, 6, https://doi.org/10.3390/d16010006 . . . . **186**

About the Editors

Srećko B. Ćurčić

Srećko B. Ćurčić, full professor, received his doctorate from the University of Belgrade - Faculty of Biology in 2005. Since 2000, he has been working at the Institute of Zoology, University of Belgrade - Faculty of Biology. His research interests include the diversity and faunistics of hypogean and soil-dwelling invertebrates; morphology, systematics, taxonomy, phylogeny, development and ecology of Coleoptera; photonic structures in insects; secretion in ground beetles; and antimicrobial action of the pygidial gland secretions of ground beetles. He is the author of more than 200 scientific articles in peer-reviewed journals in various fields of zoology. His scientific papers have been cited 1676 times in a large number of scientific publications. Prof. Ćurčić is an associate editor, subject editor, guest editor and editorial board member of a number of scientific journals.

Gordan S. Karaman

Gordan S. Karaman, academician and professor emeritus, received his doctorate from the Faculty of Science and Mathematics at the University of Belgrade in 1964. From 1963 to 1966 he worked at the Institute of Marine Biology in Kotor, from 1966 to 1990 at the Institute of Biological and Medical Research in Podgorica and from 1990 to 2004 at the Faculty of Science and Mathematics at the University of Montenegro. His research interests include the study of the zoobenthos of the Mediterranean and freshwaters; the study of cave fauna; the taxonomy, ecology and zoogeography of epigean and subterranean freshwater and marine amphipods (Crustacea) on a global scale; problems in biodiversity conservation and sustainable development; the relations between zoology and philosophy with regard to the use of taxonomic categories; and the taxonomy and distribution of ants in the Balkans. He is the author of 365 scientific articles in peer-reviewed journals in various fields of zoology. His scientific papers have been cited 5076 times in a large number of scientific publications. Dr. Karaman is a subject editor, guest editor and editorial board member of a number of scientific journals.

Preface

This Special Issue Reprint is dedicated to the study of subterranean arthropods, one of the most exotic animal groups, which at the same time are relatively little researched. As their diversity in certain regions has not been sufficiently researched, it was of interest for us to dedicate ourselves to elucidating their diversity on a global level and their taxonomy. In addition to the diversity of their fauna, one of our objectives was to study their morphological, molecular, ecological, chemoecological and behavioural diversity and, given our expertise, to consider equally the representatives of the troglofauna and the stygofauna. We hope that our Special Issue Reprint will, through leading to a better knowledge of the diversity of subterranean arthropods, highlight their great importance in the habitats they inhabit and for humans in general. It is aimed primarily at professionals involved in various aspects of the study of subterranean arthropods, as well as all nature lovers who admire this specific and fascinating group of animals.

Srećko B. Ćurčić and Gordan S. Karaman
Guest Editors

diversity

MDPI

Editorial

Recent Advances in the Diversity and Taxonomy of Subterranean Arthropods

Srećko Ćurčić [1,*] and Gordan Karaman [2]

1 Institute of Zoology, Faculty of Biology, University of Belgrade, Studentski Trg 16, 11000 Belgrade, Serbia
2 Montenegrin Academy of Sciences and Arts, Riste Stijovića 5, 81000 Podgorica, Montenegro; karaman@t-com.me
* Correspondence: srecko@bio.bg.ac.rs; Tel.: +381-11-2638-890

The subterranean fauna of arthropods is one of the richest on the planet [1]; however, it remains poorly understood. The characteristics of the subterranean environment have led hypogean arthropods to develop both morphological and physiological adaptations, such as depigmentation, complete loss of eyes, limb elongation, a slow metabolism, reduced energy expenditure, and many others [2,3]. These animals are excellent models for studying the general principles of evolution and the mechanisms of adaptation to a new environment [4,5]. Since the diversity of subterranean arthropods worldwide is still insufficiently explored and most previous studies have focused on faunal diversity [6], it is desirable to analyse further aspects of the diversity of this specific group of animals, including morphological, molecular, ecological, chemoecological, and behavioural diversity. Since the editors' expertise covers both troglobitic (associated with terrestrial subterranean habitats) and stygobitic (associated with all types of subterranean waters) arthropods [7], this Special Issue is dedicated to exploring the diversity of both groups. We hope that it will contribute to a better understanding of the diversity of subterranean arthropods across various biological disciplines, highlighting their great importance within their habitats and for humans in general [8,9].

This Special Issue provides an overview of current research on subterranean arthropods and the interactions between these specific animals and their habitats. A total of 11 papers were accepted for inclusion in this Special Issue, comprising 10 original research articles and 1 hypothesis. Four contributions deal with the diversity, phylogeny, and phylogeography of stygobitic arthropods. Another four are dedicated to the diversity and taxonomy of troglobitic representatives, while the topics of diversity of both categories of subterranean arthropods, diversity of chemicals of subterranean ground beetles, and a hypothesis on land colonisation by arthropods are each represented by one paper.

Eme et al. (contribution 1) investigated the hitherto inconclusive phylogeographic history and showed the mito-nuclear discordance of an endemic groundwater amphipod from Iceland, *Crangonyx islandicus*, which is the only known metazoan to have survived the Pleistocene under the glaciers. Double digest restriction-site associated DNA sequencing (ddRADseq) was used to clarify this mito-nuclear discordance. Various statistical methods were applied to assess sensitivity to different analytical approaches (data type, differentiation indices, and base call uncertainty). The majority of nuclear markers and methods supported ITS divergence. Nevertheless, a more complex scenario emerges, possibly involving introgression caused by male dispersal between northern sites or mitochondrial capture exacerbated by natural selection.

Balázs et al. (contribution 2) analysed 14 localities of *Niphargus* in Hungary, covering the 8 valid species of the country and including 9 previously unstudied populations.

 https://doi.org/10.3390/d17080532

Based on the sequences of three gene fragments, they reconstructed the phylogeny of the Hungarian species using maximum likelihood and Bayesian approaches. They discovered that all Hungarian *Niphargus* species are closely related to each other and that even species sampled from the same localities may belong to different clades. Some species form monophyletic clades, while others are nested in different non-Hungarian lineages. The new populations are all genetically distinct from the known species. The results suggest that the Hungarian *Niphargus* fauna has originated from seven unrelated clades, and its diversity is underestimated due to unknown populations and cryptic species.

In the study by Weber and Weigand (contribution 3), groundwater amphipods were sampled in hyporheic interstitials throughout Luxembourg and the Greater Region, and the samples were analysed using DNA barcoding. Sites characterised by gravel or coarse sand and high flow velocities of the inflowing water were the richest in species and individuals. A total of 11 species were detected, of which the *Niphargus aquilex* lineage EF of the *N. aquilex* cryptic species complex and juveniles of *N. schellenbergi* dominated, but an unknown lineage of the *N. aquilex* complex was also found. In some regions, groundwater amphipods appeared to be absent. The authors suggest that underlying sandstone formations leading to low sediment porosity may prevent physical colonisation, but also that historical water pollution may have a long-lasting effect, either through the persistence of contaminants in the sediment or through low recolonisation rates of affected populations.

Jaime et al. (contribution 4) presented an overview of the stygobitic crustaceans of the Yucatan Peninsula. They represent potential markers for water reserves, which is why the authors wanted to use them to delineate aquifers on the peninsula. As part of a literature review, 75 crustacean species from 132 subterranean systems on the peninsula were recorded together with geomorphological, hydrological, hydrogeochemical, and historical precipitation data. Fourteen UPGMA clusters were informative for mapping species composition, with the Ring of Cenotes, Caribbean Cave, and Cozumel Island delineated as consolidated aquifers. These aquifers also differ in terms of abiotic factors: the Ring of Cenotes is dominated by freshwater species, while the Caribbean Cave and Cozumel Island are characterised by marine species. The use of stygobitic crustaceans for the integrative delineation of aquifers may represent an opportunity to improve the monitoring networks of regional aquifers.

Vargovitsh (contribution 5) described two new troglomorphic springtail species from the Arrhopalitidae family that live in the Snezhnaya (=Snow) Cave System in the western Caucasus. Both are morphologically highly specialised, but have evolved into different troglobitic life forms. *Arrhopalites profundus*, with extremely long claws, a lamellar mucro, and an enlarged sensory organ on the third antennal segment, lives exclusively in hygropetric and epineustonic habitats. In contrast, *Pygmarrhopalites rystsovi*, with strongly elongated and multi-subsegmented antennae and long legs, favours open terrestrial spaces. The author discussed the troglomorphy and ecological life forms of the obligate cave-dwelling Arrhopalitidae, in general, and classified them as neustonic-hygropetric, atmobiontic, intrasubstrate, and intermediate troglomorphs.

The paper by Vesović et al. (contribution 6) gives an overview of the diversity of subterranean terrestrial arthropods in the Resava Cave in eastern Serbia, which is under protection as a natural monument and is one of the most visited caves in Serbia and its surroundings. A total of 107 arthropod species from the four main subphyla were recorded in the cave: 66 species of Hexapoda, 27 species of Chelicerata, 11 species of Myriapoda, and 3 species of Crustacea. The authors provided microhabitat descriptions and information on distribution in the cave for 4 troglobitic, 16 troglophilic, and 87 trogloxenic species. The cave was found to be relatively rich in hypogean terrestrial arthropod fauna compared to other caves in Serbia that have been biospeleologically investigated so far.

Latella and Brighenti (contribution 7) investigated the terrestrial invertebrate diversity in two ice caves (Bus delle Taccole and Caverna del Sieson) in northeastern Italy. Ice samples were taken from each explored part of the caves, which the authors used to measure the stable isotopes of oxygen and hydrogen and monitor the air temperature. The two caves had different invertebrate communities, both dominated by a combination of troglobitic and cryophilic taxa. Despite the low taxonomic richness, both caves harboured rare and endemic species, four of which are yet to be described. The isotopic signatures of the ice water differed in the different parts of the caves, suggesting that the ice formed under different climatic conditions and/or due to different frequencies of thawing/freezing.

In the article by Blasco-Aróstegui and Prendini (contribution 8), the first truly troglobitic European scorpion species, *Euscorpius studentium*, which was described a few years ago on the basis of two immature specimens, is redescribed on the basis of adult specimens. In addition, the ecological classification of all currently known subterranean scorpions of the family Euscorpiidae is updated, a key to the identification of cavernicolous scorpions occurring in the Dinaric Karst is provided, and an overview of the historical and geographical factors affecting the distribution and conservation of cave-dwelling scorpions in the Balkans is given.

In the work by Sarbu et al. (contribution 9), biological investigations were carried out in the Sulfur Cave in the Vromoner Canyon on the Greek–Albanian border, focusing on invertebrate, vertebrate, and microbial communities and investigating the structure of the subterranean food web. The invertebrate fauna includes 5 aquatic species, 25 terrestrial species, and 4 amphibiotic species. Two fish species from two families and six bat species from three families were documented. The microbial communities in the different biofilms are dominated by chemosynthetic, sulphur-oxidising microorganisms, especially filamentous bacteria. Stable isotope analysis revealed low carbon and nitrogen levels for the terrestrial and aquatic invertebrates, suggesting that the subterranean communities rely on food produced in situ by chemoautotrophic microorganisms.

Vranić et al. (contribution 10) investigated both the chemical composition of the pygidial gland secretion and the morphology of the glands in adults of the troglophilic ground beetle species *Laemostenus cavicola*. A total of seven chemical compounds were detected in the secretion mixture. Formic acid was the most dominant compound, followed by dodecyl acetate and undecane. Other chemicals were present in small amounts. The morphological structure of the pygidial glands of the studied ground beetle was compared with the structure of the glands of the related troglophilic species *Laemostenus punctatus*. In addition, the authors presented summarised data on the semiochemicals previously recorded in subterranean ground beetle species and discussed the differences in the chemical composition of secretions between and among troglobitic and troglophilic species. The results of this study show the great diversity of chemicals in the pygidial gland secretions of subterranean ground beetles.

The issue of land conquest by arthropods is discussed by Frumkin and Chipman in their paper (contribution 11), in which they offer an opinion on how arthropods could survive on land without established plants and an ozone shield. Their hypothesis is that chemolithoautotrophic cave ecosystems, which are independent of photosynthesis, may have served as a subterranean stepping stone, providing a possible explanation for the land invasion conundrum. These caves could provide abundant food and radiation protection, allowing ancient arthropods to develop strategies to adapt to new frontiers by gradually spreading from the sea into shielded cave waters, then into the hygropetric margins of cave waters, and finally to the surface.

In summary, this Special Issue contains 11 papers that provide a comprehensive study of the diversity and taxonomy of subterranean arthropods. These studies examine a variety of arthropods inhabiting different hypogean habitats, including temperate caves, ice caves, sulphur caves, groundwater, and interstitial habitats. The studies were conducted on three continents—Europe, Asia, and North America—and in all major climatic zones: temperate, polar, and tropical. The main focus was on terrestrial and aquatic arthropods, which were analysed at both species and population level, with some papers covering other invertebrate groups, vertebrates, and microbes from the same habitats in addition to arthropods. The authors of the papers used both traditional and modern data collection techniques and advanced statistical methods, as well as taxonomic, molecular, faunistic, morphological, ecological, and chemoecological methods for laboratory analyses. The results of the articles in this Special Issue described, characterised, and clarified presently little-researched topics related to subterranean arthropods. Overall, this Special Issue represents an important and encouraging contribution to the understanding of the diversity and taxonomy of subterranean arthropods on a global scale and contains valuable data that shed light on part of their biology and support the conservation of these specific organisms and their habitats.

In view of the enormous diversity of the animal group studied [10], there is still much to be discovered, examined, and clarified. Particular attention should be paid to the investigation of hotspots that have not yet been sufficiently researched (North America, Australia, Southeast Asia, Brazil, and China) [11–17]. In addition, future studies should increasingly incorporate modern techniques, such as mapping and modelling [18,19]. It is necessary to adopt an integrative, synergistic approach [20], using and combining different types of data (faunistic, morphological, molecular, environmental, etc.) to gain a more comprehensive insight into the richness and diversity of this specific group of invertebrates.

Author Contributions: Both authors managed this editorial, wrote the manuscript and contributed to the revision. All authors have read and agreed to the published version of the manuscript.

Acknowledgments: As Guest Editors of the Special Issue "Recent Advances in the Diversity and Taxonomy of Subterranean Arthropods", we would like to thank the authors whose valuable works were published in this issue and contributed to its success. Our sincere thanks also go to all the academic editors and reviewers whose voluntary work has contributed significantly to the completion of this Special Issue. Finally, we would like to thank the Serbian Ministry of Science, Technological Development and Innovation (contract number 451-03-137/2025-03/200178) for its support.

Conflicts of Interest: The authors declare no conflicts of interest.

List of Contributions:

1. Eme, D.; Westfall, K.M.; Matthíasardóttir, B.; Kristjánsson, B.K.; Pálsson, S. Contrasting phylogeographic patterns of mitochondrial and genome-wide variation in the groundwater amphipod *Crangonyx islandicus* that survived the Ice Age in Iceland. *Diversity* **2023**, *15*, 88. https://doi.org/10.3390/d15010088.
2. Balázs, G.; Borko, Š.; Angyal, D.; Zakšek, V.; Biró, A.; Fišer, C.; Herczeg, G. Not the last piece of the puzzle: *Niphargus* phylogeny in Hungary. *Diversity* **2023**, *15*, 223. https://doi.org/10.3390/d15020223.
3. Weber, D.; Weigand, A.M. Groundwater amphipods of the hyporheic interstitial: a case study from Luxembourg and the Greater Region. *Diversity* **2023**, *15*, 411. https://doi.org/10.3390/d15030411.
4. Jaime, S.; Cervantes-Martínez, A.; Gutiérrez-Aguirre, M.A.; Hernández-Flores, G.; González-Herrera, R.A.; Sánchez-Rivera, G.; Enseñat-Soberanis, F.; Delgado-Blas, V.H. Toward an integrative overview of stygobiotic crustaceans for aquifer delimitation in the Yucatan Peninsula, Mexico. *Diversity* **2025**, *17*, 77. https://doi.org/10.3390/d17020077.
5. Vargovitsh, R.S. Deep troglomorphy: new Arrhopalitidae (Collembola: Symphypleona) of different life forms from the Snezhnaya Cave System in the Caucasus. *Diversity* **2022**, *14*, 678. https://doi.org/10.3390/d14080678.

6. Vesović, N.; Deltshev, C.; Mitov, P.; Antić, D.; Stojanović, D.Z.; Stojanović, D.V.; Stojanović, K.; Božanić, M.; Ignjatović-Ćupina, A.; Ćurčić, S. The diversity of subterranean terrestrial arthropods in Resava Cave (eastern Serbia). *Diversity* **2024**, *16*, 234. https://doi.org/10.3390/d16040234.
7. Latella, L.; Brighenti, S. Exploring ice cave biodiversity in northeastern Italy. *Diversity* **2024**, *16*, 364. https://doi.org/10.3390/d16070364.
8. Blasco-Aróstegui, J.; Prendini, L. Redescription of *Euscorpius studentium* based on adult specimens; updated classification of cavernicolous Euscorpiidae; and review of cavernicolous scorpions in the Balkans. *Diversity* **2024**, *16*, 737. https://doi.org/10.3390/d16120737.
9. Sarbu, S.M.; Brad, T.; Băncilă, R.I.; Ştefan, A. Exploring biodiversity and food webs in Sulfur Cave in the Vromoner Canyon on the Greek-Albanian border. *Diversity* **2024**, *16*, 477. https://doi.org/10.3390/d16080477.
10. Vranić, S.; Vujisić, L.; Vesović, N.; Todosijević, M.; Pavićević, M.; Radović, D.; Ćurčić, S. On the diversity of semiochemicals of the pygidial gland secretions of subterranean ground beetles (Coleoptera: Carabidae). *Diversity* **2023**, *15*, 136. https://doi.org/10.3390/d15020136.
11. Frumkin, A.; Chipman, A.D. A subsurface stepping stone hypothesis for the conquest of land by arthropods. *Diversity* **2024**, *16*, 6. https://doi.org/10.3390/d16010006

References

1. Deharveng, L.; Bedos, A.; Pipan, T.; Culver, D.C. Global subterranean biodiversity: A unique pattern. *Diversity* **2024**, *16*, 157. [CrossRef]
2. Culver, D.C.; Pipan, T. *The Biology of Caves and Other Subterranean Habitats*, 2nd ed.; Oxford University Press: Oxford, UK, 2019.
3. Howarth, F.; Moldovan, O.T. The ecological classification of cave animals and their adaptations: Analysis and synthesis. In *Cave Ecology*; Moldovan, O.T., Kováč, L'., Halse, S., Eds.; Springer: Cham, Switzerland, 2018; pp. 41–67.
4. Protas, M.; Jeffery, W.R. Evolution and development in cave animals: From fish to crustaceans. *WIREs Dev. Biol.* **2012**, *1*, 823–845. [CrossRef] [PubMed]
5. Howarth, F. High-stress subterranean habitats and evolutionary change in cave-inhabiting arthropods. *Am. Nat.* **1993**, *142*, S65–S77. [CrossRef] [PubMed]
6. Sollai, G.; Giglio, A.; Giulianini, P.G.; Crnjar, R.; Solari, P. Topic: Arthropod biodiversity: Ecological and functional aspects. *Insects* **2024**, *15*, 766. [CrossRef] [PubMed]
7. Sket, B. Can we agree on an ecological classification of subterranean animals? *J. Nat. Hist.* **2008**, *42*, 1549–1563. [CrossRef]
8. Sket, B. Subterranean (hypogean) habitats in karst and their fauna. In *The Wetland Book*; Finlayson, C., Milton, G., Prentice, R., Davidson, N., Eds.; Springer: Dordrecht, The Netherlands, 2017; pp. 1–14.
9. Mammola, S.; Malumbres-Olarte, J.; Vaccarelli, I.; Nanni, V.; Bellvert, A.; Jarić, I. On art, science, and the conservation of subterranean ecosystems. *Subterr. Biol.* **2025**, *51*, 1–19. [CrossRef]
10. Wynne, J.J. (Ed.) *Cave Biodiversity: Speciation and Diversity of Subterranean Fauna*; John Hopkins University Press: Baltimore, MD, USA, 2022.
11. Culver, D.C.; Deharveng, L.; Pipan, T.; Bedos, A. An overview of subterranean biodiversity hotspots. *Diversity* **2021**, *13*, 487. [CrossRef]
12. Bregović, P.; Zagmajster, M. Understanding hotspots within a global hotspot—Identifying the drivers of regional species richness patterns in terrestrial subterranean habitats. *Insect Conserv. Biodivers.* **2016**, *9*, 268–281. [CrossRef]
13. Christman, M.C.; Doctor, D.H.; Niemiller, M.L.; Weary, D.J.; Young, J.A.; Zigler, K.S.; Culver, D.C. Predicting the occurrence of cave-inhabiting fauna based on features of the Earth surface environment. *PLoS ONE* **2016**, *11*, e0160408. [CrossRef] [PubMed]
14. Guzick, M.T.; Austin, A.D.; Cooper, S.J.B.; Harvey, M.S.; Humphreys, W.F.; Bradford, T.; Eberhard, S.M.; King, R.A.; Leijs, R.; Muirhead, K.A.; et al. Is the Australian subterranean fauna uniquely diverse? *Invert. Syst.* **2010**, *24*, 407–418. [CrossRef]
15. Souza Silva, M.; Ferreira, R.L. The first two hotspots of subterranean biodiversity in South America. *Subterr. Biol.* **2016**, *19*, 1–21. [CrossRef]
16. Trajano, E.; Gallão, J.E.; Bichuette, M.E. Spots of high diversity of troglobites in Brazil: The challenge of measuring subterranean diversity. *Biodivers. Conserv.* **2016**, *25*, 1805–1828. [CrossRef]
17. Latella, L. Biodiversity: China. In *Encyclopedia of Caves*, 3rd ed.; White, W.B., Culver, D.C., Pipan, T., Eds.; Academic Press: London, UK, 2019; pp. 127–135.
18. Zagmajster, M.; Christman, M.C. Mapping subterranean biodiversity. In *Encyclopedia of Caves*, 3rd ed.; White, W.B., Culver, D.C., Pipan, T., Eds.; Academic Press: London, UK, 2019; pp. 678–685.

19. Mammola, S.; Leroy, B. Applying species distribution models to caves and other subterranean habitats. *Ecography* **2017**, *40*, 1–14. [CrossRef]
20. Gibson, L.; Humphreys, W.F.; Harvey, M.; Hyder, B.; Winzer, A. Shedding light on the hidden world of subterranean fauna: A transdisciplinary research approach. *Sci. Total Environ.* **2019**, *684*, 381–389. [CrossRef] [PubMed]

diversity MDPI

Article

Contrasting Phylogeographic Patterns of Mitochondrial and Genome-Wide Variation in the Groundwater Amphipod *Crangonyx islandicus* That Survived the Ice Age in Iceland

David Eme [1,2,*], Kristen M. Westfall [1,3], Brynja Matthíasardóttir [1,4], Bjarni Kristófer Kristjánsson [5] and Snæbjörn Pálsson [1,*]

1 Department of Life and Environmental Sciences, University of Iceland, Askja, Sturlugata 7, 101 Reykjavik, Iceland
2 RiverLy Research Unit, National Research Institute for Agriculture Food and Environment (INRAE), 5 rue de la Doua, 69100 Villeurbanne, France
3 Fisheries and Oceans Canada, Pacific Biological Station, Nanaimo, BC V9T 6N7, Canada
4 National Institutes of Health, NHGRI, 10 Center Drive, Bethesda, MD 20892, USA
5 Department of Aquaculture and Fish Biology, Hólar University, Háeyri 1, 550 Sauðárkrókur, Iceland
* Correspondence: david.eme@inrae.fr (D.E.); snaebj@hi.is (S.P.)

Abstract: The analysis of phylogeographic patterns has often been based on mitochondrial DNA variation, but recent analyses dealing with nuclear DNA have in some instances revealed mito-nuclear discordances and complex evolutionary histories. These enigmatic scenarios, which may involve stochastic lineage sorting, ancestral hybridization, past dispersal and secondary contacts, are increasingly scrutinized with a new generation of genomic tools such as RADseq, which also poses additional analytical challenges. Here, we revisited the previously inconclusive phylogeographic history, showing the mito-nuclear discordance of an endemic groundwater amphipod from Iceland, *Crangonyx islandicus*, which is the only metazoan known to have survived the Pleistocene beneath the glaciers. Previous studies based on three DNA markers documented a mitochondrial scenario with the main divergence occurring between populations in northern Iceland and an ITS scenario with the main divergence between the south and north. We used double digest restriction-site-associated DNA sequencing (ddRADseq) to clarify this mito-nuclear discordance by applying several statistical methods while estimating the sensitivity to different analytical approaches (data-type, differentiation indices and base call uncertainty). A majority of nuclear markers and methods support the ITS divergence. Nevertheless, a more complex scenario emerges, possibly involving introgression led by male-biased dispersal among northern locations or mitochondrial capture, which may have been further strengthened by natural selection.

Keywords: introgression; incomplete lineage sorting; selection; subglacial refugia; populations; genomics; groundwater; RADseq

1. Introduction

An understanding of the evolutionary histories of species, including geographic population structure, degree of connectivity/or isolation, past demographics or migration routes, offers crucial knowledge about species and is the main goal of phylogeography [1–4]. Phylogeographic studies have traditionally relied on small fragments of mitochondrial DNA (mtDNA) [2,3]. The predominance of mitochondrial markers was predicated by a well-developed procedure to obtain data and by properties that allowed the inference of relationships between closely related taxa such as fast mutation rate, low or absent recombination rate, small population size and quasi-neutral evolution [2,5], but see [3]. However, the mitochondrial genome is only a single marker that can have a singular evolutionary history [1,6]. The analysis of variation within the nuclear genome is thus needed in order to obtain a better estimate of the history of the populations studied [1,6]. Such studies have in some cases revealed divergent phylogeographic scenarios than inferred from mtDNA, i.e., mito-nuclear discordance conundrums [1,6], and raised questions about the inferences of the underlying evolutionary processes [1,6,7]. Two studies in the highly dynamic Icelandic aquatic ecosystem showed that an endemic groundwater amphipod species, *Crangonyx islandicus* Svavarsson and Kristjánsson, 2006, presented a mito-nuclear discordant phylogeographic pattern interrogating its evolutionary history and population structure [8,9]. As the observed discordance was limited to a single nuclear marker [9], sampling a larger proportion of the nuclear genome is warranted to assess the robustness of the mito-nuclear discordance and the evolutionary history of *C. islandicus*. However, it poses additional analytical challenges when considering a large diversity of nuclear markers.

Mito-nuclear discordance has been found in several species and can be driven both by stochastic and introgression processes [1,6,7,10–12]. Stochasticity can lead to incomplete lineage sorting (ILS) among genes due to genetic drift [1], which implies the persistence of ancestral polymorphism and in different genealogies among genes [1,6,7]. Different genealogies can, in addition, be maintained by different modes of inheritance and ploidy (e.g., bi-parental inheritance for nuclear DNA vs. maternal inheritance for mitochondrial DNA). Incomplete lineage sorting usually does not leave a particular pattern, and frequencies among different patterns should be homogeneous. On the contrary, introgression implies hybridization among individuals coming from distinct isolated populations through secondary contact allowed by the removal of the dispersal barrier. Introgression should be more prominent among taxa with closer relationships than among distantly related species [13]. Mito-nuclear discordance resulting from introgression can be driven by adaptation or sex-biased mechanisms such as sex-biased dispersal, hybrid survival and mate choice [6,7]. Purifying or balancing selection, augmented by selection on linked sites due to reduced or no recombination, may in addition lead to deviations in genetic patterns among different genomic regions such as in mitochondria, sex- and autosomal chromosomes (e.g., [14]), and such impacts can be stronger in small populations where mutations with weak fitness effects can accumulate [15]. As mito-nuclear discordance can strongly affect the inferences about the underlying evolutionary processes [1,6,7], implications for conservation measures may need to be revisited as the population structure, gene flow and effective population size inferred from a single or few markers may not represent the divergence history among populations but a singular pattern [16]. Revising mitochondrial phylogeographic patterns with nuclear DNA input is thus crucial to clarify the relationships of closely related taxa and the population processes (drift, mutation, migration, selection) that shape genetic diversity as a whole.

Abundant polymorphic nuclear markers can now be obtained for non-model organisms through reduced representation genome sequencing methods such as restriction-site-associated DNA sequencing (RADseq) [17,18]. Restriction-site-associated DNA sequencing has been successfully used to infer recent population structure [19,20], population history [21,22], resolve enigmatic phylogenetic relationships between species [23–25] and to assess the extent of hybridization and introgression between species [26]. However, two points need to be considered when analyzing RADseq data to infer phylogeographic scenarios. First, RADseq loci present a large diversity of markers with different modes of inheritance (mitochondrial vs. nuclear) and differences within the genome such as variation in mutation rates (e.g., in conserved coded regions and repetitive neutral regions), recombination rates and selection pressure [27]. Consequently, the genetic diversity among loci may be highly variable, especially when the full DNA sequence of different loci is considered instead of unlinked single nucleotide polymorphisms (SNPs) per locus. Statistics that are used to infer putative population structure or population parameters can be differently sensitive to this variability in genetic diversity [28–31]. For instance, the differentiation indices F-statistics [32] and Gst [33], contrary to the D statistic [29], are sensitive to high levels of heterozygosity and can thus be misleading to reveal population structure, but still offer a good estimate of the demographic history [34]. Care is thus needed when interpreting commonly reported population genetic summary statistics as the type of genetic information retained (full sequence, haplotype or unlinked SNPs) may amplify variability in genetic diversity among loci. Second, there is a trade-off between the number of individuals sampled and the accuracy of the genotype assigned by the sequencing read depth at each locus [35–37]. For a given sequencing effort, higher sample size increases estimates of allele frequencies within a population and the probability of detecting rare alleles, but at the cost of decreasing sequencing read depth across loci, which may not be sufficient to remove sequencing errors and potentially biasing downstream summary statistics [38–40]. Recently, several methods have been developed to overcome this issue taking into account the base call uncertainty [38,39,41–44]. Despite the fact that these methods are still in their infancy, they offer the opportunity to evaluate the robustness of the summary statistics using classic base calling for low-coverage datasets [44].

In this study, we revisit the phylogeographic history, showing the mito-nuclear discordance of *C. islandicus* using mitochondrial and RADseq data, while estimating the sensitivity of different analytical approaches (data-type, differentiation indices and base call uncertainty). *Crangonyx islandicus* is one of the two groundwater amphipod species endemic to Iceland that are found in springs in lava fields throughout the volcanic zone of the country [8,45,46]. It is the only metazoan species known to have survived beneath the glaciers during the Pleistocene, likely in fissures along the tectonic plate boundary in Iceland [8] (Figure 1a). A phylogeographic study based on its 16S and COI mitochondrial genes revealed a clear pattern of population divergence of five mitochondrial monophyletic clades, hereafter referred to as the mitochondrial scenario (Figure 1b) (clades AA′, BC, D, E and F). The differentiation of the clades followed geographical separation within Iceland; clades AA′, BC and D in southern Iceland diverged from each other from 0.4 to approximately 1 Myr ago, from E about 1.3 Myr ago and clade F, located in northeastern Iceland, presenting a putative cryptic species, diverged approximately 4.8 Myr ago from the others [8]. However, the variation in the nuclear internal transcribed spacer regions (ITS) 1 and 2 showed a different phylogeographic pattern potentially driven both by concerted and divergent evolution, hereafter referred to as the ITS scenario (Figure 1b) [9]. Contrary to the mitochondrial markers, ITS1 showed the oldest split between locations from northern Iceland (E and F) and those from southern Iceland, which share a common duplicated ITS1 region of 269 base pairs. Contrary to the mitochondrial DNA, ITS do not support the presence of cryptic species [9]. Further nuclear information is needed to evaluate the discrepancy between the patterns obtained with these two markers.

Here, we used the mitochondrial and double digest RADseq (ddRADseq) sequence data of the individuals sampled at four locations belonging to three mitochondrial clades (AA′, E and F) previously identified by Kornobis et al. [8] to revisit the phylogeography of *C. islandicus* and test the likelihood of the mitochondrial vs. ITS scenario. First, to evaluate the mitochondrial scenario proposed by Kornobis et al. [8], we performed a Bayesian structured coalescent analysis [47] with the subset of three clades to reconstruct the phylogeographic history of the mitochondrial DNA. Second, we used ddRADseq data to test population structure using different data types (SNP, haplotypes, sequences), differentiation indices, including indices taking into account the base call uncertainty and multivariate and Bayesian multi-coalescence analyses. Finally, using ddRADseq data we evaluated the level of incongruence among loci, and tested the amount of admixture between sampling locations caused by introgression or ILS.

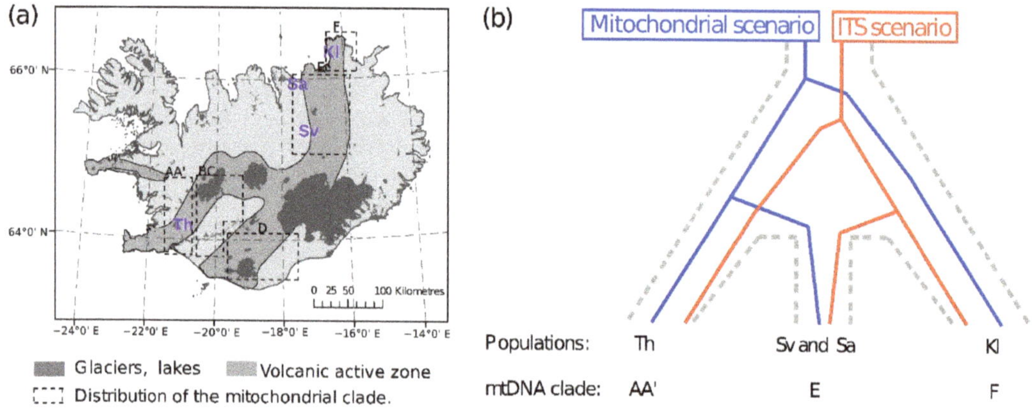

Figure 1. Sampling sites, distribution of mitochondrial DNA (mtDNA) lineages and comparison of divergence in mtDNA and in the internal transcribed spacer region (ITS). (**a**) Map of the sampling sites (in purple) and schematic distribution, denoted with squares and capital letters, of the mitochondrial clades (AA′, BC, D, E and F) previously defined [8]. (**b**) Two alternative scenarios of the population relationships proposed in the literature [8,9] showing mito-nuclear discordance. The grey dotted line represents the undefined "species tree". Th: Lake Thingvallavatn, Sa: Sandur Adaldal, Sv: Lake Svartárvatn, Kl: Klapparós.

2. Materials and Methods

2.1. Sampling

Fifty-nine individuals of *C. islandicus* were sampled in four springs in 2013, representing three distinct mitochondrial clades [8]: clade AA′ in Lake Thingvallavatn (Th: southern Iceland), clade E at Sandur Adaldal (Sa) and in Lake Svartárvatn (Sv) (northern Iceland) and clade F in Klapparós (Kl: northeastern Iceland) (Figure 1a, Supplementary Materials Table S1.1). We used dip nets after applying electricity with electric fishing gear to sample specimens near groundwater springs. Specimens were fixed in 96% ethanol and stored at −20 °C, or frozen with dry ice and stored at −80 °C.

2.2. Molecular Techniques

Total genomic DNA was extracted from the fifty-nine individuals using a standard phenol-chloroform protocol [48].

Mitochondrial DNA: To validate the mitochondrial pattern observed for these locations by Kornobis et al. [8], 21 out of the 59 individuals were newly sequenced for the CO1 and 16S genes in this study and these sequences were completed by 37 individuals previously sampled in the immediate surroundings of these four locations [8] for a total of 58 individuals used in the mitochondrial analysis (see Supplementary Materials, Table S1.2). For PCR amplifications of the CO1 and 16S genes for 21 new individuals, we followed the same protocol as described in Kornobis et al. [8], and the sequencing was performed in both directions using the Sanger method on a Genetic Analyser (3500xL Applied Biosystem, Waltham, MA, USA).

Restriction-site-associated DNA sequencing: A double digest restriction-site associated DNA sequencing (ddRADseq) library was constructed from 59 individuals, using modified protocols from [49,50]. Total genomic DNA (100–500 ng) was sequentially digested using the restriction endonucleases Sau3AI (1U) and ApeKI (2U), respectively, each for 4 h at the manufacturer's (NEB) recommended temperatures in NEB Buffer 4. Digested DNA (100 ng) was ligated to adapters (sequences in [51]) containing unique combinatorial barcodes (16 unique 5 bp barcodes for ApeKI adapters and 5 unique 6 bp barcodes for Sau3AI adapters) for each individual (barcode and adapter sequences in Supplementary

Materials S2) using T4 DNA ligase (NEB) in supplied buffer at 21 °C for 4 h. Ligation reactions contained a 6:1 molar excess of adapter to fragmented DNA, calculated using the mean fragment size determined from an agarose gel. Ligated DNA was pooled and purified using magnetic beads (Macherey-Nagel NGS clean-up and size selection) following the manufacturer's protocol. Size selection of ligated DNA fragments was performed on a Pippin Prep (Sage Science, Beverly, MA, USA) with 2% ethidium-free agarose gels and external size standard. The narrow range setting included a mean fragment size of 350 bp ± 18 bp. The eluate was split among eight PCR reactions and amplified using the primers and PCR conditions as described in Elshire et al. [51]. Each PCR reaction had a total volume of 25 µL containing 1× OneTaq Master Mix with Standard Buffer (NEB), 0.5 mM each primer and 8 µL template DNA. Polymerase chain reaction products were pooled and purified by magnetic beads before quantification using a SYBR Gold fluorometric assay (protocol in Supplementary Materials S3). The library was prepared for sequencing following the manufacturer's instructions with a final concentration of 38 nM. The library was sequenced on an Illumina HiSeq2500 using the Illumina TruSeq kit (2 × 125 bp). To increase coverage of the data, the library was resequenced twice on an Illumina MiSeq2000 using the MiSeq Reagent Kit v2 (2 × 150 bp).

2.3. *Bioinformatic Pipeline*

2.3.1. Single Nucleotide Polymorphisms, Haplotypes and Sequence Datasets Base-Called with pyRAD

The process_radtag.pl command in STACKS v1.3 program [50] was used to demultiplex the paired-end libraries from ddRadseq. Reads with uncalled and low quality (phred score < 10) bases were excluded, the remaining reads trimmed to 100 bp and barcodes and adapters removed. The reads of the HiSeq run and the two MiSeq runs were assembled. The reads were assembled de novo into loci using pyRAD v1.3.1 [52], as the alignment clustering approach allows for the presence of indels to improve the identification of homology among divergent taxa. The restriction recognition overhang sites were removed. Low quality converted base calls below a score of 20 in pyRAD were considered as N and reads with more than 10 Ns were removed. A threshold of 80% was used to cluster the reads within individuals [25]. Heterozygosity and error rate were estimated in pyRAD using the maximum likelihood formula proposed by Lynch [53]. We used a statistical base call with a minimum depth coverage of five and then, due to low-coverage data, we generated two additional datasets using a consensus base call with a minimum depth coverage of three (C3) and two (C2) to retrieve additional loci. We removed the consensus sequences with more than 10 Ns. To exclude potential paralogs, the consensus sequences with more than two alleles after error correction and 10 heterozygous sites were removed (step 5 in pyRAD). Then, the consensus sequences were clustered across individuals at 80% similarity [24] and aligned (step 6 in pyRAD). Finally, we retained all the loci present at least in 50% of the individuals (n = 30) to generate the output formats of the C2 and C3 datasets (step 7 in pyRAD).

Single nucleotide polymorphism analyses were performed using one SNP per locus. The SNPs were selected either randomly (C2SNPr, C3SNPr, see Table 1) or selecting the most informative SNP ((C2SNPb, C3SNPb) as the one with the lowest number of missing data (individuals) and in the case of equal amount of missing data one was chosen randomly). The SNP selections were performed in R [54] from the .vcf output file provided by pyRAD. Haplotype analyses were performed using the information of all the SNPs present in a locus. We used the .alleles output file provided by pyRAD to generate fasta alignments for each locus. Then, we generated the haplotype files in Fstat formats removing the sequences with more than 25% ambiguous position (i.e., second read missing) and considering indel as a 5th state (Cov2H5th, Cov3H5th) or as missing data (Cov2Hmis, Cov3Hmis) using the haplotype function of the R package haplotypes [55]. Finally, all the haplotype datasets were also derived in their sequence equivalent datasets considering nucleotide information (Table 1).

To test the influence of low-frequency alleles, potential remaining paralogs and loci detected under selection, we considered a "brute" dataset and a "clean" dataset, where we filtered out the loci in low frequency, considered as potential paralogs and under selection. We removed all the loci with a minimum allele frequency (MAF) below 5% [19,56] and those with an observed heterozygosity above 0.5, which can be considered as potential paralogs [19,57]. To remove loci detected under selection, we ran Bayescan v2.01 [58], considered as one of the most robust programs, using the differentiation approach [59]. To decrease the high false discovery rate, we followed Lotterhoos and Whitlock [60] using a prior odd = 10,000. For each SNP and haplotypes dataset, we performed two Bayescan runs using 50,000 MCMC generations preceded by 20 pilot runs of 5000 generations. The program Blastn v2.4.0 in NCBI (https://blast.ncbi.nlm.nih.gov/Blast.cgi?PROGRAM=blastn&PAGE_TYPE=BlastSearch&LINK_LOC=blasthome (accessed on 16 March 2016)) was used to infer the role of loci putatively under selection through search in GenBank. Loci identified to be under selection were further analyzed by designing primers in sequences under selection and by amplifying and sequencing from up to 79 individuals per loci from 14 sites (see Supplementary Materials S4 for additional details) from the species range [8]. The population patterns of these markers were then compared to the overall pattern obtained from this study and previous studies [8,9].

2.3.2. Genotype Likelihood Datasets with ANGSD

As low-coverage datasets are prone to base call uncertainty, which can bias downstream analyses [39,40], we applied recent methods that by-pass the base call step and estimate the genotype likelihood [38] to compute almost unbiased summary statistics such as Fst-like [39]. Such an approach implemented in ANGSD [43] and other related software [41] requires converting the original fastq files into BAM files. As no reference genome was available for *C. islandicus* or its closely related taxa, we generated two mock reference genomes for each dataset (cov2 and cov3) using a consensus sequence with a threshold of 95% similarity among all the individuals of a concatenated sequence of all the selected loci assembled by PyRAD (see previous paragraph). The demultiplexed fastq files of each individual were aligned separately against these mock reference genomes in Bowtie2 [61] and only the aligned paired-end reads were retained and converted into SAM files. SAMTOOLS [62] was used to sort the reads and convert the files in BAM format for the two datasets (cov2GL and cov3GL). ANGSD v0.9.1 was used to perform quality filtering of the reads. Bad reads with SAM score > 255 (-remove_bad 1) and reads with multiple best hits (-unique_only 1) were removed with a minimum mapping quality of 30 (-minMapQ 30) and a minimum base quality of 20 (-minQ 20). Sites likely to be polymorphic with a p-value less than 10^{-6} (-SNP_pval 1e-6) and with a minimum coverage of 2 and a maximum coverage of 20 for at least 50% of the individuals were retained. To test the effect of low-frequency sites, we also considered a "brute" dataset and a "clean" dataset, where the sites with a MAF < 0.05 were removed. For all these datasets, the genotype likelihood function provided by SAMTOOLS [62] and implemented in ANGSD (-GL 1) was applied.

2.4. Mitochondrial Analysis

All the CO1 and 16S sequences were aligned by Clustalo, implemented in Seaview v.4.2.12 [63] and checked by eye. A Bayesian structured coalescent tree was reconstructed using MultiTypeTree (MTT) [47] in BEAST2 v. 2.3.1 [64] to estimate the tree topology, robustness, divergence time, migration rate among the three mitochondrial clades defined by Kornobis et al. [8] (clades AA', E and F) as well as to assess their effective population size. We estimated the best substitution model for the CO1 (TN93) and 16S (HKY+I) genes according to the Akaike criterion corrected for a small sample size (AICc) with MrAIC.pl v1.4.6 [65]. Two MTT runs were performed considering independent substitution and clock models for CO1 and 16S but with a fixed tree topology because both belong to the same linkage group. To estimate the divergence date, we used the same set-up as in

Kornobis et al. [8], applying a strict molecular clock, divergence rates of 1.4–2.6% per million years for CO1 and 0.53–2.2% for 16S and using a uniform distribution prior. For the substitution models, we used the default priors, while for the population size and migration rate prior we used a Lognormal distribution (mean = 0, sd = 2). The two independent Monte Carlo Markov Chains (MCMCs) were run for 200 million generations, sampled every 20,000 generations. The first 10% of each chain was discarded as burn-in, and then both chains were combined using LogCombiner. Finally, we checked the convergence of the combined chain (ESS > 200) with TRACER V1.6 [66] and extracted the maximum clade credibility tree (MCCT) using TreeAnnotator, which was displayed with FigTree v1.4.2 (available at http://tree.bio.ed.ac.uk/software/figtree/, accessed on 26 March 2016).

Table 1. Datasets used in the present study and the number of markers (i.e., SNPs or loci) retained after filtering (Nb. Mark.). Dataset Name: abbreviations of the name of datasets, Data Type: the type of genetic data involved for each locus (i.e., SNP, haplotype or DNA sequence), Min. Cov: the minimal coverage retained for a locus, Cl. Data: using a brute (br) dataset coming from pyRAD or a cleaned (cl) dataset removing loci with a minimum allele frequency (MAF) below 5%, or with an observed heterozygosity above 0.5 or detected under selection by Bayescan [58]. Nb. Mark: number of DNA markers retained in the analysis.

Dataset Name	Data Type	Min. Cov.	Cl. Data	Nb. Mark.
C2SNPr_br	SNP (random SNP)	2	br	326
C2SNPb_br	SNP (best SNP)	2	br	326
C2SNPr_cl	SNP (random SNP)	2	cl	230
C2SNPb_cl	SNP (best SNP)	2	cl	196
C2Hmiss_br	Haplotype (indel missing)	2	br	295
C2H5th_br	Haplotype (indel 5th state)	2	br	313
C2Hmiss_cl	Haplotype (indel missing)	2	cl	202
C2H5th_cl	Haplotype (indel 5th state)	2	cl	228
C3SNPr_br	SNP (random SNP)	3	br	108
C3SNPb_br	SNP (best SNP)	3	br	108
C3SNPr_cl	SNP (random SNP)	3	cl	90
C3SNPb_cl	SNP (best SNP)	3	cl	73
C3Hmiss_br	Haplotype (indel missing)	3	br	103
C3H5th_br	Haplotype (indel 5th state)	3	br	109
C3Hmiss_cl	Haplotype (indel missing)	3	cl	71
C3H5th_cl	Haplotype (indel 5th state)	3	cl	83
C2Seqmiss_br	Sequences the same as C2Hmiss_br	2	br	295
C2Seqmiss_cl	Sequences the same as C2Hmiss_cl	2	cl	202
C2Seq5th_br	Sequences the same as C2H5th_br	2	br	313
C2Seq5th_cl	Sequences the same as C2H5th_cl	2	cl	228
C3Seqmiss_br	Sequences the same as C3Hmiss_br	3	br	103
C3Seqmiss_cl	Sequences the same as C3Hmiss_cl	3	cl	71
C3Seq5th_br	Sequences the same as C3H5th_br	3	br	109
C3Seq5th_cl	Sequences the same as C3H5th_cl	3	cl	83
C2GLr_br	Genotype likelihood	2	br	310 [b]
C2GLr_cl	Genotype likelihood	2	Cl [a]	300 [b]
C3GLr_br	Genotype likelihood	3	br	103 [b]
C3GLr_cl	Genotype likelihood	3	Cl [a]	103 [b]

[(a)] Loci with a MAF below 5% were removed. [(b)] Maximum number of markers used in pairwise comparisons.

2.5. Restriction-Site-Associated DNA Sequencing Analysis

2.5.1. Population Differentiation Estimates among Sampling Locations

Pairwise population differentiation was estimated using five different metrics: G_{ST}, G''_{ST} [30], D_{ST} [29], Φ_{ST} and F_{ST}-like statistics based on genotype likelihood [39]. G_{ST}, G''_{ST} and D_{ST} were estimated for SNP and haplotype datasets, and Φ_{ST} was estimated on the sequence datasets. We use G_{ST}, G''_{ST} and D_{ST} due to their different sensitivity to the mutation rate, level of heterozygosity and population size [30,34]. G''_{ST}, the standardized measure of G_{ST} taking into account bias of sample size, and D_{ST} are less sensitive to the level of heterozygosity (genetic diversity). D_{ST} [29] is more sensitive to the variation of mutation rate between loci but it is considered as a better measure of differentiation in a wide range of circumstances because it is less sensitive to the population size [29–31,34]. To test the statistical significance of the differentiation among populations, we used Fisher's exact test, implemented in the mmod R package [67] using 1000 permutations for each locus. Then, the *p*-values of each locus were combined following Fisher's method [68] and corrected using Holm's formula for multiple testing [69]. We also computed the pairwise genetic distances between populations by taking differences between sequences into account (Φ_{ST}), using K80 (Kimura 2 parameters) and TN93 [70] substitution models. As both models provided similar results (R^2 = 0.99), only the K80 model Φ_{ST}s were retained. Statistical significance of the pairwise Φ_{ST}s was evaluated using 1000 permutations for each locus and the *p*-values were combined using Fisher's method [68] and corrected for multiple testing according to Holm's formula [69]. Pairwise G_{ST}, G''_{ST} and D_{ST}s were estimated using the mmod R package, while Φ_{ST} was computed with the pegas R package [71]. The base call uncertainty caused by the low-coverage dataset was taken into account by computing a Fst-like statistic proposed by Fumagalli et al. [39], based on genotype likelihood implemented in ANGSD using realSFS tools.

The differences between various pairwise differentiation indices and different datasets in Table 1 were summarized by calculating distances and visualized with a metric multidimensional scaling approach in R. The distances were calculated as 1-*r*, where *r* is the Pearson correlation coefficient based on the pairwise distances among populations computed among indices/datasets.

2.5.2. Congruence Tests among Loci

An empirical approach was used to estimate the proportion of loci supporting the mitochondrial scenario (Klapparós, considered as the most divergent population, was used as an outgroup) vs. the ITS scenario (Lake Thingvallavatn as an outgroup) using G_{ST}, G''_{ST}, D_{ST} and Φ_{ST} as differentiation measure among sampling locations. The difference in the extent of genetic differentiation between the outgroup and the other populations obtained for the two scenarios was tested using a non-parametric Wilcoxon test in R.

2.5.3. Population Structure and Relationship among Groups

Population structure was further inferred with both SNPs and haplotype datasets using the discriminant analysis of principal components (DAPC [72]) implemented in the adegenet v.2.0.1 R package [73]. The DAPCs were performed both considering the best clustering scheme inferred by the sequential K-means approach and using the sampling sites as a prior. The choice of the optimal number of clusters (K), within a range of 2 to 10, was performed according to the lowest score of the BIC criteria. The optimal number of retained principal components was performed using a cross validation procedure [72].

The statistical support of the number of independent lineages (or populations) and their phylogenetic relationships was estimated by performing a Bayesian multi-coalescence analysis using unlinked bi-allelic SNP with SNAPP [74] with a Bayes Factors Delimitation (BDF*) procedure [75,76], implemented in BEAST2. SNAPP estimates the species tree using unlinked bi-allellic SNPs while by-passing the estimate of multiple gene trees. It allows estimating the statistical support of the tree topology and posterior probability of the present and ancestral effective population size assuming constant population size for each branch of the tree and sudden changes [74]. For computational tractability, five individuals were selected per location with the least number of missing bi-allelic SNPs for the datasets with a minimum coverage of 2 and 3, respectively, and after removing loci in low frequency and potential paralogs. Three models were compared, M1: the four sampling locations as distinct entities [(Sv,Sa,Th,Kl)], M2: the mitochondrial scenario with three distinct entities [(((Sv,Sa),Th), Kl)] and M3: the ITS scenario with two distinct entities (the populations in the north vs. Thingvallvatn in the south [((Sv,Sa,Kl), Th)]. For each model, two MCMCs of

2,000,000 iterations sampled every 100 iterations were performed. The first 10% of each chain was discarded and the remainder of both chains was combined using LogCombiner. The convergence of the parameters was checked with TRACER (ESS > 200).

The statistical supports of the three models were estimated with the Bayes Factors (BF), using a path-sampling approach to estimate the marginal likelihoods [76]. Two independent runs were conducted for each model with 20 steps of path sampling with 100,000 MCMC iterations per step. We sampled the chain every 100 iterations after discarding the first 1000 steps as pre-burn-in and then the first 10% as burn-in. The average marginal likelihood (ML) of the two runs of each model was calculated and the formula BF = 2*($ML_{model1} - ML_{model2}$) was used to compare the two models. According to the scale of Kass and Raftery [77], the support for model 1 is considered to be strong for BF > 2, very strong for BF > 6 and decisive for BF > 10.

2.5.4. Introgression Analysis

The four-taxon D statistic, also called ABBA-BABA test [78,79], was used to disentangle the contribution of ILS and introgression (admixture) caused by ancestral hybridization among the four putative populations. If "A" is an outgroup/ancestral allele and "B" a derived allele, the D statistic represents the occurrence of two discordant patterns, ABBA and BABA, in a four-taxon tree topology [(((P1,P2),P3),O)], with P1, P2 and P3 as clades under investigation and O as an outgroup. Under ILS caused by stochastic sorting of ancestral polymorphism, frequencies of both patterns are expected to be similar, while an excess of one of those patterns may arise if introgression occurs between P3 and P2 or P1 [24,79]. D is positive when P2 exchanges with P3 and D is negative when P1 exchanges with P3. We used the SNP frequency of heterozygous site to compute the D statistic considering two alternative scenarios: the mitochondrial scenario tests the introgression between Thingvallavatn (P3) and Svartárvatn (P1) or Sandur (P2) using Klapparós as an outgroup, and the ITS scenario tests the introgression between Klapparós (P3) and Svartárvatn (P1) or Sandur (P2) using Thingvallavatn as an outgroup. Switching the position of Svartárvatn and Sandur did not qualitatively affect the results. To obtain the most complete data matrix, we selected the five most informative individuals per population (maximizing the locus coverage among populations) for two datasets (with a minimal coverage of 2 and 3, respectively) and after removing low-frequency data (Cov2Hmiss_cl and, Cov3Hmiss_cl datasets). For the ingroup taxa, we selected randomly one individual for each of P1, P2 and P3 populations among the five individuals retained per population, while pooling the five individuals of the outgroup to estimate the SNP frequency [26]. For each of these combinations of individuals (i.e., 125 replicates in total), we computed the D statistic. To assess the significance of the D statistic over replicates, we followed the procedure of Eaton et al. [26]. For each replicate, we performed 200 bootstrap iterations over loci with replacement to estimate the standard deviation of the D statistic and convert the observed D statistic as a Z-score. The significance of the Z-score was assessed by a conservative p-value at alpha = 0.01 after a Holm correction for multiple testing over replicates. A D statistic with an absolute Z-score value above 3 is generally considered as significant.

3. Results

3.1. Mitochondrial Results

The MultiTypeTree analysis based on the two mitochondrial genes (CO1 and 16S), including twenty-one new individuals, perfectly supports the previously described mitochondrial scenario [8]. The MCCT showed four well-supported monophyletic groups (node pp = 0.99–1, Figure 2a). The individuals from Klapparós belong to the clade F, the individuals of the two northern locations (Sandur and Lake Svartárvatn) belong without sub-geographical structure to the clade E, and the individuals sampled in Lake Thingvallavatn belong to the clade AA′. The confidence interval of the date of divergence between the clade F (95% HPD = 2.3–4.8 Myr) and all remaining clades (95% HPD = 0.7–1.8 Myr) did not overlap and supports a clear ancestral divergence of the mitochondrial genes between the northeastern population (Clade F) and a clade formed by the ancestors of the populations in the north (clade E) and the south (clades A and A′, Figure 2a). There was no evidence of recent mitochondrial gene flow among the mitochondrial clades, i.e., the confidence intervals of the posterior distribution of the migration rates always overlapped 0. Assuming a constant mutation rate for all the mitochondrial clades, the posterior distributions of θ showed that the Thingvallavatn population (Clade A-A′) tends to have a higher effective population size (θ_{median} = 0.49, 95% HPD = 0.23–0.87) than that from Klapparós (θ_{median} = 0.14, 95% HPD = 0.03–0.35) and the northern population (both samples of the clade E, Sandur and Svartárvatn, were considered as one population according to the absence of geographical structure, θ_{median} = 0.07, 95% HPD = 0.01–0.18, Figure 2b).

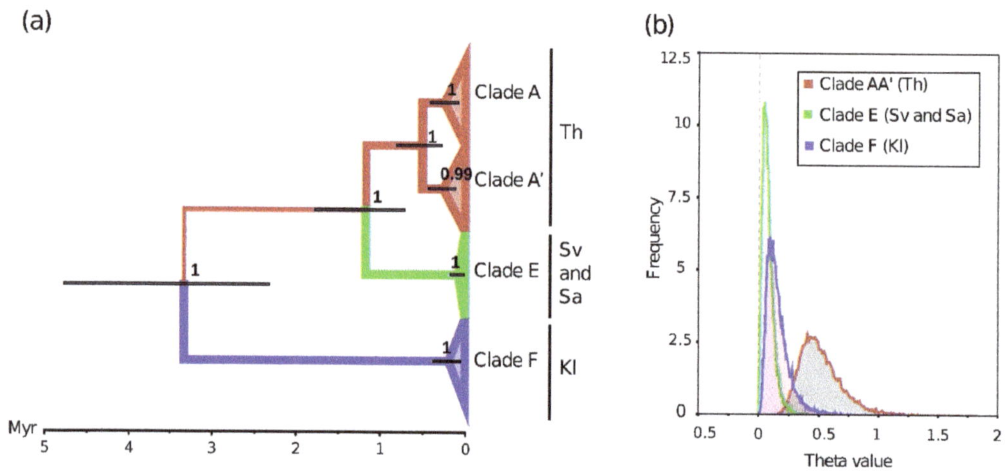

Figure 2. Divergence of mitochondrial genotypes of *Crangonyx islandicus* from four geographic locations. (**a**) Maximum clade credibility tree (MCCT) and (**b**) posterior distribution of θ ($\theta = Ne^{*}\mu$) parameters inferred by MultiTypeTree [47] in BEAST2 for four locations of *Crangonyx islandicus* using CO1 and 16S mitochondrial genes belonging to three mitochondrial clades, previously identified by [8]. Color branches of the MCCT represent the most likely localization of the ancestors of the four populations inferred by MultiTypeTree. Theta = Ne*μ, Ne = effective population size, μ = mutation rate. Th: Lake Thingvallavatn, Sv: Lake Svartárvatn, Kl: Klapparós, Sa: Sandur.

3.2. Restriction-Site-Associated DNA Sequencing Results

3.2.1. Description of the Datasets

The difference in the minimal coverage used, from two to three, had a major effect on the total number of de novo loci and SNPs retained in the analysis by pyRAD shifting from 335 to 123 loci, respectively, of which 323 and 108 were polymorphic, respectively. The coverage cut-off also reduced the dataset from 3539 SNPs with two to 835 SNPs with three. The haplotype datasets contained more alleles when indels were considered as 5th state compared to when they were considered as missing data (C2, missing = 4.5 ± 3.3; 5th = 6.5 ± 4.9; paired Wilcoxon test $p < 0.001$: C3, missing = 3.8 ± 3.2; 5th = 5.5 ± 4.7; paired Wilcoxon test $p < 0.001$). No unlinked SNPs were identified as putatively under selection by Bayescan, while haplotype datasets revealed seven loci under putative selection using a minimum coverage of two, regardless of whether the indels are considered as missing data or not. For a minimum coverage of three, we detected five and six loci under putative selection when indels were considered as missing data or as 5th state, respectively. Among the loci detected as putatively under selection, the Blastn approach was able to retrieve three annotated DNA fragments with an E-value statistic below 0.001: a histone H3 nuclear gene, a mitochondrial NADH dehydrogenase subunit 5 and another mitochondrial fragment overlapping the genes ND2 and tRNA-Trp (Supplementary Materials Table S5.1). The NADH dehydrogenase subunit 5 was found only with the dataset with a minimal coverage of two. The proportion of loci in low frequency (with MAF < 0.05), under putative selection and with observed heterozygosity above 0.5 was 29.6 ± 9.7% for the SNPs and 28.4 ± 3.6% for the haplotypes. These data were removed when considering the "clean" datasets.

3.2.2. Differentiation among Populations

A large majority (i.e., 92%) of the datasets and the differentiation metrics supported the ITS scenario (Figure 3). The Φ_{ST}, in particular, was insensitive to the datasets used and supported systematically the ITS scenario (Figure 3). For all datasets, the significance of the pairwise G_{ST}, G''_{ST} and D_{ST} between populations was always supported ($p < 0.001$), even after p-value correction for multiple testing among populations. However, all the pairwise Φ_{ST} between the southern (Thingvallavatn) and the northern populations (Sandur, Svartárvatn and Klapparós) were significant at $\alpha = 0.05$ but not among the northern populations (except between Sandur and Svartárvatn, with a minimum coverage of three, Table S5.2).

All unlinked SNP datasets supported the ITS scenario regardless of the differentiation metrics, t coverage or the presence/absence of the loci in low-frequency or under selection (Figure 3, Table S5.2). Similarly, regardless of the dataset used, the F_{ST}-like estimates based on genotype likelihood taking into account the base call uncertainty always supported the ITS scenario. However, minimal coverage had a stronger effect on the differentiation among populations than the effect of removing the loci under low frequency at a given coverage (Figure 3). Furthermore, all haplotype datasets with a minimal coverage of three supported the ITS scenario regardless of the differentiation metrics used. However, with a coverage of two, considering the presence of indels as informative or not, removing loci in low frequency and under selection affected the differentiation metrics variably, as well as the inferred scenario. Two approaches showed opposite results. First, G_{ST}, G''_{ST} and D_{ST} supported the ITS scenario when indels were considered as missing data and all the loci were retained, while in the second approach, considering indels as a 5th state and removing the loci in low frequency and under selection, these three metrics supported the mitochondrial scenario. For the latter approach, 68% of the loci removed supported the ITS scenario. Finally, two haplotype datasets (C2Hmiss_cl and C2H5th_br) showed discordant results when using different differentiation metrics, with the G_{ST} supporting the mitochondrial scenario, while the G''_{ST} and D_{ST} supported the ITS scenario. Such discrepancies between the G_{ST} and the two other differentiation metrics might be explained by the greater sensitivity of G_{ST} to the significant increase in the number of alleles and the heterozygosity within subpopulation (Hs) for all the loci supporting the discordant scenario compared to all the loci supporting the concordant scenario (Cov2H5th_br: nb loci discordant = 22, Wilcoxon test $P_{Hs(global)} = 0.024$, $P_{nbAlleles} < 0.001$; Cov2Hmiss_cl: nb loci discordant = 7, Wilcoxon test $P_{Hs(global)} = 0.005$, $P_{nbAlleles} = 0.012$).

Globally, and for each pairwise population comparison, the G''_{ST} showed a higher absolute differentiation value and D_{ST} showed the lowest absolute values, but all showed a strong correlation (r = 0.88–1, Supplementary Materials Figures S5.1 and S5.2). All the differentiation metrics also supported lower absolute value with a clear decrease in the variability among loci for the pairwise differentiation between Sandur vs. Svartárvatn (two northern populations), and between Sandur and Klapparós (Figures S5.1 and S5.2).

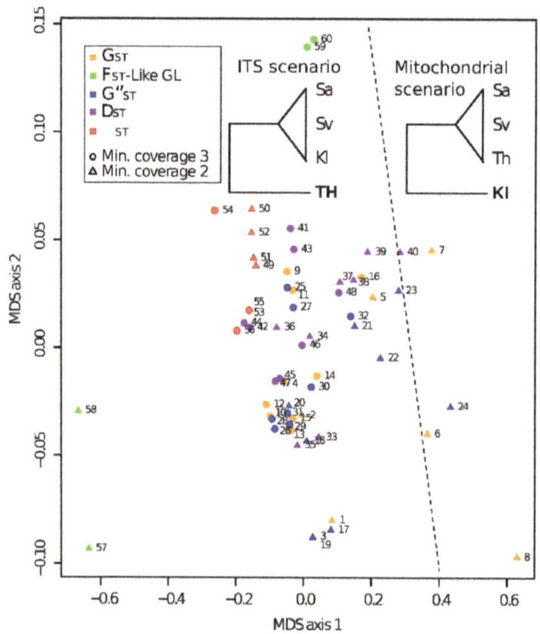

Figure 3. Multidimensional scale plot of the correlations between five differentiation indices based on the divergence among the four sampling locations of *Crangonyx islandicus* and various RADseq datasets. Numbers of RADseq datasets are reported in Table S5.2. Trees present the two alternative scenarios based on the mtDNA and ITS genetic variation. Kl: Klapparós, Sa: Sandur, Sv: Lake Svartárvatn, Th: Lake Thingvallavatn.

Two of the three loci detected under putative selection, the mtDNA NADH dehydrogenase subunit 5 and histone 3 loci, were successfully amplified and sequenced from 73 and 79 individuals, respectively, from 14 sites representing the five mitochondrial clades within Iceland (see Supplementary Materials Table S4.1). The NADH sequences of 620 bp showed the same phylogeographic pattern as found earlier by Kornobis et al. [8], based on COI and 16S, with distinct monophyletic mtDNA lineages in the southwest (A, $n = 33$), south (B/C, $n = 7$) and southeast (D, $n = 25$), which were more similar to each other than to the samples from the north (E, $n = 8$, Supplementary Materials Figure S5.3). Unfortunately, no NADH sequences were obtained from the population from Klapparós in the northeast (F, Table S4.1). The variation in histone 3 was confined to a single SNP (A/G), which was variable in the RADseq data with fixed differences between southern and northern Iceland, with the G allele present only in the population from Lake Thingvallavatn, while the A allele was fixed in the populations from Svartárvatn, Sandur and Klapparós (Figure S5.4). Partial histone H3 sequences (314 bp) from 79 individuals belonging to mtDNA lineages A–F showed similar results with only the G allele in the southwest (clade A, A') and the south (clade B), a large majority of G (44 G and 4 A) in the southeast (clade D) and a huge majority of A (56 A and 4 G) in the north (clade E, F).

3.2.3. Congruence Tests among Loci

Considering all datasets, 57.4 ± 5.0% of the loci on average supported the ITS scenario rather than the mitochondrial scenario for the G_{ST}, G''_{ST} and D_{ST}. This proportion increased to 74.9 ± 4.0% on average when considering the Φ_{ST}. However, no significant difference was found for the extent of differentiation between the two scenarios for any of the datasets and differentiation metrics (Wilcoxon test $p > 0.1$).

3.2.4. Population Structure and Relationship among Groups Using SNP and Haplotype Frequencies

The K-means clustering approach, based on the RADseq data, retained two clusters systematically, except for the haplotype datasets using a minimal coverage of two and after removing the loci in low frequency and under selection. For these two exceptions, the BIC inferred by sequential K-means never reached clear minima and supported the maximum number of clusters allowed (i.e., 10). On average, the first two axes of the DAPC retained 18.3 ± 7.2% of the total variance. The clustering segregated the individuals into two groups, Thingvallavatn (southwestern Iceland) vs. the three populations from northern Iceland. The assignment plot shows that the level of admixture is rather weak between these two groups, and three to five individuals present in Thingvallavatn can be considered admixed depending on the dataset (Figure 4a–c). Regardless of the dataset, considering sampling location as the a priori number of clusters showed higher levels of admixture, especially among the three northern samples (Figure 4b–d).

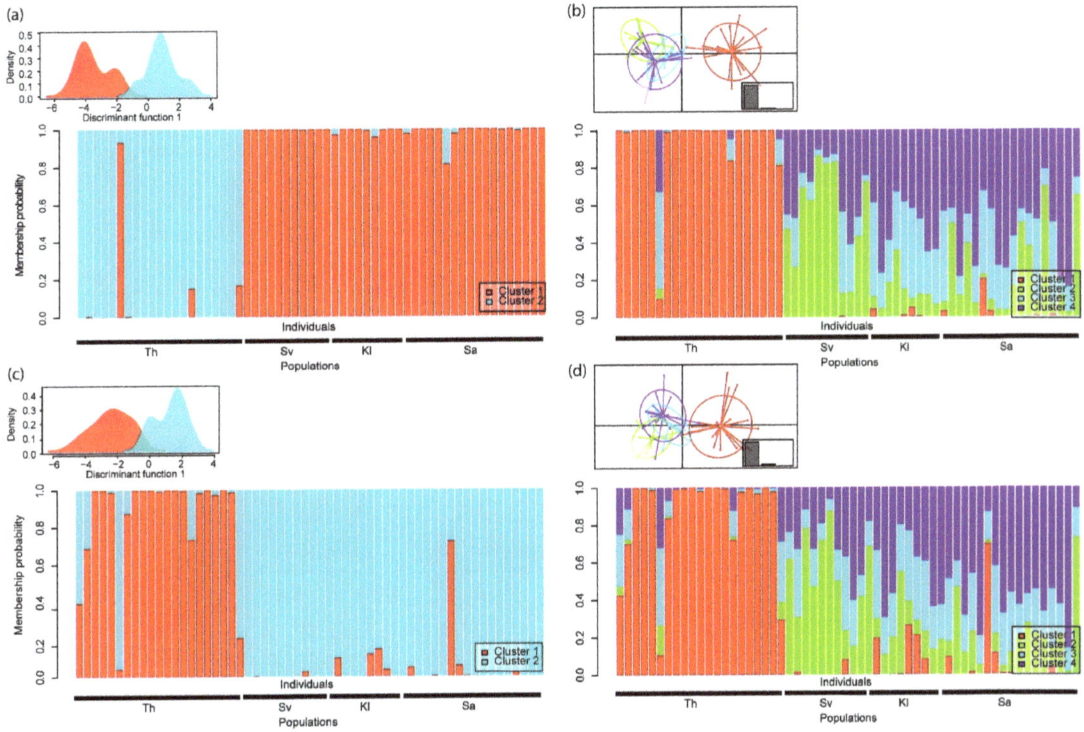

Figure 4. Discriminant analysis of principal components (DAPC) of RADseq data of four sampling locations of *Crangonyx islandicus*. The analysis is based on haplotypes, indels as missing data and with a minimum coverage of three (C3Hmiss dataset). The plots in (**a**,**b**) display the results of the C3Hmiss dataset before removing the loci in low frequency (n = 103) and after (n = 71): (**c**,**d**). The plots in (**a**,**c**) display the results of the optimal number of clusters inferred by a sequential K-means approach according to BIC criteria. The plots in (**b**,**d**) display the results of four sampling locations used as a priori number of clusters. The scatter plot of the DAPC is displayed as small plots and presents the main discriminants axis (1 axis for (**a**,**c**), and 2 axes for (**b**,**d**)), the ellipses represent the inertia of the clusters and the dots represent individuals. The barplots show the assignment probability of each individual to the different clusters. Th: Lake Thingvallavatn, Sv: Lake Svartárvatn, Kl: Klapparós, Sa: Sandur.

The Bayes Factor Delineation (BFD*) approach based on the bi-allelic datasets with the five most informative individuals per location favored the model of four distinct populations (M1), showing a strong support against M2 (three populations with the mitochondrial scenario) and a very strong support against M3 (two populations with the ITS scenario) for both a coverage of two and three (Table 2). However, the topology of the most supported four-population species tree supported clearly the ITS scenario, with the northern populations more closely related together than with the Thingvallavatn population (Figure 5 and Figure S5.5). The posterior probability of the node supporting the northern clade is high (C3SNPr dataset $p = 0.97$, $n = 88$; C2SNPr dataset $p = 1$, $n = 223$), but the node support of the clade E (from Sandur vs. Svartárvatn) is very weak (C3SNPr dataset $p = 0.77$, $n = 88$; C2SNPr dataset $p = 0.59$, $n = 223$). However, the confidence intervals of the date divergence between Thingvallavatn and the three other northern populations overlap, showing a large variance in the coalescence process (Figure 5a). The posterior distributions of θ of the populations overlapped as well and showed similar θ_{median} between 0.10 and 0.13 (Thingvallavatn $\theta_{median} = 0.13$, 95% HPD = 0.08–0.19; Klapparós $\theta_{median} = 0.10$, 95% HPD = 0.05–0.16; Sandur $\theta_{median} = 0.12$, 95% HPD = 0.07–0.19; Svartárvatn $\theta_{median} = 0.11$, 95% HPD = 0.06–0.17, see Figure 5b).

Table 2. Bayes Factors (BF) of the Bayes Factor Delineation (BFD*) approach [75,76] testing the support of three models of population structure of *Crangonyx islandicus* using RADseq data on unlinked bi-allelic SNPs after removing low-frequency SNPs considered as potential paralogs or putatively under selection. We used the scale of Kass and Raftery [77] to evaluate the support of the best model. The support for model 1 is strong for BF > 2, very strong for BF > 6 and decisive for BF > 10, with BF computed using the formula BF = 2*($ML_{model1} - ML_{model2}$), with ML for marginal likelihood expressed in log scale. SD: Standard deviation.

Dataset	Minimum Coverage	Number of Unlinked Bi-Allelic SNPs	Models	Averaged Marginal Likelihood among Runs (SD)	Test	BF
C3SNPr	3	88	1 (4 populations)	−549.91 (0.09)	M1 vs. M2	9.84
			2 (3 populations mitochondrial scenario)	−554.83 (0.20)	M2 vs. M3	71.54
			3 (2 populations ITS scenario)	−590.61 (0.16)	M1 vs. M3	81.38
C2SNPr	2	223	1 (4 populations)	−1245.15 (0.14)	M1 vs. M2	179.02
			2 (3 populations mitochondrial scenario)	−1334.59 (0.09)	M2 vs. M3	332.53
			3 (2 populations ITS scenario)	−1500.91 (0.02)	M1 vs. M3	511.55

3.2.5. Introgression Tests

The introgression tests based on the D statistic showed a comparable amount of introgression among the northern populations of C. islandicus (from Sandur and Svartárvatn) with either the southern (from Thingvallavatn) or the northeastern population (from Klapparós). A test based on the conservative dataset, with a minimal coverage of three (including 21 to 45 loci) and the mitochondrial scenario with Klapparós population as an outgroup, supported an introgression between Thingvallavatn and Sandur populations, in the genetic combinations of 34 individuals for [Sv,Sa,Th][Kl] but 28 individuals when the structure of the ingroup in the test was changed [Sa,Sv,Th][Kl] (Table 3). Only one combination supported an introgression between Thingvallavatn and Svartárvatn populations. The tests based on the ITS scenario (i.e., with the Thingvallavatn population as an outgroup) showed a similar support of introgression among the northern populations with 28 or 25 combinations of individuals, indicating an introgression between Svartárvatn and Klapparós populations, but only one combination between Sandur and Klapparós populations (Table 3). However, when considering additional loci with a minimal coverage of two (between 62 and 128 loci were used for comparisons), a small number of combinations of individuals supported an introgression between Thingvallavatn and Sandur (five), Thingvallavatn and Svartárvatn (four or five), Klapparós and Svartárvatn (five or seven) and Klapparós and Sandur populations (three or four, Table 3).

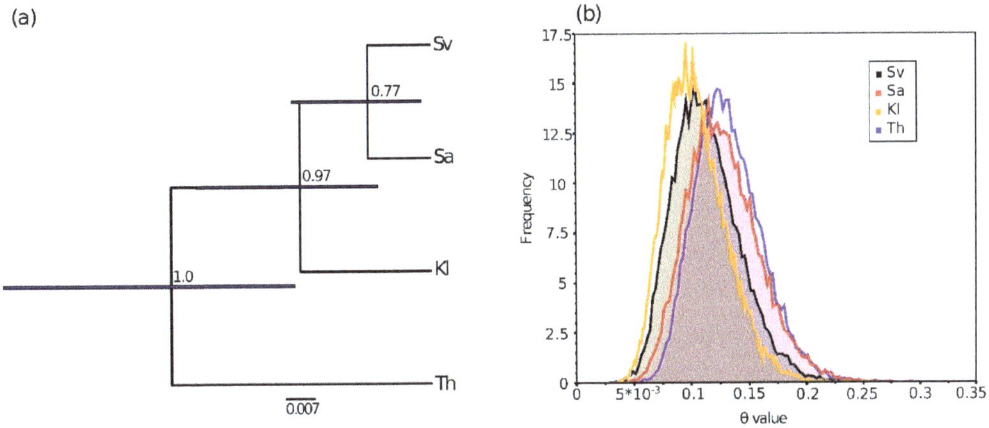

Figure 5. Divergence of RADseq genotypes of *Crangonyx islandicus* from the four sampling locations. (**a**) Maximum clade credibility tree (MCCT) and (**b**) posterior distribution of theta (θ = Ne*μ) parameters inferred by SNAPP [73] according to the Bayes Factor Delineation's (BFD* [75,76]) best scenario for the four populations using 88 unlinked bi-allelic SNPs dataset (C3SNPr). Node number on the MCCT displays the posterior probability of the node, and the dark blue node bar indicates the 95% HPD interval of the node age. Th: Lake Thingvallavatn, Sv: Lake Svartárvatn, Kl: Klapparós, Sa: Sandur.

Table 3. Results of the four-taxon D statistic testing for introgression among sampling locations of *Crangonyx islandicus* using RADseq data. P1, P2 and P3 are the populations under investigation for introgression and O is the outgroup. Range Z is the range of the Z value, upper values deviated significantly from 0 ($p < 0.01$). NSign/Ntot for positive D reports the number of significant replicates (combination of individuals) over the total number of tests for positive D statistic indicating an introgression between P2 and P3. NSign/Ntot for negative D indicates an introgression between P1 and P3. Range NbLoci is the range of loci used among replicates to test the D statistic. Th: Lake Thingvallavatn, Sv: Lake Svartárvatn, Kl: Klapparós, Sa: Sandur. Min. cov. presents the minimum coverage of datasets 3 (Cov3Hmiss_cl) and 2 (Cov2Hmiss_cl).

Min Cov.	Test	P1	P2	P3	O	Range Z	NSign/Ntot for Positive D	NSign/Ntot for Negative D	Range NbLoci
3	Mito.1	Sv	Sa	Th	Kl	0–10.1	28/125	0/125	33–45
3	Mito.2	Sa	Sv	Th	Kl	0–10.1	1/125	34/125	33–45
3	ITS.1	Sv	Sa	Kl	Th	0–10.1	28/125	0/125	21–43
3	ITS.2	Sa	Sv	Kl	Th	0–8.2	25/125	1/125	21–43
2	Mito.1	Sv	Sa	Th	Kl	0–7.7	5/125	4/125	101–128
2	Mito.2	Sa	Sv	Th	Kl	0–8.2	5/125	5/125	101–128
2	ITS.1	Sv	Sa	Kl	Th	0–6.4	7/125	3/125	62–128
2	ITS.2	Sa	Sv	Kl	Th	0–6.0	5/125	4/125	62–128

4. Discussion

The current study showed the robustness of the mito-nuclear discordance of the phylogeographic history of *C. islandicus*, as newly sequenced individuals from previously described populations for mitochondrial DNA confirmed the previous mitogenomic findings by Kornobis et al. [8], while a multitude of nuclear genomic markers sequenced through ddRADseq provided further support to the ITS scenario [9]. A structured coalescent analysis using the mitochondrial 16S and CO1 sequences from the same four populations assessed with ddRADseq data confirmed the previous mitochondrial pattern and supports a clear pre-Ice Age divergence between the northeastern Icelandic (from Klapparós) population and the northern and southwestern populations. Despite the use of multiple ddRADseq analytical approaches designed to overcome potential misinterpretations, including different data types (SNPs, haplotypes, DNA sequences), different dataset configurations (brute vs. clean), base call uncertainty and various metrics, most approaches (differentiation indices, DAPC, BFD*) were congruent and supported the ITS scenario, with the main divergence occurring between the northern vs. southern populations. We showed that a higher proportion of the RADseq loci followed the ITS scenario compared to the mitochondrial scenario, which was still strongly recovered by a small proportion of RADseq loci. The admixture revealed by DAPC was higher among northern populations than between southern and northern populations (admixture was found for several individuals though), which globally provides support to the ITS scenario. A specific approach used to disentangle the contribution to ILS and introgression/ancestral hybridization related to these discordant genealogies showed a comparable introgression signal between populations in northern (from Sandur and Svartárvatn) and northeastern Iceland (from Klapparós) and between the population in the south (from Thingvallavatn) and the ones from the north. Overall, despite the ITS scenario receiving greater support from the RADseq data, our results suggest a complex scenario involving not only incomplete lineage sorting but also possibly introgression led by male-biased dispersal among northern locations (but see the discussion below) or mitochondrial capture and then the accelerated isolation of the mitochondrial DNA among the northeastern, southern and southwestern populations possibly strengthened by selection, a scenario that would require additional research to be fully tested.

4.1. Methodological Uncertainties with RADseq Data

As a genome reduction approach, RADseq analysis assesses variation from small genomic fragments sampled from throughout the genome without any a priori information about their mode of inheritance, location, substitution rate or recombination rate. Overall, the RADseq loci may present an important variability among all these characteristics that may affect commonly used summary statistics in population genetics when population differentiations and other population parameters are estimated. Our results reveal no systematic bias for a particular differentiation metric; however, they confirm the potential higher sensitivity of G_{ST} to loci with high genetic diversity and may thus affect the estimate of population structure as the proportion of such markers increases (cf. [30]). Only five out of sixty statistics calculated for the subdivision of the populations supported the mitochondrial scenario; the other supported the ITS scenario. Of the five statistics supporting the mtDNA, three were based on GST, which does not take into account the dependency of the statistic on the overall variation as G''_{ST} and D_{ST} do, which may thus give a biased overall estimate. Furthermore, these five methods are based on the lower coverage datasets (minimal coverage two), which included a higher number of loci but at the price of higher base call uncertainty, and four of them included indels as the 5th state. Higher coverage and omitting indels is a more conservative approach, as increasing coverage offers a more robust estimate of the DNA sequence polymorphisms and indels can lead to problems in alignment and may reflect errors in sequencing [80]. Thus, those five approaches (dataset and metrics) supporting the mtDNA-scenario may be clearly suboptimal to assessing the population structure. Specific methods accounting for uncertainty related to low-coverage data (regardless of the minimal coverage) based on a maximum likelihood base call estimate (ANGSD) confirmed the ITS population structure scenario, also inferred by most classic approaches not accounting for such a bias.

Methods relying on genotype likelihood instead of SNP calling to infer population statistics offer great promise to considerably decrease the bias caused by base call uncertainty, even for very low-coverage datasets (i.e., x2) but remain limited to estimating basic summary statistics [27,39,41,43]. However, our results suggested that the potential noise brought by low-coverage data may not be strong enough to blur the main phylogeographic signal. These results are in agreement with the findings of several studies [36,37], reporting that when facing the trade-off of sampling more individuals per population or increasing the per sample depth coverage to infer population parameters, the former offers the best strategies because each sequence read from a new individual brings more information than additional reads (considering that the reads with a low coverage are discarded for the analyses) already present in the pool [36]. In addition, the trade-off between the number of individuals per population vs. the depth coverage of the DNA loci can be further tangled by the genome size of the organism [81]. Indeed, organisms with large genomes may increase the number of restriction sites and thus DNA loci, which overall limit the number of read copies per loci.

Although the genome size of *C. islandicus* is currently unknown, we suspect it could be large, which could partly explain the low coverage of our dataset. Indeed, crustacean and amphipod species in particular are prone to gigantism in their genome size, especially species supposedly with low metabolic rates [82] present in cold water environments such as the Arctic [83,84] and deep waters from Lake Baikal [85], which share many similarities with the groundwater habitats of *C. islandicus* [8,45,46]. Moreover, groundwater species tend to have larger genome sizes than their surface sister species due to the reduction in effective population size [86].

When the information in a DNA sequence was reduced to a haplotype or a single SNP per locus, the results did not show any systematic bias toward a particular scenario. However, the use of substitution models including the DNA sequence of the fragments such as in the Φ_{ST}-index, showed consistently robust patterns regardless of coverage (2 or 3), or when removing loci in low frequency, potential paralogs or loci under selection. Further, our results confirm the increase in power to detect candidate loci showing the signature of selection when using haplotype data inferred from the full sequence of the RADseq loci in comparison with the use of a single SNP per locus [87,88]. Considering indels as a fifth state when building the haplotypes might provide additional power to further detect loci under selection, as suggested by our results.

4.2. The Phylogeographic Pattern of *Crangonyx islandicus*

We have shown here that different analyses based on the extended mtDNA dataset and a larger nuclear dataset (ddRADseq) than in [8,9] confirmed the different mitochondrial and nuclear phylogeographic patterns in *C. islandicus*. Mito-nuclear discordance among structured populations or closely related taxa has been detected in several taxonomic groups [1,6,7], including a groundwater species complex belonging to genus *Crangonyx Bate*, 1859 [89]. Disentangling the phylogeographic scenario and its causes remains extremely difficult, as many mechanisms can lead to the same patterns. However, from the previous and present results it is clear that this groundwater species diverged within Iceland through the Pleistocene, when Iceland was repeatedly covered by glaciers [46,90]. The main split observed for most genomic markers, the ITS scenario, between southern and northern Iceland, reflects different watersheds separated by the interior highlands of the country, whereas the earliest split in the mtDNA was between the northern and northeastern areas followed by the subsequent structuring of the southeast, south and southwest of the country from an ancestor of the northern population. We propose below tentative and non-mutually exclusive phylogeographic scenarios implying several mechanisms commonly proposed in the literature to explain such complex mito-nuclear discordant patterns [6,7] with some explanation about their likelihood.

A first scenario, based only on stochastic forces driven by a strong incomplete lineage sorting between the mitochondrial and most nuclear markers, sounds appealing at first but it seems relatively unlikely. Indeed, mito-nuclear discrepancies may be caused by the lower population effective size (Ne) of the mitogenome (1/4 of the nuclear), which makes it more likely to be quickly reciprocally monophyletic (shorter coalescence times due to faster loss of ancestral polymorphisms) [6]. Mitochondrial DNA has thus been referred to as a 'leading' indicator, rather than 'lagging' as for nuclear markers, which more often show paraphyletic or polyphyletic relationships because nuclear markers take more time to coalesce [1]. In addition to a less effective population size (Ne) in mtDNA, background selection or hitch-hiking might further reduce Ne or the variation within populations, e.g. [13], and thus to increased population differentiation, as estimated with F_{ST}, which is not yet apparent with nuclear DNA, especially if the ancestral population had a large Ne. The decoupling of the mitochondrial vs. nuclear substitution rates has been reported in many phyla [91], but a recent study showed a lower nuclear substitution rate for subterranean organisms than for surface organisms due to a lower metabolic rate and reduced genome replication rate, while the substitution rate of the mitochondrial genome remains unaffected [92]. This would further increase the differences in substitution rates between the mitochondrial and nuclear genomes for subterranean taxa. The populations from the north (Sandur and Svartárvatn) and northeast (Klapparós), currently located on each side of the volcanic active zone characterized by a high geothermal area and volcanic activity north of Lake Mývatn, may share a common origin further south in the interior of the country. The divergence between these northern and northeastern populations might be too recent for the nuclear DNA that has not yet achieved a complete lineage sorting (i.e., ILS) and showed strong admixture, while the mitochondrial genome reaches reciprocal monophyly much more quickly.

Even though ILS may have played a role in the mito-nuclear discordance, we propose a second scenario involving introgression among populations previously isolated due to the opening of dispersal corridors that may not be conducive anymore. Such a scenario might be especially prevalent in the interior of the country, where geographical distances are shorter and individuals from populations characterized by different mtDNA lineages are more likely to meet. Furthermore, clear phylogeographic mito-nuclear discordant patterns usually rule out the ILS scenario [6]. Indeed, it is difficult to explain the strong and discordant phylogeographic structure among the populations, especially the main and deep split of the nuclear DNA between the northern and southern populations (i.e., Thingvallavatn) without involving introgression.

The DAPC admixture plots and introgression test results supported a certain amount of ancestral hybridization, especially between the populations in the north (from Sandur and Svartárvatn) and the northeast (from Klapparós), but also between the populations in the south (Thingvallavatn) and the north, which implies the presence of secondary contacts [13,26,79] and thus the opening of dispersal corridors. The two events may though have led to different incongruences between the mtDNA and the nuclear genomes. The discordant pattern between the northern and northeastern populations suggests that introgression occurred mostly on the nuclear genome and not on the mitochondrial genome that shows a clear phylogeographic pattern of reciprocal monophyly. This might have resulted from male-biased dispersal, but no males have been identified within the species [44] and the sex differences may be subtle. The mtDNA from the Klapparós population might also have been captured and replaced by a distinct unsampled ancestral population.

In the study by Kornobis et al. [8], the largest variation was observed at higher altitudes, suggesting these sites were the sources for populations at lower altitudes (<100 m), which were below sea level about 10,000 years ago (see references in [8]). Both mtDNA and RADseq data showed a close genetic proximity among the two northern populations from Sandur in Adaldal and Lake Svartárvatn, which may be explained by the recent colonization of Sandur, which is a low altitude area that was submerged by sea at the end of the last glacial period of the Ice Age about 10 kyr ago. Sandur is likely to have been colonized from the areas around lakes (Svartárvatn and/or Mývatn) further south. The sites are linked today by rivers and major lava fields, which ran to the north along the valleys Bárdardalur (9500 and 9000 years ago) and Laxárdalur (5000 and 2200 years ago) [93]. It is though noteworthy that samples from the interior of Iceland, where the different evolutionary lineages are separated by shorter distances and where crossing may have occurred, are scarce. The clustering of the H3 gene from the small sample from the highlands, Kjölur (mtDNA C), with southeast Iceland (mtDNA D) indicates that reproduction between the two lineages may have happened there. A similar scenario might have happened south of Lake Mývatn, where the ancestral admixture of samples now inhabiting the sampled sites in northern and northeastern Iceland might have occurred. Although the divergence in mtDNA between samples from different locations follows well the geographic distances and geological features, there are some exceptions that point to the fact that dispersal can occur via rivers on the surface or along lava fields, whose topology and thus migratory routes are likely to have changed considerably over time during the different glacial and interglacial periods of Iceland, subsidence, post-glacial rebound and volcanic activities. One example is the distribution of the BC mtDNA haplotypes, which are found in separate regions of the volcanic zones in southern Iceland [8].

Ancient introgression followed by isolation and the differentiation of the mtDNA lineages among northern populations offers an appealing scenario for the split among northern populations but it does not yet explain why nuclear DNA showed the main and deep split between the northern and southern populations (i.e., from Lake Thingvallavatn). Interestingly, both the mtDNA and RADseq data showed that the southwestern population from Thingvallavatn harbored the highest theta (a proxy of Ne assuming a constant mutation rate), especially for the mtDNA, which was almost four and seven times higher than the theta of the populations from the northeast (Klapparós) and the north (Sandur and Svartárvatn), respectively. Populations with the highest Ne or genetic diversity usually suggest a long presence of a large population (center of diversification) or they can also correspond to a secondary contact zone among previously isolated populations [94,95]. Southwestern Iceland (i.e., from Thingvallavatn) represents a large area of potential habitats for *C. islandicus,* which may have maintained several populations through time [8]. The genetic diversity of the Lake Thingvallavatn population, and thus its Ne, may have been fed by regular connections and secondary contacts with peripheral populations belonging to the mitochondrial clades A, A′ and B [8]. Those peripheral populations might be regularly connected and isolated depending on the opening/closing of fissures due to geothermal and tectonic activities in the regions [8]. The long presence of the southwestern population, and possibly northeastern populations, could explain the deep divergence in the nuclear genome. Moreover, the closer similarity of the mtDNA clades in northern and southern populations compared to the one in the northeast could also indicate an ancestral hybridization, where mtDNA was more likely to introgress or be captured by one lineage [6] but has then diverged more rapidly than the nuclear genome due to smaller Ne, which could be strengthened by selection [7,96].

5. Conclusions

Well-supported inferences about the phylogeographic patterns of populations rely on good knowledge about the genomic and geographic variation within species. Here we show that the genome-wide markers obtained with RADseq analysis provide greater support to previous results on the nuclear ITS region [9] rather than the results from the mitochondrial analyses, albeit a large variation among loci. The main genetic divergence within *C. islandicus* follows the split between northern and southern Iceland, but the analyses indicate the evidence of introgression between the main evolutionary lineages within the country. The analysis of whole genome variation could further give insights into the evolutionary histories of these populations. Although RADseq analyses provide information from a large number of markers, information on their location within the genome could allow further inferences, which could take into account the possible impacts of varying effective size and selection due to variation in ploidy, structural variation or linkage. Furthermore, increased geographic sampling, especially from the interior of

Iceland, may reveal more mixed populations and thus insight into the history of the populations that have been studied and are mostly found in the lowland areas of the country. Overall, our study paved the way to combine large numbers of both nuclear and mitochondrial markers to disentangle species' evolutionary history and revisit previous mitochondrial phylogeographic scenarios, especially among closely related species, as a more complex picture may emerge.

Supplementary Materials: The following supporting information can be downloaded at: https://www.mdpi.com/article/10.3390/d15010088/s1, Supplementary Materials S1—Sampling information; Table S1.1: Sampling information on 59 individuals of *Crangonyx islandicus* obtained for RADseq analyses and to complement mitochondrial analyses; Table S1.2: Information on 37 individuals of *Crangonyx islandicus* extracted from Kornobis et al. [1] used to complete the MultiTypeTree structured coalescent analysis. Supplementary Materials S2—Adapter and barcode sequences; Supplementary Materials S3—Protocol of the SYBR gold fluorometric assay for dsDNA quantitation; Supplementary Materials S4—Additional analysis of the NADH and H3 loci showed indication of selection in RADseq analysis [97–99]. Table S4.1: Samples obtained for sanger sequencing for NADH dehydrogenase subunit 5 and the histone 3 (H3), which showed indication of selection in RADseq analysis. Figure S4.1: Sampling sites of *Crangonyx islandicus* in Iceland. Further description of the sites is given in Table S4.1. Dark grey zone presents the volcanic zone while glaciers are in light grey. Table S4.2: Primers used for amplification of mtDNA NADH dehydrogenase subunit 5 and the histone exon 3 (H3). The sequence length of NADH5 was 760 bp and of H3 was 314 bp. Supplementary Materials S5—Additional results. Table S5.1: Blastn results of the RADseq loci detected under selection by Bayescan. Table S5.2: Overall pairwise differentiation measures among populations of *Crangonyx islandicus* inferred using RADseq data allowing the correspondence between the labels in Figure 3 and the type of data used. Bold values show significant *p*-value at $\alpha = 0.05$ pairwise G_{ST}, G''_{ST}, D_{ST} and Φ_{ST} among populations using a combined Fisher's *p*-value among loci corrected for a multiple testing using Holm's formula. Clean_Dataset: using a brute (br) dataset coming from pyRAD or a cleaned (cl) dataset removing loci with a Minimum Allele Frequency (MAF) below 5%, with an observed heterozygosity above 0.5 and detected under selection by Bayescan. Th: Lake Thingvallavatn, Sv: Lake Svartárvatn, Kl: Klapparós, Sa: Sandur. Figure S5.1: Global and pairwise differentiation values of each RADseq locus among the four populations of *Crangonyx islandicus*. D_{ST}, G_{ST} and G''_{ST} are computed using haplotype information of a dataset considering indels as missing data and a minimal coverage of the RADSeq loci of 3 (C3Hmiss dataset). The Φ_{STs} were computed using the sequences of the same dataset. The dataset C3Hmiss before (a) ($n = 103$), and after (b) removing the loci in low frequency, considered as potential paralogs or under selection ($n = 71$). Th: Lake Thingvallavatn, Sv: Lake Svartárvatn, Kl: Klapparós, Sa: Sandur. The black bands display the median distance, the box represents the interquartile distance, the whiskers are long up to 1.5 the interquartile distance while the black dots represent the outliers. Figure S5.2: Global and pairwise differentiation values of each RADseq locus among the four populations of *Crangonyx islandicus*. D_{ST}, G_{ST} and G''_{ST} are computed using haplotype information of a dataset considering indels as missing data and a minimal coverage of the RADSeq loci of 2 (Cov2Hmiss dataset). The Φ_{STs} were computed using the sequences of the same dataset. The dataset Cov2Hmiss before (a) ($n = 295$), and after (b) removing the loci in low frequency, considered as potential paralogs or under selection ($n = 202$). Th: Lake Thingvallavatn, Sv: Lake Svartárvatn, Kl: Klapparós, Sa: Sandur. The black bands display the median distance, the box represents the interquartile distance, the whiskers are long up to 1.5 interquartile distance while the black dots represent the outliers. Figure S5.3: Phylogenetic tree of 73 individuals of *Crangonyx islandicus* based on 620 bp of the partial mtDNA NADH dehydrogenase subunit 5 gene detected as putatively under selection by Bayescan. This maximum likelihood tree was built with Phyml [98] using a GTR evolution model. Tips represent the different haplotypes and tip labels show the label of the site location as described in Table S4.1 and the number of individuals (i.e., frequencies) belonging to a given haplotype is provided within brackets. The branch support (aLRT) displayed at nodes was obtained with the difference in likelihoods with and without the respective branch. Geographic origin is presented with vertical bars using the following letters: SV: southwestern Iceland, S: southern Iceland, SA: southeastern Iceland and N: northern Iceland. Site G is from Kjölur, central Iceland (see Figure S4.1). Figure S5.4: Phylogenetic tree of 79 individuals of *Crangonyx islandicus* based on 374 bp of partial histone 3 (H3) gene detected as putatively under selection by Bayescan. This maximum likelihood tree was built with Phyml [98] using a GTR evolution model. Tips represent the different haplotypes and tip labels show the label of the site location as described in Table S4.1 and the number of individuals (i.e., frequencies) belonging to a given haplotype is provided within brackets. The branch support (aLRT) displayed at nodes was obtained with the difference in likelihoods with and without the respective branch. Geographic origin is presented with vertical bars using the following letters: SV: southwestern Iceland, S: southern Iceland, SA: southeastern Iceland and N: northern Iceland. Figure S5.5: (**a**) Maximum clade credibility tree (MCCT) and (**b**) posterior distribution of theta (θ = Ne*μ) parameters inferred by SNAPP [74] according to the Bayes Factor Delineation's (BFD* [76]) best scenario for the four populations of *Crangonyx islandicus* using 223 unlinked bi-allelic SNP dataset (C2SNPr). Node number on the MCCT displays the posterior probability of the node and the dark blue node bar presents the 95% HPD interval of the node age.

Author Contributions: D.E. and S.P. performed the conceptualization and designed the study, B.K.K., S.P. and D.E. performed the sampling and the morphological identification of the individuals, K.M.W., D.E. and B.M. performed the laboratory work, D.E., K.M.W. and S.P. defined the methodology, D.E. and S.P. analyzed the data, D.E. performed the visualization, D.E. and S.P. led the writing, S.P. secured the funding. All authors have read and agreed to the published version of the manuscript.

Funding: The study was supported by Research Grant No. 130244-05, funded by the Icelandic Research Fund.

Data Availability Statement: Publicly available datasets were analyzed in this study. The data and parameter files, as well as R scripts and bash codes used to perform the analyses of the mitochondrial and the RADseq data, are available at: the Zenodo Digital Repository https://doi.org/10.5281/zenodo.7506345. The RADseq data are available in NCBI with the BioProject number PRJNA917071. The new CO1, 16S, NADH5 and H3 DNA sequences have been deposited in GenBank and accession numbers are available in Supplementary Materials Tables S1.1 and S4.1.

Acknowledgments: We would like to thank Christophe J. Douady, Tristan Lefébure and Agnes-Katharina Kreiling for their help in the field.

Conflicts of Interest: The authors declare no conflict of interest.

References

1. Zink, R.M.; Barrowclough, G.F. Mitochondrial DNA under siege in avian phylogeography. *Mol. Ecol.* **2008**, *17*, 2107–2121. [CrossRef] [PubMed]
2. Avise, J.C. Phylogeography: Retrospect and prospect. *J. Biogeogr.* **2009**, *36*, 3–15. [CrossRef]
3. Galtier, N.; Nabholz, B.; Glémin, S.; Hurst, G.D.D. Mitochondrial DNA as a marker of molecular diversity: A reappraisal. *Mol. Ecol.* **2009**, *18*, 4541–4550. [CrossRef] [PubMed]
4. Garrick, R.C.; Bonatelli, I.A.S.; Hyseni, C.; Morales, A.; Pelletier, T.A.; Perez, M.F.; Rice, E.; Satler, J.D.; Symula, R.E.; Thomé, M.T.C.; et al. The evolution of phylogeographic data sets. *Mol. Ecol.* **2015**, *24*, 1164–1171. [CrossRef] [PubMed]
5. Avise, J.C.; Arnold, J.; Ball, R.M.; Bermingham, E.; Lamb, T.; Neigel, J.E.; Reeb, C.A.; Saunders, N.C. Intraspecific phylogeography: The mitochondrial DNA bridge between population genetics and systematics. *Annu. Rev. Ecol. Evol. Syst.* **1987**, *18*, 489–522. [CrossRef]
6. Toews, D.P.L.; Brelsford, A. The biogeography of mitochondrial and nuclear discordance in animals. *Mol. Ecol.* **2012**, *21*, 3907–3930. [CrossRef]
7. Bonnet, T.; Leblois, R.; Rousset, F.; Crochet, P.A. A reassessment of explanations for discordant introgressions of mitochondrial and nuclear genomes. *Evolution* **2017**, *71*, 2140–2158. [CrossRef]
8. Kornobis, E.; Pálsson, S.; Kristjánsson, B.K.; Svavarsson, J. Molecular evidence of the survival of subterranean amphipods (Arthropoda) during Ice Age underneath glaciers in Iceland. *Mol. Ecol.* **2010**, *19*, 2516–2530. [CrossRef]
9. Kornobis, E.; Pálsson, S. Discordance in variation of the ITS region and the mitochondrial COI gene in the subterranean amphipod *Crangonyx islandicus*. *J. Mol. Evol.* **2011**, *73*, 34–44. [CrossRef]
10. Funk, D.J.; Omland, K.E. Species-level paraphyly and polyphyly: Frequency, causes, and consequences, with insights from animal mitochondrial DNA. *Annu. Rev. Ecol. Evol. Syst.* **2003**, *34*, 397–423. [CrossRef]
11. McCracken, K.G.; Sorenson, M.D. Is homoplasy or lineage sorting the source of incongruent mtDNA and nuclear gene trees in the stiff-tailed ducks (*Nomonyx-Oxyura*)? *Syst. Biol.* **2005**, *54*, 35–55. [CrossRef] [PubMed]
12. Petit, R.J.; Excoffier, L. Gene flow and species delimitation. *Trends Ecol. Evol.* **2009**, *24*, 386–393. [CrossRef] [PubMed]
13. Pease, J.B.; Hahn, M.W. Detection and polarization of introgression in a five-taxon phylogeny. *Syst. Biol.* **2015**, *64*, 651–662. [CrossRef] [PubMed]
14. Charlesworth, B.; Jensen, J.D. Effects of selection at linked sites on patterns of genetic variability. *Annu. Rev. Ecol. Evol. Syst.* **2021**, *52*, 177–197. [CrossRef]

15. Palsson, S.; Pamilo, P. The effects of deleterious mutations on linked neutral variation in small populations. *Genetics* **1999**, *153*, 475–483. [CrossRef]
16. Moritz, C.; Potter, S. The importance of an evolutionary perspective in conservation policy planning. *Mol. Ecol.* **2013**, *22*, 5969–5971. [CrossRef]
17. Davey, J.W.; Hohenlohe, P.A.; Etter, P.D.; Boone, J.Q.; Catchen, J.M.; Blaxter, M.L. Genome-wide genetic marker discovery and genotyping using next-generation sequencing. *Nat. Rev. Genet.* **2011**, *12*, 499–510. [CrossRef]
18. Andrews, K.; Good, J.; Miller, M.R.; Luikart, G.; Hohenloe, P.A. Harnessing the power of RADseq for ecological and evolutionary genomics. *Nat. Rev. Genet.* **2016**, *17*, 81–92. [CrossRef]
19. Benestan, L.; Gosselin, T.; Perrier, C.; Sainte-Marie, B.; Rochette, R.; Bernatchez, L. RAD genotyping reveals fine-scale genetic structuring and provides powerful population assignment in a widely distributed marine species, the American lobster (*Homarus americanus*). *Mol. Ecol.* **2015**, *24*, 3299–3315. [CrossRef]
20. Pavey, S.A.; Gaudin, J.; Normandeau, E.; Dionne, M.; Castonguay, M.; Audet, C.; Bernatchez, L. RAD sequencing highlights polygenic discrimination of habitat ecotypes in the panmictic American eel. *Curr. Biol.* **2015**, *25*, 1666–1671. [CrossRef]
21. Emerson, K.J.; Merz, C.R.; Catchen, J.M.; Hohenlohe, P.A.; Cresko, W.A.; Bradshaw, W.E.; Holzapfel, C.M. Resolving postglacial phylogeography using high-throughput sequencing. *Proc. Natl. Acad. Sci. USA* **2010**, *107*, 16196–16200. [CrossRef]
22. Catchen, J.; Bassham, S.; Wilson, T.; Currey, M.; O'Brien, C.; Yeates, Q.; Cresko, W.A. The population structure and recent colonization history of Oregon threespine stickleback determined using restriction-site associated DNA-sequencing. *Mol. Ecol.* **2013**, *22*, 2864–2883. [CrossRef] [PubMed]
23. Eaton, D.A.R.; Ree, R.H. Inferring phylogeny and introgression using RADseq data: An example from flowering plants (*Pedicularis*: Orobanchaceae). *Syst. Biol.* **2013**, *62*, 689–706. [CrossRef] [PubMed]
24. Hipp, A.L.; Eaton, D.A.R.; Cavender-Bares, J.; Fitzek, E.; Nipper, R.; Manos, P.S. A framework phylogeny of the American oak clade based on sequenced RAD data. *PLoS ONE* **2014**, *9*, e93975. [CrossRef] [PubMed]
25. Takahashi, T.; Nagata, N.; Sota, T. Application of RAD-based phylogenetics to complex relationships among variously related taxa in a species flock. *Mol. Phylogenet. Evol.* **2014**, *80*, 137–144. [CrossRef] [PubMed]
26. Eaton, D.A.R.; Hipp, A.L.; González-Rodríguez, A.; Cavender-Bares, J. Historical introgression among the American live oaks and the comparative nature of tests for introgression. *Evolution* **2015**, *69*, 2587–2601. [CrossRef]
27. Nevado, B.; Ramos-Onsins, S.E.; Perez-Enciso, M. Resequencing studies of non-model organisms using closely related reference genomes: Optimal experimental designs and bioinformatics approaches for population genomics. *Mol. Ecol.* **2014**, *23*, 1764–1779. [CrossRef]
28. Hedrick, P.W. A standardized genetic differentiation measure. *Evolution* **2005**, *59*, 1633–1638.
29. Jost, L. GST and its relatives do not measure differentiation. *Mol. Ecol.* **2008**, *17*, 4015–4026. [CrossRef]
30. Meirmans, P.G.; Hedrick, P.W. Assessing population structure: FST and related measures. *Mol. Ecol. Resour.* **2011**, *11*, 5–18. [CrossRef]
31. Verity, R.; Nichols, R.A. What is genetic differentiation, and how should we measure it—G_{ST}, D, neither or both? *Mol. Ecol.* **2014**, *23*, 4216–4225. [CrossRef] [PubMed]
32. Weir, B.S.; Cockerham, C.C. Estimating F-statistics for the analysis of population structure. *Evolution* **1984**, *38*, 1358–1370. [PubMed]
33. Nei, M. Analysis of gene diversity in subdivided populations. *Proc. Natl. Acad. Sci. USA* **1973**, *70*, 3321–3323. [CrossRef] [PubMed]
34. Whitlock, M.C. G_{ST} and D do not replace F_{ST}. *Mol. Ecol.* **2011**, *20*, 1083–1091. [CrossRef]
35. Crawford, J.E.; Lazzaro, B.P. Assessing the accuracy and power of population genetic inference from low-pass next-generation sequencing data. *Front. Genet.* **2012**, *3*, 66. [CrossRef]
36. Buerkle, C.A.; Gompert, Z. Population genomics based on low coverage sequencing: How low should we go? *Mol. Ecol.* **2013**, *22*, 3028–3035. [CrossRef]
37. Fumagalli, M. Assessing the effect of sequencing depth and sample size in population genetics inferences. *PLoS ONE* **2013**, *8*, e79667. [CrossRef]
38. Nielsen, R.; Korneliussen, T.; Albrechtsen, A.; Li, Y.; Wang, J. SNP calling, genotype calling, and sample allele frequency estimation from new-generation sequencing data. *PLoS ONE* **2012**, *7*, e37558. [CrossRef]
39. Fumagalli, M.; Vieira, F.G.; Korneliussen, T.S.; Linderoth, T.; Huerta-Sánchez, E.; Albrechtsen, A.; Nielsen, R. Quantifying population genetic differentiation from next-generation sequencing data. *Genetics* **2013**, *195*, 979–992. [CrossRef]
40. Han, E.; Sinsheimer, J.S.; Novembre, J. Characterizing bias in population genetic inferences from low-coverage sequencing data. *Mol. Biol. Evol.* **2014**, *31*, 723–735. [CrossRef]
41. Fumagalli, M.; Vieira, F.G.; Linderoth, T.; Nielsen, R. NgsTools: Methods for population genetics analyses from next-generation sequencing data. *Bioinformatics* **2014**, *30*, 1486–1487. [CrossRef] [PubMed]
42. Korneliussen, T.S.; Moltke, I.; Albrechtsen, A.; Nielsen, R. Calculation of Tajima's D and other neutrality test statistics from low depth next-generation sequencing data. *BMC Bioinform.* **2013**, *14*, 289. [CrossRef] [PubMed]
43. Korneliussen, T.S.; Albrechtsen, A.; Nielsen, R. OpenaAccess ANGSD: Analysis of next generation sequencing data. *BMC Bioinform.* **2014**, *15*, 356. [CrossRef] [PubMed]

44. Vieira, F.G.; Lassalle, F.; Korneliussen, T.S.; Fumagalli, M. Improving the estimation of genetic distances from next-generation sequencing data. *Biol. J. Linn. Soc.* **2016**, *117*, 139–149. [CrossRef]
45. Svavarsson, J.; Kristjánsson, B.K. *Crangonyx islandicus* sp. nov., a subterranean freshwater amphipod (Crustacea, Amphipoda, Crangonyctidae) from springs in lava fields in Iceland. *Zootaxa* **2006**, *1365*, 1–17. [CrossRef]
46. Kristjánsson, B.K.; Svavarsson, J. Subglacial refugia in Iceland enabled groundwater amphipods to survive glaciations. *Am. Nat.* **2007**, *170*, 292–296. [CrossRef]
47. Vaughan, T.G.; Kühnert, D.; Popinga, A.; Welch, D.; Drummond, A.J. Efficient Bayesian inference under the structured coalescent. *Bioinformatics* **2014**, *30*, 2272–2279. [CrossRef]
48. Hillis, D.M.; Moritz, C.; Mable, B.K. *Molecular Systematics*; Sinauer Associates: Sunderland, MA, USA, 1996.
49. Elshire, R.J.; Glaubitz, J.C.; Sun, Q.; Poland, J.A.; Kawamoto, K.; Buckler, E.S.; Mitchell, S.E. A robust, simple genotyping-by-sequencing (GBS) approach for high diversity species. *PLoS ONE* **2011**, *6*, e19379. [CrossRef]
50. Peterson, B.K.; Weber, J.N.; Kay, E.H.; Fisher, H.S.; Hoekstra, H.E. Double Digest RADseq: An inexpensive method for de novo SNP discovery and genotyping in model and non-model species. *PLoS ONE* **2012**, *7*, e37135. [CrossRef]
51. Catchen, J.M.; Amores, A.; Hohenlohe, P.; Cresko, W.; Postlethwait, J.H. Stacks: Building and genotyping loci de novo from short-read sequences. *G3 Genes Genomes Genet.* **2011**, *1*, 171–182. [CrossRef]
52. Eaton, D.A.R. PyRAD: Assembly of de novo RADseq loci for phylogenetic analyses. *Bioinformatics* **2014**, *30*, 1844–1849. [CrossRef] [PubMed]
53. Lynch, M. Estimation of nucleotide diversity, disequilibrium coefficients, and mutation rates from high-coverage genome-sequencing projects. *Mol. Biol. Evol.* **2008**, *25*, 2409–2419. [CrossRef] [PubMed]
54. R Core Team. *R: A Language and Environment for Statistical Computing*; R Foundation for Statistical Computing: Vienna, Austria, 2015; Available online: https://www.R-project.org/ (accessed on 1 November 2015).
55. Aktas, C. Haplotypes: Haplotype Inference and Statistical Analysis of Genetic Variation. R Package Version 1.0. 2015. Available online: https://CRAN.R-project.org/package=haplotypes (accessed on 1 November 2015).
56. Roesti, M.; Hendry, A.P.; Salzburger, W.; Berner, D. Genome divergence during evolutionary diversification as revealed in replicate lake-stream stickleback population pairs. *Mol. Ecol.* **2012**, *21*, 2852–2862. [CrossRef] [PubMed]
57. Hohenlohe, P.A.; Amish, S.J.; Catchen, J.M.; Allendorf, F.W.; Luikart, G. Next-generation RAD sequencing identifies thousands of SNPs for assessing hybridization between rainbow and westslope cutthroat trout. *Mol. Ecol. Resour.* **2011**, *11*, 117–122. [CrossRef]
58. Foll, M.; Gaggiotti, O. A genome-scan method to identify selected loci appropriate for both dominant and codominant markers: A Bayesian perspective. *Genetics* **2008**, *180*, 977–993. [CrossRef]
59. De Mita, S.; Thuillet, A.C.; Gay, L.; Ahmadi, N.; Manel, S.; Ronfort, J.; Vigouroux, Y. Detecting selection along environmental gradients: Analysis of eight methods and their effectiveness for outbreeding and selfing populations. *Mol. Ecol.* **2013**, *22*, 1383–1399. [CrossRef]
60. Lotterhos, K.E.; Whitlock, M.C. Evaluation of demographic history and neutral parameterization on the performance of FST outlier tests. *Mol. Ecol.* **2014**, *23*, 2178–2192. [CrossRef]
61. Langmead, B.; Salzberg, S. Fast gapped-read alignment with Bowtie 2. *Nat. Meth.* **2012**, *9*, 357–359. [CrossRef]
62. Li, H. A statistical framework for SNP calling, mutation discovery, association mapping and population genetical parameter estimation from sequencing data. *Bioinformatics* **2011**, *27*, 2987–2993. [CrossRef]
63. Gouy, M.; Guindon, S.; Gascuel, O. Sea view version 4: A multiplatform graphical user interface for sequence alignment and phylogenetic tree building. *Mol. Biol. Evol.* **2010**, *27*, 221–224. [CrossRef]
64. Bouckaert, R.; Heled, J.; Kühnert, D.; Vaughan, T.; Wu, C.H.; Xie, D.; Suchard, M.A.; Rambaut, A.; Drummond, A.J. BEAST 2: A software platform for Bayesian evolutionary analysis. *PLoS Comput. Biol.* **2014**, *10*, e1003537. [CrossRef] [PubMed]
65. Nylander, J.A.A. *MrAIC. pl. Program Distributed by the Author*; Evolutionary Biology Centre, Uppsala University: Uppsala, Sweden, 2004.
66. Rambaut, A.; Suchard, M.A.; Xie, D.; Drummond, A.J. Tracer, version 1.6; 2014. Available online: http://tree.bio.ed.ac.uk/software/tracer/ (accessed on 26 March 2016).
67. Winter, D.J. MMOD: An R library for the calculation of population differentiation statistics. *Mol. Ecol. Resour.* **2012**, *12*, 1158–1160. [CrossRef]
68. Sokal, R.R.; Rohlf, F.J. *Biometry: The Principles and Practice of Statistics in Biological Research*; W.H. Freeman: San Francisco, CA, USA, 1994; 880p.
69. Holm, S. A simple sequentially rejective multiple test procedure. *Scand. J. Stat.* **1979**, *6*, 65–70.
70. Tamura, K.; Nei, M. Estimation of the number of nucleotide substitutions in the control region of mitochondrial DNA in humans and chimpanzees. *Mol. Biol. Evol.* **1993**, *10*, 512–526.
71. Paradis, E. Pegas: An R package for population genetics with an integrated-modular approach. *Bioinformatics* **2010**, *26*, 419–420. [CrossRef] [PubMed]
72. Jombart, T.; Devillard, S.; Balloux, F. Discriminant analysis of principal components: A new method for the analysis of genetically structured populations. *BMC Genet.* **2010**, *11*, 94. [CrossRef] [PubMed]
73. Jombart, T. Adegenet: A R package for the multivariate analysis of genetic markers. *Bioinformatics* **2008**, *24*, 1403–1405. [CrossRef]
74. Bryant, D.; Bouckaert, R.; Felsenstein, J.; Rosenberg, N.A.; Roychoudhury, A. Inferring species trees directly from biallelic genetic markers: Bypassing gene trees in a full coalescent analysis. *Mol. Biol. Evol.* **2012**, *29*, 1917–1932. [CrossRef]

75. Grummer, J.A.; Bryson, R.W.; Reeder, T.W. Species delimitation using Bayes factors: Simulations and application to the *Sceloporus scalaris* species group (Squamata: Phrynosomatidae). *Syst. Biol.* **2014**, *63*, 119–133. [CrossRef]
76. Leache, A.D.; Fujita, M.K.; Minin, V.N.; Bouckaert, R.R. Species delimitation using genome-wide SNP data. *Syst. Biol.* **2014**, *63*, 534–542. [CrossRef]
77. Kass, R.E.; Raftery, A.E. Bayes Factors. *J. Am. Stat. Assoc.* **1995**, *90*, 773–795. [CrossRef]
78. Green, R.E.; Krause, J.; Briggs, A.W.; Maricic, T.; Stenzel, U.; Kircher, M.; Patterson, N.; Li, H.; Zhai, W.; Fritz, M.H.; et al. A draft sequence of the Neandertal genome. *Science* **2010**, *328*, 710–722. [CrossRef] [PubMed]
79. Durand, E.Y.; Patterson, N.; Reich, D.; Slatkin, M. Testing for ancient admixture between closely related populations. *Mol. Biol. Evol.* **2011**, *28*, 2239–2252. [CrossRef] [PubMed]
80. Albers, C.A.; Lunter, G.; MacArthur, D.G.; McVean, G.; Ouwehand, W.H.; Durbin, R. Dindel: Accurate indel calls from short-read data. *Genome Res.* **2011**, *21*, 961–973. [CrossRef]
81. Herrera, S.; Reyes-Herrera, P.H.; Shank, T.M. Predicting RAD-seq marker numbers across the eukaryotic tree of life. *Genome Biol. Evol.* **2015**, *7*, 3207–3225. [CrossRef]
82. Alfsnes, K.; Leinaas, H.P.; Hessen, D.O. Genome size in arthropods; different roles of phylogeny, habitat and life history in insects and crustaceans. *Ecol. Evol.* **2017**, *7*, 5939–5947. [CrossRef]
83. Rees, D.J.; Dufresne, F.; Glémet, H.; Belzile, C. Amphipod genome sizes: First estimates for Arctic species reveal genomic giants. *Genome* **2007**, *50*, 151–158. [CrossRef]
84. Dufresne, F.; Jeffery, N. A guided tour of large genome size in animals: What we know and where we are heading. *Chrom. Res.* **2011**, *19*, 925–938. [CrossRef]
85. Jeffery, N.W.; Yampolsky, L.; Gregory, T.R. Nuclear DNA content correlates with depth, body size, and diversification rate in amphipod crustaceans from ancient Lake Baikal, Russia. *Genome* **2017**, *60*, 303–309. [CrossRef]
86. Lefébure, T.; Morvan, C.; Malard, F.; François, C.; Konecny-Dupré, L.; Guéguen, L.; Weiss-Gayet, M.; Seguin-Orlando, A.; Ermini, L.; Sarkissian, C.D.; et al. Less effective selection leads to larger genomes. *Genome Res.* **2017**, *27*, 1016–1028. [CrossRef]
87. Fariello, M.I.; Boitard, S.; Naya, H.; San Cristobal, M.; Servin, B. Detecting signatures of selection through haplotype differentiation among hierarchically structured populations. *Genetics* **2013**, *193*, 929–941. [CrossRef]
88. Leitwein, M.; Duranton, M.; Rougemont, Q.; Gagnaire, P.A.; Bernatchez, L. Using haplotype information for conservation genomics. *Trends Ecol. Evol.* **2020**, *35*, 245–258. [CrossRef] [PubMed]
89. Cannizzaro, A.G.; Balding, D.; Lazo-Wasem, E.A.; Sawicki, T.R. A new species rises from beneath Florida: Molecular phylogenetic analyses reveal cryptic diversity among the metapopulation of *Crangonyx hobbsi* Shoemaker, 1941 (Amphipoda: Crangonyctidae). *Org. Div. Evol.* **2020**, *20*, 387–404. [CrossRef]
90. Geirsdottir, Á.; Miller, G.H.; Axford, Y.; Ólafsdóttir, S. Holocene and latest Pleistocene climate and glacier fluctuations in Iceland. *Quat. Sci. Rev.* **2009**, *28*, 2107–2118. [CrossRef]
91. Allio, R.; Donega, S.; Galtier, N.; Nabholz, B. Large variation in the ratio of mitochondrial to nuclear mutation rate across animals: Implications for genetic diversity and the use of mitochondrial DNA as a molecular marker. *Mol. Biol. Evol.* **2017**, *34*, 2762–2772. [CrossRef] [PubMed]
92. Saclier, N.; François, C.M.; Konecny-Dupré, L.; Lartillot, N.; Guéguen, L.; Duret, L.; Malard, F.; Douady, C.J.; Lefébure, T. Life history traits impact the nuclear rate of substitution but not the mitochondrial rate in isopods. *Mol. Biol. Evol.* **2018**, *35*, 2900–2912. [CrossRef] [PubMed]
93. Hjartarson, Á. Víðáttumestu hraun Íslands (The largest lava of Iceland). *Náttúrufræðingurinn* **2011**, *81*, 37–48.
94. Hewitt, G.M. Genetic consequences of climatic oscillations in the Quaternary. *Philos. Trans. R. Soc. Lond. Ser. B Biol. Sci.* **2004**, *359*, 183–195. [CrossRef]
95. Provan, J.; Bennett, K.D. Phylogeographic insights into cryptic glacial refugia. *Trends Ecol. Evol.* **2008**, *23*, 564–571. [CrossRef]
96. Bernardo, P.H.; Sánchez-Ramírez, S.; Sánchez-Pacheco, S.J.; Alvares-Castaneda, S.T.; Aguileira-Miller, E.F.; Mendez-de la Cruz, F.R.; Murphy, R.W. Extreme mito-nuclear discordance in a peninsular lizard: The role of drift, selection, and climate. *Heredity* **2019**, *123*, 359–370. [CrossRef]
97. Stephens, M.; Smith, N.; Donnelly, P. A new statistical method for haplotype reconstruction from population data. *Am. J. Hum. Gen.* **2001**, *68*, 978–989. [CrossRef] [PubMed]
98. Guindon, S.; Dufayard, J.F.; Lefort, V.; Anisimova, M.; Hordijk, W.; Gascuel, O. New algorithms and methods to estimate maximum-likelihood phylogenies: Assessing the performance of PhyML.3.0. *Syst. Biol.* **2010**, *59*, 307–321. [CrossRef] [PubMed]
99. Paradis, E.; Claude, J.; Strimmer, K. APE: Analyses of phylogenetics and evolution in R language. *Bioinformatics* **2004**, *20*, 289–290. [CrossRef] [PubMed]

diversity

MDPI

Article

Not the Last Piece of the Puzzle: *Niphargus* Phylogeny in Hungary

Gergely Balázs [1,2,*], Špela Borko [3], Dorottya Angyal [4], Valerija Zakšek [3], Anna Biró [1,2,5], Cene Fišer [3] and Gábor Herczeg [1,2]

1 ELKH-ELTE-MTM Integrative Ecology Research Group, 1117 Budapest, Hungary
2 Department of Systematic Zoology and Ecology, Institute of Biology, ELTE Eötvös Loránd University, 1117 Budapest, Hungary
3 SubBio Lab, Department of Biology, Biotechnical Faculty, University of Ljubljana, 1000 Ljubljana, Slovenia
4 Calle 31 #436A x 36 y 36A Colonia Jesús Carranza, Mérida 97109, Mexico
5 Doctoral School of Biology, Institute of Biology, ELTE Eötvös Loránd University, 1117 Budapest, Hungary
* Correspondence: gergely.balazs@ttk.elte.hu

Abstract: The Palaearctic genus *Niphargus* is a promising model system to understand subterranean fauna genesis in Europe. The Pannonian Plain (mainly covered by Hungary) in Central Europe, once being the area of the Paratethys, is a key area for *Niphargus* diversification. However, our knowledge on Hungarian species of *Niphargus* is primarily based on sporadic taxonomical works from the pre-molecular era. Here, we studied 14 localities, covering the eight valid Hungarian species of *Niphargus* and including nine previously unstudied populations. Based on sequences of three gene fragments, we reconstructed their phylogeny using maximum likelihood and Bayesian approaches. We found that not all Hungarian species of *Niphargus* are closely related, and even species sampled at the same localities can belong to different clades. Some Hungarian species form monophyletic clades, while others are nested in various non-Hungarian lineages. The new populations are all genetically distinct from the known species. Our results suggest that the Hungarian *Niphargus* fauna has originated from seven unrelated clades and its diversity is underestimated due to unknown populations and cryptic species. The detection of genetically distinct species of *Niphargus* from non-carbonate regions calls for further research efforts. The high diversity and the number of putative new species in the *N. tatrensis* clade warrants further, high-resolution phylogenetic studies.

Keywords: *Niphargus*; Amphipoda; subterranean fauna; phylogeny

1. Introduction

The genus *Niphargus* SCHIØDTE, 1849 (Crustacea: Amphipoda) is the most diverse subterranean amphipod genus [1]. This taxon with more than 400 known and numerous undescribed species is the most salient element of the subterranean fauna in the Western Palaearctic [2,3]. The genus is an important model system that may help answer several questions addressing the origin of subterranean fauna in Europe. *Niphargus* originated in an area that nowadays belongs to Western Europe, and subsequently dispersed eastward, with at least five radiations [4,5]. High species richness of the genus can be attributed to speciation that followed the colonisation of new ecological niches [6], but also to fragmented karstic systems that enabled speciation within the same environment in micro-allopatry [7], eventually resulting in numerous morphologically indistinguishable or very similar species, called cryptic species [8,9]. Bursts of speciation can be tentatively attributed to the colonisation of the novel regions, such as karstic masses that emerged from ancient seas, or to major geological events that reshuffled dispersal routes [10].

Accurate reconstruction of the speciation and biogeography of the genus, however, depends on completeness of the taxonomic inventory. The taxonomic structure of the genus is far from resolved. The results of the attempted large-scale taxonomic revisions of the genus evidently showed [4,11,12] that smaller-scale local revisions are needed [13], as they can point out taxonomic uncertainties, so revealing important taxa missed in large-scale studies. Molecular studies imply that on average every described species in fact comprises two-three cryptic, yet undescribed species, which often show extremely limited distribution [11]. Sampling effort is highly uneven across the range of the genus. While some regions have been thoroughly sampled, some countries remained generally underexplored, or were not explored with molecular methods. Insufficient sampling is particularly critical in regions with turbulent geological history that in the past acted as major dispersal routes or speciation centres. Such are the Pannonian lowlands and some karstic regions within it, which belonged to the ancient Paratethys. The current area of Hungary occupies the central area of the former Pannonian Sea, which was part of the ancient Paratethys. The regression of the Paratethys prompted speciation in surface waters [14–16] and after the Pannonian Sea got isolated from it and became the shallow Pannonian Lake [17], it most likely remained a dispersal route for aquatic animals [18].

Research on *Niphargus* in Hungary had its peak in the first half of the 20th century when most of the taxonomic works along with species descriptions were published e.g., [19–22]. In the following decades, the number of publications declined and only one species description was published [23]. During this period, studies on *Niphargus* were flourishing at the international level and our knowledge about the genus increased considerably along with the number of species and the number of morphological traits used for species identification [24]. A previous study based on literature data and morphotaxonomy [25] showed that although there are more than 20 species of *Niphargus* reported from Hungary [22,26], only eight are currently valid: *N. tatrensis* (WRZESNIOWSKY, 1890); *N. molnari* MÉHELŸ, 1927; *N. aggtelekiensis* DUDICH, 1932; *N. hrabei* S. KARAMAN, 1932; *N. valachicus* DOBREANU and MANOLACHE, 1933; *N. gebhardti* SCHELLENBERG, 1934; *N. hungaricus* MÉHELŸ, 1937; *N. forroi* G. KARAMAN, 1986. Most of the taxonomic work on Hungarian species of *Niphargus* had been done prior to genetic methods becoming available. Therefore, the insights into genetic diversity and phylogenetic structure of the Hungarian species are limited to a small number of species [15,27,28].

In the present paper, we studied the phylogenetic relationships of the known Hungarian taxa of *Niphargus* (including six subterranean and two surface-dwelling species). We also included numerous new samples representing previously unknown Hungarian populations. We were particularly interested in whether (i) the Hungarian species of *Niphargus* are closely related, or belong to phylogenetically distinct clades, (ii) the clades incorporating the Hungarian species/populations are exclusive to Hungary, or the Hungarian species are members of clades with larger distribution and (iii) the new populations belong to known species or are genetically distinct from them.

2. Materials and Methods

2.1. Sample Collection

Most of the valid Hungarian species (*N. molnari*, *N. gebhardti*, *N. aggtelekiensis*, *N. hungaricus*, *N. forroi*) were described from Hungarian type localities. To secure species identity, we sampled at type localities of the valid species where it was possible (Figure 1 and Table 1). In the case of two out of the above five species, sampling at the type locality was not possible. The type locality of *N. hungaricus* is the Jávor Spring (KőszegMts) [29], but due to changes in the morphology of the spring orifice, the local population became physically inaccessible. Therefore, we collected samples from the closest known location, the Borha Valley mine tunnel situated 300 m from the spring. The type locality of *N. molnari* is the Mánfai-kőlyuk Cave where the species is not present anymore, most probably due to artificial changes in connection with the utilisation of the cave as a water reservoir [30]. Therefore, a sample was collected from the closest known location, the Abaligeti Cave [27], which is 7.5 km away. In the case of *N. aggtelekiensis*, besides from the type locality (Baradla-Domica Cave System), we also collected samples from the hydrologically separate, yet geographically close Rákóczi No. 1 Cave. In the case of the two surface-dwelling species that have wide distribution ranges (*N. hrabei*, described from Slovakia [31], and *N. valachi-*

cus, described from Romania [32]) we collected samples from known Hungarian locations (one sample per species). According to previous studies, these species show very limited genetic variation throughout their range [15,28], therefore we assumed that the sampling locality does not affect our results. The occurrence of *N. tatrensis* in Hungary is doubtful [25]. In this case we included a sample from the Kecske-lyuk Cave in the Bükk Mts, which was identified as *N. tatrensis* based on morphological characters. We also included samples from two undescribed, but morphologically distinct species known only from the Molnár János Cave. The sampling of valid Hungarian species and the undescribed species inhabiting the Molnár János Cave was carried out between 2013 and 2016 (for details see [15,25,27,30,33]). Ever since the beginning of studies on *Niphargus* in Hungary [34], the research focus was always on caves, thus almost all the species of *Niphargus* described from Hungary (with the sole exception of *N. hungaricus*) are known from karstic areas. To get a more detailed view on the *Niphargus* fauna of the country, we carried out extensive sampling in non-karstic mountainous areas too. Between 2016 and 2022, we visited over 80 mines and springs in the Visegrád Mts, Börzsöny Mts, Mátra Mts, Balaton Uplands, Bakony Mts and Zemplén Mts. This effort resulted in sites with new findings of *Niphargus* out of which—based on morphological traits and/or preliminary genetic data—we detected putative new species to science. Samples representing these putative new species to science (from Dömös, Vasbánya Spring, Kánya Spring, Werbőczy Spring, Gejzír Spring 1, Gejzír Spring 2) were also included in the present analysis. Collected adults were stored in 96% ethanol at 10 °C until DNA extraction.

Figure 1. Sampling locations in Hungary (coloured circles) and the samples from surrounding regions (white circles). Colour codes for the Hungarian samples are the same as in Figures 2 and 3, representing clades revealed by this study.

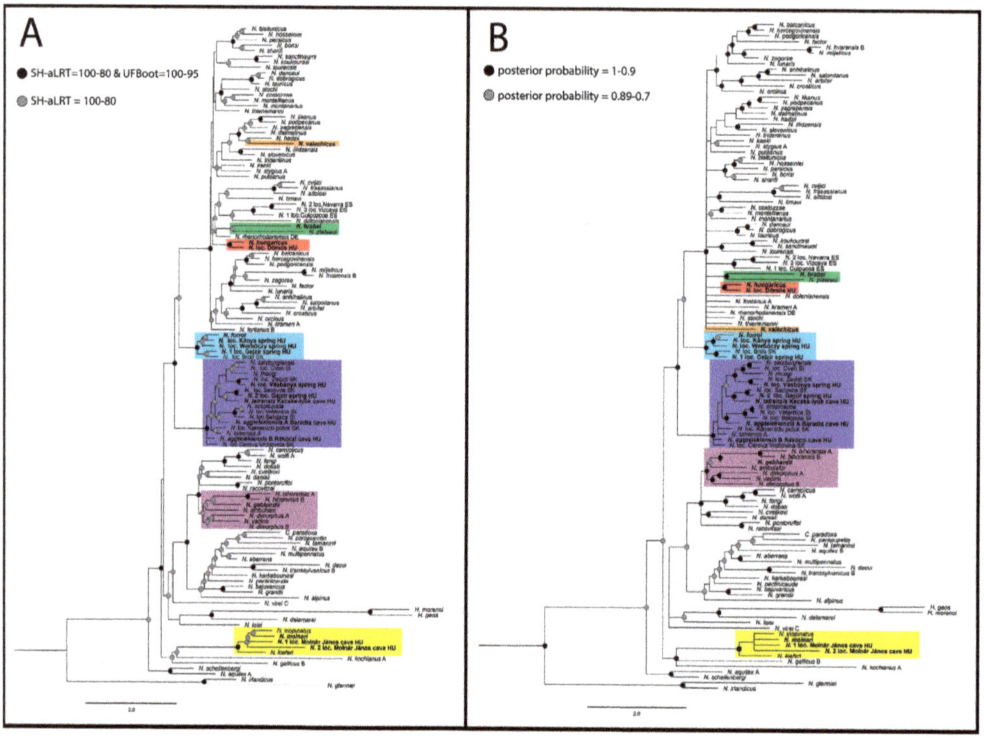

Figure 2. Phylogenetic relationships of the species in the genus *Niphargus* inferred from maximum likelihood (**A**) and Bayesian (**B**) approaches, including ultrafast bootstrap (UFBoot), Shimodaria–Hasewaga approximate likelihood ratio test (SH–aLRT) and Bayesian posterior probability support values. Outgroup species have been removed. The specimens from Hungary are in bold and the clades including specimens from Hungary are colour coded using the same colours as on Figures 1 and 3.

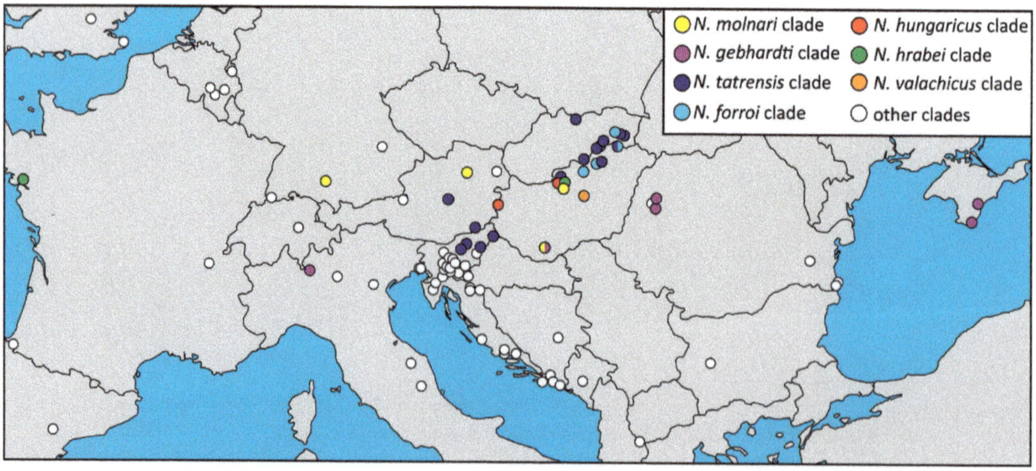

Figure 3. Geographic distribution of the clades that include Hungarian species. Colour codes are the same as in Figures 1 and 2.

Table 1. Sampling locations of Hungarian taxa of *Niphargus*.

Sample Name	Sampling Location	Region, Settlement	Latitude (N)	Longitude (E)
Niphargus aggtelekiensis A Baradla cave	Baradla-Domica Cave System *	Aggtelek Karst, Aggtelek	48.4831	20.5440
Niphargus aggtelekiensis B Rákóczi cave	Rákóczi No. 1 Cave	Aggtelek Karst, Bódvarákó	48.5208	20.7489
Niphargus forroi	Diabáz Cave *	Bükk Mts, Bánkút	48.0955	20.4822
Niphargus gebhardti	Abaligeti Cave *	Mecsek Mts, Abaliget	46.1374	18.1158
Niphargus hrabei	Göd	Pest Plain, Göd	47.7159	19.1409
Niphargus hungaricus	Borha Valley, mine tunnel	Kőszegi Mts, Kőszeg	47.351	16.4843
Niphargus molnari	Abaligeti Cave	Mecsek Mts, Abaliget	46.1374	18.1158
Niphargus tatrensis Kecske-lyuk cave	Kecske-lyuk Cave	Bükk Mts, Alsóhámor	48.1175	20.6316
Niphargus valachicus	Farmos	Pest Plain, Farmos	47.3608	19.8269
Niphargus 1 loc. Molnár János cave	Molnár János Cave	Buda Thermal Karst, Budapest	47.5181	19.0358
Niphargus 2 loc. Molnár János cave	Molnár János Cave	Buda Thermal Karst, Budapest	47.5181	19.0358
Niphargus loc. Vasbánya spring	Vasbánya Spring	Börzsöny Mts, Szokolya	47.8850	19.0368
Niphargus loc. Dömös	Dömös, mine tunnel	Visegrád Mts, Dömös	47.7548	18.9087
Niphargus loc. Kánya spring	Kánya Spring	Mátra Mts, Galyatető	47.9268	19.9131
Niphargus loc. Werbőczy spring	Werbőczy Spring	Mátra Mts, Galyatető	47.9207	19.9167
Niphargus 1 loc. Gejzír spring	Gejzír Spring	Zemplén Mts, Telkibánya	48.4820	21.3584
Niphargus 2 loc. Gejzír spring	Gejzír Spring	Zemplén Mts, Telkibánya	48.4820	21.3584

Type localities are marked with "*".

2.2. DNA Extraction, PCR, and Analysis

We extracted genomic DNA with QIAamp DNA Micro Kit (Qiagen, Hilden, Germany) or GenElute Mammalian Genomic DNA Miniprep Kit (Sigma Aldrich, St. Louis, MO, USA) according to the manufacturer's specification using the pereiopods and antennae of the animals. We amplified the fragments of three genes: two fragments of 28S rRNA gene, the histone 3 subunit A (H3), and a fragment of mitochondrial cytochrome oxidase I (COI). The fragments of 28S rRNA gene were amplified using primers 28Slev2, 28Sdes2 [35], 28Slev3, and 28Sdes5, and the corresponding internal primers [36]. The histone H3 gene was amplified using primers H3AFR and H3AR2 [37], and COI was amplified using primers LCO1490 and HCO 2198 [38] applying PCR cycler settings as described in Angyal et al. [27]. The PCR products were purified using Roche High Pure Purification Kit (Merck, Darmstadt, Germany) or Exonuclease I and Fast AP Thermosensitive Alkaline Phosphatase (Thermo Fisher Scientific, Waltham, MA, USA) and sequenced at Macrogen Europe (Amsterdam, Netherlands) or at the Molecular Taxonomy Laboratory of the Hungarian Natural History Museum (Budapest, Hungary) using the same primers as for amplification. The chromatograms were assembled and edited in Geneious 11.0.3 (Biomatters Ltd., Auckland, New Zealand). To identify the phylogenetic relationships of the species of *Niphargus* from Hungary we compiled a dataset comprising 121 specimens (including *Carinurella paradoxa* SKET, 1964, *Haploginglymus geos* JURADO-RIVERA, 2017, and *Haploginglymus morenoi* IANNILLI, MINELLI and RUFFO, 2009) that well cover the phylogenetic diversity of *Niphargus* in the entire distribution range [3,39]. *Pseudoniphargus gorbeanus* NOTENBOOM, 1986 was used as an outgroup [5]. The collected material is deposited at the Department of Systematic Zoology and Ecology, Eötvös Loránd University, Budapest, Hungary. The list of the species studied, origin of samples, and the GenBank accession numbers are available in Supplementary Material, Table S1.

We aligned the sequences in MAFFT 7.388 [40] using the E-INS-I algorithm with the scoring matrix 1PAM/k = 2 and with the highest gap penalty. Alignments were concatenated, partitioned by gene and codon position, and the optimal substitution model for each partition was chosen using Partition Finder 2 [41,42] under the corrected Akaike information criterion (AICc). The optimal substitution models were GTR+I+Γ for the second and third codon position of COI and the first and third codon position of 28S 22 and H3; SYM+I+Γ for the first codon position of 28S 35 and COI and JC+I for the second codon position of H3. We reconstructed the phylogenetic relationships with Bayesian inference (BI) in MrBayes v3.2.6 [43], where we used the optimal substitution models inferred with Partition Finder; and the maximum likelihood (ML) method in IQ-TREE 2.2.0 [44], where we used built-in automatic best fit substitution model search [45].

A Bayesian MCMC tree search was run for 20 million generations with two independent runs with four chains for each run. Trees were sampled each 2000th generation. After reaching the stationary phase, the first 25% of trees were discarded as burn-in, and from the remaining trees we calculated the 50% majority rule consensus tree. The ML phylogenetic analysis was run using simultaneous optimisation of substitution models, with ultrafast bootstrap approximation (UFBoot; [46]). Support values were calculated using the Shimodaira–Hasegawa approximate likelihood ratio test (SH-aLRT) approach [47]. Phylogenetic analyses were run on the CIPRES Science Gateway (https://www.phylo.org; [48]).

3. Results

Phylogenetic relationships of the species of *Niphargus* from Hungary showed that most are not closely related and belong to different clades, sometimes even if they are from the same locality (Figure 2). Some of the Hungarian species form monophyletic clades, while others are nested in various non-Hungarian lineages. The structure of the ML (Figure 2A) and Bayesian tree (Figure 2B) differs considerably reflecting weak support among the well-supported clades. Sister relationships among the species are generally well-supported and congruent in both trees. The structural differences and uncertainties are most probably the results of a lack of genetic information at certain splits, which were reconstructed differently (albeit with no support) in both approaches [49]. Although the phylogenetic position of larger clades that include Hungarian species remain unclear, some cautious conclusions can be drawn based on our results.

Below we overview the results of phylogenetic analysis clade by clade. The first clade comprises *N. molnari* from the Mecsek Mts, two undescribed species from the Molnár János Cave, Budapest and two species hitherto not recorded from Hungary, *Niphargus kieferi* SCHELLENBERG, 1936 and *N. inopinatus* SCHELLENBERG, 1932.

Niphargus gebhardti, another species from the Mecsek Mts, is nested within a clade which includes species scattered from Italy to the Crimean Peninsula.

Five samples, (*N. tatrensis* from Kecske-lyuk Cave (Bükk Mts), Gejzír Spring 2 (Zemplén Mts), Vasbánya Spring (Börzsöny Mts), and two samples of *N. aggtelekiensis*) are part of the *N. tatrensis* clade together with many other specimens from Slovenia, Austria and Slovakia, including the relatively recently described species from the *N. tatrensis* species group (*Niphargus scopicauda* FIŠER, COLEMAN, ZAGMAJSTER, ZWITTNIG, GERECKE and SKET, 2010, *N. salzburgensis* SCHELLENBERG, 1935, *N. moogi* STOCH, CHRISTIAN and FLOT, 2020). The specimen identified as *N. tatrensis* living in the Kecske-lyuk Cave is clearly separated from *N. tatrensis* from its type locality (labelled as *N. tatrensis* A), indicating that the Hungarian population belongs to a new putative species. The position of *N. aggtelekiensis* from the type locality (*N. aggtelekiensis* A Baradla cave) further confirms the valid status of the species. Interestingly, our other *N. aggtelekiensis* sample from the Rákóczi No. 1 Cave shows clear genetic differentiation from the topotype. The Gejzír Spring 2 and the Vasbánya Spring samples are both from taxa that require further investigations.

Four samples, (*N. forroi*, Gejzír Spring 1 (Zemplén Mts), Kánya Spring (Mátra Mts), and Werbőczy Spring (Mátra Mts)) form a well-defined and separate clade together with a taxon from Southeast Slovakia (*N.* loc. Brdo). Genetic distinctness of the three Hungarian populations indicate that each of them represents most likely a new putative species.

Niphargus hungaricus and the sample from Dömös mine tunnel (Visegrád Mts) form a well-supported monophyletic clade, although its position in the entire phylogeny remains uncertain, hence the sister clades and the origin of the clade cannot be defined.

The position of the two surface-dwelling species, *N. hrabei* and *N. valachicus*, proves their different origins, which is in accordance with the results of previous studies [15,28].

4. Discussion

The result of the first genetic insights into Hungarian species of *Niphargus* brought up several important conclusions. First, the Hungarian *Niphargus* fauna is phylogenetically diverse. Species originated in seven unrelated clades. Most of these clades show a west-east distribution. This pattern was probably shaped by the geological history of the Paratethys (*N. tatrensis*, *N. gebhardti* and *N. molnari* clades, see [50] and [5] for comparisons). Second, the diversity of *Niphargus* in Hungary is underestimated. We detected almost twice as many (putative) species of *Niphargus* than known from the most recent checklist [25]. Some species show distinct morphology (e.g., species from the Molnár János Cave), but some are morphologically similar (e.g., *N. aggtelekiensis* from the Rákóczi No. 1 Cave) and should be taxonomically evaluated in focused molecular studies. Third, the detection of *Niphargus* in non-carbonate regions calls for additional sampling. We have no samples from riverine interstitials, a habitat type that should be systematically studied in the future. Moreover, sampling in carbonate and non-carbonate regions—despite our effort—might be far from exhaustive and should be intensified to properly define species distributions and their population structure.

Phylogenetic analysis with the present data has certain limitations. The position of the clades with Hungarian taxa of *Niphargus* is in many cases weakly supported and in this respect, considerable differences can be observed between the ML and Bayesian trees. Consequently, we cannot define the absolute position of the Hungarian taxa within the genus. Our phylogeny trees contain weak support values in many cases, especially at basal splits. We can think of various reasons behind the low values. In general, we can assume that by increasing the number of samples and the number of genes, better results can be expected. This is only true if the markers are representing similar phylogeny, and there is no considerable difference in the speed of genetic changes between lineages [51–53]. Moreover, by including more genes, the amount of phylogenetically noninformative information could increase by increasing the chance for inclusion of genes under environmental selection. In such case, the analysis would result in well-supported, but incorrect trees [54]. Although it is possible that we could obtain better results by using other or additional markers, our options were rather limited as the markers we could use were strongly determined by the already available sequences for comparison. More even sampling distribution is also a good way to improve tree reliability [55,56]. Unfortunately, in the case of subterranean species, this possibility is severely constrained due to the limited number of known and accessible locations. Despite these limitations, the composition of the clades with Hungarian samples are well supported in most cases and consistent on both the ML and Bayesian trees. Taken together, we can state that even though the results of our analysis cannot be used to draw unquestionable conclusions about the phylogeny of the studied species, they can provide solid background for further, more specific studies.

The clade including *N. molnari* and the two species from the Molnár János Cave contains two other species. *N. kieferi* is a widely distributed species that lives in groundwater habitats in Germany and France, while *N. inopinatus* also lives in non-karstic habitats from Germany to Slovakia [57] (Figure 3). Both species have wide distribution ranges and were found in interstitial habitats in the Danube basin and could be expected in Hungary as well.

Niphargus gebhardti is clustered together with species from Crimea (*N. dimorphus* Birstein, 1961, *N. vadimi* Birstein, 1961), Romania (*N. bihorensis* Schellenberg, 1940), and Italy (*N. ambulator* G. Karaman, 1975) that form a well-defined clade. Interestingly, although the clade covers a wide geographic range, the occurrence of its members is limited to a narrow latitudinal range between 44° and 46° (Figure 3). The sister clade to the *N. gebhardti* clade, containing species from Slovenia, Bulgaria, Romania, and Iran is also scattered along a west-east line. The distribution of the species in the *N. gebhardti* clade and the species in the sister clade suggest the North Dinaric origin of the lineage.

The clade which contains samples identified as *N. aggtelekiensis* and *N. tatrensis* from Hungary exhibits many uncertainties, yet its members are all from the Western and Eastern Carpathians (Figure 3). *Niphargus aggtelekiensis* from the type locality is assuredly separated from other taxa, but the other supposed *N. aggtelekiensis* sample from the nearby Rákóczi No 1 Cave also appears to be genetically distinct. While the geographical distance between the two locations is only 15 km, the genetic distance is quite substantial (16% uncorrected distance for COI). Although genetic distance on a mitochondrial marker alone is not sufficient to draw any firm conclusions, this result certainly calls for further investigation. The positions of the relatively recently identified species from the *N. tatrensis* group (*N. scopicauda, N. moogi, N. salzburgensis*) are not surprising and are in accordance with previous studies [46,54]. Although the phylogenetic information provided by this study on the other Hungarian samples in the *N. tatrensis* clade is not sufficient to make secure taxonomic conclusions, we believe that some remarks are worth mentioning. While the taxonomic status of *N. tatrensis* from the Bükk Mts is unclear, we can state that the population living in the Kecske-lyuk Cave does not belong to the *N. tatrensis*. The samples from Gejzír Spring, Zemplén Mts (Gejzír 2) and Vasbánya Spring, Börzsöny Mts most probably also represent putative new species. The sample from Vasbánya Spring is genetically almost identical to *Niphargus* found near Bratislava, West Slovakia [58]. In general, many unidentified taxa can be found in the *N. tatrensis* clade. Their relative position to the topotype suggests that some Northern and Western Carpathian samples included in our analysis might represent additional putative new species. Taken together, we can say that while the Hungarian samples turned out to be important pieces in the *N. tatrensis* clade, unfortunately, our results have only further complicated the long-lasting problematic taxonomy of the group [50,59]. Based on our results, we think that the solution to overcome the difficulties is to increase the resolution with denser sampling in smaller geographical areas and to use higher number of genetic markers in future studies.

The *N. forroi* clade appears to be well-defined on both trees. It contains *N. forroi* from the type locality and four putative new species. It is worth mentioning that the geographical distance between Kánya Spring and Werbőczy Spring is only 700 m, yet the two samples show genetic distinctness. In one of his works, Méhelÿ [22] mentioned a *Niphargus* species he found in the Mátra Mts and named it *N. matrensis*. It is possible that one of the taxa revealed in our study refers to this species, yet, as the author did not provide a morphological description and precise type locality, the species has to be treated as *species inquirenda*. While relationships of taxa within the *N. forroi* clade are not well-supported, the clade itself is clearly monophyletic. Bearing in mind that *N. forroi* is a relatively small species typically found in infiltrating waters in the caves of the Bükk Mts and that the other Hungarian specimens from the clade are also small and elongated and from non-karstic springs, it is plausible that this lineage contains "small-pore" species (for additional information on morphotypes of *Niphargus* see [60]). Members of the *N. tatrensis* clade are relatively big and typically found in cave streams. Therefore, we can assume niche partitioning, where species from the *N. tatrensis* and *N. forroi* clades can be found together, as in the caves of the Bükk Mts and in the subterranean aquatic habitat sampled via the Gejzír Spring, Zemplén Mts.

Niphargus hungaricus and the sample from Dömös mine tunnel (Visegrád Mts) form a well-supported monophyletic clade. Both sampling locations are situated in non-karstic areas. The low number of species in the clade can be explained by the insufficient sampling of *Niphargus* in the non-karstic areas. Our results suggest that the Danube River can act as a strong dispersal barrier, as previously it was also suggested by Fišer et al. [59]. While numerous members of the *N. tatrensis* clade can be found north and east of the Danube River, the lineage is completely missing from Transdanubian locations (Figure 3). Likewise, members of the *N. hungaricus* clade can only be found in the Transdanubian region. This pattern is the most obvious in the case of the Dömös and Vasbánya Spring sample pair, as the Börzsöny and the Visegrád Mts are actually two parts of a single geological unit divided by the Danube River [61].

The closest relative of *N. hrabei* is the widely distributed *N. plateaui* CHEVREUX, 1901 from France. This result is consistent in both trees and in accordance with a previous study [28]. In the case of these two species, the clade is rather weakly defined and can only be outlined on the ML tree. Based on the sister species found on the ML tree, we can cautiously assume the western and eastern expansion of the clade from Mediterranean refugia.

The position of *N. valachicus* on the Bayesian tree does not provide any information on its relatives. Based on the sister species revealed by the ML tree, we can suggest its northern

Dinaric origin, which is in line with the theory proposed by Copilaș-Ciocianu et al. [28], even though the sister species are only partially overlapping.

Based on the results, we can draw some conclusions regarding the *Niphargus* fauna of different regions in Hungary. The two species inhabiting the caves of the Mecsek Mts, *N. gebhardti* and *N. molnari*, are from different lineages and their co-occurrence is most likely a result of independent colonisation. The taxa found in the North Hungarian Mountains (except the Visegrád Mts) are from two major lineages—the *N. tatrensis* and the *N. forroi* clades. The high number of taxa in the *N. tatrensis* clade is probably a result of an expansion followed by allopatric speciation. In a previous study focusing on the phylogeny of the *N. tatrensis* group, multiple secondary contacts were also proposed based on the difference of the variability of COI and ITS sequences [50]. The two undescribed species of *Niphargus* from the Molnár János Cave are obviously closely related, but due to the relatively weak support values it is not clear whether sympatric or allopatric speciation is more likely.

While Hungary is not renowned for its extensive karstic areas, picturesque limestone caves or rich subterranean fauna [62], this study has shown that the subterranean fauna of this country is understudied. This is rather unfortunate, given that the geographic position of Hungary falls within the area of the ancient Paratethys. Moreover, during the regression of the Paratethys, the Pannonian Sea and later the Pannonian Lake with their occasionally decreased salinity [17] could serve as a dispersal route for freshwater animal species. Indeed, the results show that species found across the karstic and non-karstic regions of Hungary may improve the biogeography of some clades and may be of utmost importance in future reconstruction analyses. We hope this preliminary study might prompt future research on the subterranean fauna of the country.

Supplementary Materials: The following supporting information can be downloaded at https://www.mdpi.com/article/10.3390/d15020223/s1. Table S1: Data on the taxa, sample locations and sequences used in this study.

Author Contributions: Conceptualization, G.B. and D.A.; methodology, G.B., D.A., A.B., Š.B. and V.Z.; validation, G.B. and Š.B.; formal Analysis, Š.B.; resources, G.B., G.H. and C.F.; data curation, G.B., D.A., A.B. and Š.B.; writing—original draft preparation, G.B., C.F. and A.B.; writing—review and editing, G.B., A.B., Š.B., V.Z., G.H. and C.F.; visualization, G.B.; supervision, G.H.; project administration, G.H.; funding acquisition, G.H. and C.F. All authors have read and agreed to the published version of the manuscript.

Funding: This research was supported by the joint Hungarian–Slovenian grant for international cooperation (Hungarian National Research, Development and Innovation Fund: #SNN-125627; Slovenian Research Agency: #N1-0069 and #N1-0096) and the Research Core Funding (Slovenian Research Agency: #P1-0184, J1-2464). GB was supported by the ÚNKP-20-4 New National Excellence Program of the Ministry of Innovation and Technology, Hungary from the source of the National Research, Development and Innovation Found.

Data Availability Statement: The data presented in this study are available on request from the corresponding author.

Acknowledgments: We are truly grateful for all the cavers who helped our work in the field. We would also like to thank the people in SubBio Lab (Ljubljana, Slovenia) and Bracken–Grissom Lab (Florida, USA) for their help with this study. Sampling would not be possible without the help of colleagues and officials, namely Péter Borza (Institute of Aquatic Ecology, Budapest, Hungary), Péter Gruber (Aggtelek National Park, Jósvafő, Hungary), Gergely Ferenczy (Bükk National Park, Miskolc, Hungary) and Csaba Egri (Ministry of Agriculture, Budapest, Hungary).

Conflicts of Interest: The authors declare no conflict of interest.

References

1. Väinölä, R.; Witt, J.D.S.; Grabowski, M.; Bradbury, J.H.; Jazdzewski, K.; Sket, B. Global diversity of amphipods (Amphipoda; Crustacea) in freshwater. *Hydrobiologia* **2008**, *595*, 241–255. [CrossRef]
2. Sket, B. High biodiversity in hypogean waters and its endangerment—The situation in Slovenia, the Dinaric karst, and Europe. *Crustaceana* **1999**, *72*, 767–779. [CrossRef]
3. Esmaeili-Rineh, S.; Sari, A.; Delić, T.; Moškrič, A.; Fišer, C. Molecular phylogeny of the subterranean genus *Niphargus* (Crustacea: Amphipoda) in the Middle East: A comparison with European niphargids. *Zool. J. Linn. Soc.* **2015**, *175*, 812–826. [CrossRef]
4. McInerney, C.E.; Maurice, L.; Robertson, A.L.; Knight, L.R.F.D.; Arnscheidt, J.; Venditti, C.; Dooley, J.S.G.; Mathers, T.; Matthijs, S.; Eriksson, K.; et al. The ancient Britons: Groundwater fauna survived extreme climate change over tens of millions of years across NW Europe. *Mol. Ecol.* **2014**, *23*, 1153–1166. [CrossRef] [PubMed]
5. Borko, Š.; Trontelj, P.; Seehausen, O.; Moškrič, A.; Fišer, C. A subterranean adaptive radiation of amphipods in Europe. *Nat. Commun.* **2021**, *12*, 3688. [CrossRef] [PubMed]
6. Borko, Š.; Collette, M.; Brad, T.; Zakšek, V.; Flot, J.F.; Vaxevanopoulos, M.; Sarbu, S.; Fišer, C. Amphipods in a Greek cave with sulphidic and non-sulphidic water: Phylogenetically clustered and ecologically divergent. *Syst. Biodivers.* **2019**, *17*, 558–572. [CrossRef]
7. Orsini, L.; Vanoverbeke, J.; Swillen, I.; Mergeay, J.; De Meester, L. Drivers of population genetic differentiation in the wild: Isolation by dispersal limitation, isolation by adaptation and isolation by colonization. *Mol. Ecol.* **2013**, *22*, 5983–5999. [CrossRef]
8. Lefébure, T.; Douady, C.J.; Gouy, M.; Trontelj, P.; Briolay, J.; Gibert, J. Phylogeography of a subterranean amphipod reveals cryptic diversity and dynamic evolution in extreme environments. *Mol. Ecol.* **2006**, *15*, 1797–1806. [CrossRef]
9. Trontelj, P.; Fišer, C. Cryptic species diversity should not be trivialised. *Syst. Biodivers.* **2009**, *7*, 1–3. [CrossRef]
10. Delić, T.; Stoch, F.; Borko, Š.; Flot, J.F.; Fišer, C. How did subterranean amphipods cross the Adriatic Sea? Phylogenetic evidence for dispersal–vicariance interplay mediated by marine regression–transgression cycles. *J. Biogeogr.* **2020**, *47*, 1875–1887. [CrossRef]
11. Eme, D.; Zagmajster, M.; Delić, T.; Fišer, C.; Flot, J.F.; Konecny-Dupré, L.; Pálsson, S.; Stoch, F.; Zakšek, V.; Douady, C.J.; et al. Do cryptic species matter in macroecology? Sequencing European groundwater crustaceans yields smaller ranges but does not challenge biodiversity determinants. *Ecography* **2018**, *41*, 424–436. [CrossRef]
12. Borko, Š.; Altermatt, F.; Zagmajster, M.; Fišer, C. A hotspot of groundwater amphipod diversity on a crossroad of evolutionary radiations. *Divers. Distrib.* **2022**, *28*, 2765–2777. [CrossRef]
13. Fišer, C.; Alther, R.; Zakšek, V.; Borko, Š.; Fuchs, A.; Altermatt, F. Translating *Niphargus* barcodes from Switzerland into taxonomy with a description of two new species (Amphipoda, Niphargidae). *Zookeys* **2018**, *760*, 113–141. [CrossRef]
14. Copilaș-Ciocianu, D.; Petrusek, A. The southwestern Carpathians as an ancient centre of diversity of freshwater gammarid amphipods: Insights from the *Gammarus fossarum* species complex. *Mol. Ecol.* **2015**, *24*, 3980–3992. [CrossRef] [PubMed]
15. Copilaș-Ciocianu, D.; Fišer, C.; Borza, P.; Balázs, G.; Angyal, D.; Petrusek, A. Low intraspecific genetic divergence and weak niche differentiation despite wide ranges and extensive sympatry in two epigean *Niphargus* species (Crustacea: Amphipoda). *Zool. J. Linn. Soc.* **2017**, *181*, 485–499. [CrossRef]
16. Mamos, T.; Wattier, R.; Burzyński, A.; Grabowski, M. The legacy of a vanished sea: A high level of diversification within a European freshwater amphipod species complex driven by 15 My of Paratethys regression. *Mol. Ecol.* **2016**, *25*, 795–810. [CrossRef]
17. Kázmér, M. Birth, life and death of the Pannonian Lake. *Palaeogeogr. Palaeoclimatol. Palaeoecol.* **1990**, *79*, 171–188. [CrossRef]
18. Rendoš, M.; Delić, T.; Copilaș-Ciocianu, D.; Fišer, C. First insight into cryptic diversity of a Caucasian subterranean amphipod of the genus *Niphargus* (Crustacea: Amphipoda: Niphargidae). *Zool. Anz.* **2021**, *290*, 1–11. [CrossRef]
19. Dudich, E. Új rákfajok Magyarország faunájában. Neue Krebstiere in der Fauna Ungarns. *Arch. Balaton.* **1927**, *1*, 343–387.
20. Dudich, E. Biologie der Aggteleker Tropfsteinhöhle "Baradla" in Ungarn. *Spaläolog. Monogr.* **1932**, *13*, 1–246.
21. Méhelÿ, L. *Új férgek és rákok a Magyar faunában. Neue Würmer und Krebse aus Ungarn*; Held: Budapest, Hungary, 1927; p. 19.
22. Méhelÿ, L. *A Niphargus kutatás új útjai. Neue Wege der Niphargus-Forschung*; Stephaneum: Budapest, Hungary, 1941; p. 36.
23. Karaman, G.S. One new species of family Niphargidae (Gammaridea), *Niphargus forroi* sp. n., from Hungary. *Acta Zool. Acad. Sci. Hung.* **1986**, *32*, 61–72.
24. Fišer, C.; Trontelj, P.; Luštrik, R.; Sket, B. Towards a unified taxonomy of *Niphargus* (Crustacea: Amphipoda): A review of morphological variability. *Zootaxa* **2009**, *2061*, 1–22. [CrossRef]
25. Balázs, G.; Angyal, D.; Kondorosy, E. *Niphargus* (Crustacea: Amphipoda) species in Hungary: Literature review, current taxonomy and the updated distribution of valid taxa. *Zootaxa* **2015**, *3974*, 361–376. [CrossRef] [PubMed]
26. Muskó, I. Occurrence of Amphipoda in Hungary since 1853. *Crustaceana* **1994**, *66*, 144–152. [CrossRef]
27. Angyal, D.; Balázs, G.; Zakšek, V.; Krízsik, V.; Fišer, C. Redescription of two subterranean amphipods *Niphargus molnari* Méhelÿ, 1927 and *Niphargus gebhardti* Schellenberg, 1934 (Amphipoda, Niphargidae) and their phylogenetic position. *ZooKeys* **2015**, *509*, 53–85. [CrossRef] [PubMed]
28. Copilaș-Ciocianu, D.; Fišer, C.; Borza, P.; Petrusek, A. Is subterranean lifestyle reversible? Independent and recent large-scale dispersal into surface waters by two species of the groundwater amphipod genus *Niphargus*. *Mol. Phylogenet. Evol.* **2018**, *119*, 37–49. [CrossRef]
29. Méhelÿ, L. *Niphargus hungaricus*, ein neuer Amphipode aus Ungarn. *Zool. Anz.* **1937**, *120*, 117–119.

30. Angyal, D.; Balázs, G. Distinguishing characters of *Niphargus gebhardti* Schellenberg, 1934 and *Niphargus molnari* Méhelÿ, 1927 (Crustacea: Amphipoda): A clarification. *Opus. Zool.* **2013**, *44*, 3–8.
31. Karaman, S. Beitrag zur Kenntnis der Süsswasser-Amphipoden (Amphipoden unterirdischer Gewässer). *Prirodoslovne razprave.* **1932**, *2*, 179–232.
32. Dobreanu, E.; Manolache, C. Beitrag zur Kenntnis der Amphipodenfauna Rumäniens. *Notat. Biol.* **1933**, *1*, 103–108.
33. Pérez-Moreno, J.L.; Balázs, G.; Wilkins, B.; Herczeg, G.; Bracken-Grissom, H.D. The role of isolation on contrasting phylogeographic patterns in two cave crustaceans. *BMC Evol. Biol.* **2017**, *17*, 247. [CrossRef] [PubMed]
34. Frivaldszky, J. Adatok a magyarhoni barlangok faunájához. *Math.Természettud. Közl.* **1865**, *3*, 17–53.
35. Verovnik, R.; Sket, B.; Trontelj, P. The colonization of Europe by the freshwater crustacean *Asellus aquaticus* (Crustacea: Isopoda) proceeded from ancient refugia and was directed by habitat connectivity. *Mol. Ecol.* **2005**, *14*, 4355–4369. [CrossRef] [PubMed]
36. Fišer, C.; Zagmajster, M.; Zakšek, V. Coevolution of life history traits and morphology in female subterranean amphipods. *Oikos* **2013**, *122*, 770–778. [CrossRef]
37. Colgan, D.; McLauchlan, A.; Wilson, G.; Livingston, S.P.; Edgecombe, G.D.; Macaranas, J.; Cassis, G.; Gray, M. Histone H3 and U2 snRNA DNA sequences and arthropod molecular evolution. *Aust. J. Zool.* **1998**, *46*, 419–437. [CrossRef]
38. Folmer, O.; Black, M.; Hoeh, W.; Lutz, R.; Vrijenhoek, R. DNA primers for amplification of mitochondrial cytochrome c oxidase subunit I from diverse metazoan invertebrates. *Mol. Mar. Biol. Biotechnol.* **1994**, *3*, 294–299.
39. Jurado-Rivera, J.A.; Álvarez, G.; Caro, J.A.; Juan, C.; Pons, J.; Jaume, D. Molecular systematics of *Haploginglymus*, a genus of subterranean amphipods endemic to the Iberian Peninsula (Amphipoda: Niphargidae). *Contrib. Zool.* **2017**, *86*, 239–260. [CrossRef]
40. Katoh, K.; Standley, D.M. MAFFT Multiple Sequence Alignment Software Version 7: Improvements in Performance and Usability. *Mol. Biol. Evol.* **2013**, *30*, 772–780. [CrossRef]
41. Lanfear, R.; Frandsen, P.B.; Wright, A.M.; Senfeld, T.; Calcott, B. PartitionFinder 2: New Methods for Selecting Partitioned Models of Evolution for Molecular and Morphological Phylogenetic Analyses. *Mol. Biol. Evol.* **2016**, *34*, 772–773. [CrossRef]
42. Guindon, S.; Dufayard, J.F.; Lefort, V.; Anisimova, M.; Hordijk, W.; Gascuel, O. New algorithms and methods to estimate maximum-likelihood phylogenies: Assessing the performance of PhyML 3.0. *Syst. Biol.* **2010**, *59*, 307–321. [CrossRef]
43. Ronquist, F.; Teslenko, M.; van der Mark, P.; Ayres, D.L.; Darling, A.; Höhna, S.; Larget, B.; Liu, L.; Suchard, M.A.; Huelsenbeck, J.P. MrBayes 3.2: Efficient Bayesian phylogenetic inference and model choice across a large model space. *Syst. Biol.* **2012**, *61*, 539–542. [CrossRef] [PubMed]
44. Nguyen, L.T.; Schmidt, H.A.; von Haeseler, A.; Minh, B.Q. IQ-TREE: A fast and effective stochastic algorithm for estimating maximum likelihood phylogenies. *Mol. Biol. Evol.* **2015**, *32*, 268–274. [CrossRef] [PubMed]
45. Kalyaanamoorthy, S.; Minh, B.Q.; Wong, T.K.F.; Von Haeseler, A.; Jermiin, L.S. ModelFinder: Fast model selection for accurate phylogenetic estimates. *Nat. Methods* **2017**, *14*, 587–589. [CrossRef] [PubMed]
46. Minh, B.Q.; Nguyen, M.A.T.; von Haeseler, A. Ultrafast approximation for phylogenetic bootstrap. *Mol. Biol. Evol.* **2013**, *30*, 1188–1195. [CrossRef] [PubMed]
47. Anisimova, M.; Gil, M.; Ois Dufayard, J.F.; Dessimoz, C.; Gascuel, O. Survey of branch support methods demonstrates accuracy, power, and robustness of fast likelihood-based approximation schemes. *Syst. Biol.* **2011**, *60*, 685–699. [CrossRef]
48. Miller, M.A.; Pfeiffer, W.; Schwartz, T. Creating the CIPRES Science Gateway for inference of large phylogenetic trees. In Proceedings of the Gateway Computing Environments Workshop (GCE), New Orleans, LA, USA, 14 November 2010; pp. 1–8. [CrossRef]
49. Som, A. Causes, consequences and solutions of phylogenetic incongruence. *Brief. Bioinform.* **2014**, *16*, 536–548. [CrossRef]
50. Stoch, F.; Christian, E.; Flot, J.-F. Molecular taxonomy, phylogeny and biogeography of the *Niphargus tatrensis* species complex (Amphipoda, Niphargidae) in Austria. *Org. Divers. Evol.* **2020**, *20*, 701–722. [CrossRef]
51. Felsenstein, J. Cases in which parsimony or compatibility methods will be positively misleading. *Syst. Biol.* **1978**, *27*, 401–410. [CrossRef]
52. Philippe, H.; Derelle, R.; Lopez, P.; Pick, K.; Borchiellini, C.; Boury-Esnault, N.; Vacelet, J.; Renard, E.; Houliston, E.; Quéinnec, E.; et al. Phylogenomics revives traditional views on deep animal relationships. *Curr. Biol.* **2009**, *19*, 706–712. [CrossRef]
53. Philippe, H.; Brinkmann, H.; Lavrov, D.V.; Littlewood, D.T.J.; Manuel, M.; Wörheide, G.; Baurain, D. Resolving difficult phylogenetic questions: Why more sequences are not enough. *PLoS Biol.* **2011**, *9*, e1000602. [CrossRef]
54. Phillips, M.J.; Delsuc, F.; Penny, D. Genome-scale phylogeny and the detection of systematic biases. *Mol. Biol. Evol.* **2004**, *21*, 1455–1458. [CrossRef]
55. Heath, T.; Hedtke, S.; Hillis, D. Taxon sampling and the accuracy of phylogenetic analyses. *J. Syst. Evol.* **2008**, *46*, 239–257. [CrossRef]
56. Nabhan, A.R.; Sarkar, I.N. The impact of taxon sampling on phylogenetic inference: A review of two decades of controversy. *Brief. Bioinform.* **2011**, *13*, 122–134. [CrossRef] [PubMed]
57. Hudec, I.; Mock, A. *Niphargus plurispinosus* sp. n. (Crustacea, Amphipoda), a stygophile and hypotelminorheic representative from Central Europe. *Subterr. Biol.* **2014**, *13*, 65–87. [CrossRef]
58. Stloukal, E.; Kuzl, D. First record of niphargid amphipods in region of Bratislava (western Slovakia). *Folia faun. Slov.* **2015**, *20*, 157–162.

59. Fišer, C.; Coleman, C.O.; Zagmajster, M.; Zwittnig, B.; Gerecke, R.; Sket, B. Old museum samples and recent taxonomy: A taxonomic, biogeographic and conservation perspective of the *Niphargus tatrensis* species complex (Crustacea: Amphipoda). *Org. Divers. Evol.* **2010**, *10*, 5–22. [CrossRef]
60. Trontelj, P.; Blejec, A.; Fišer, C. Ecomorphological convergence of cave communities. *Evolution* **2012**, *66*, 3852–3865. [CrossRef]
61. Korpás, L. (Ed.) *Explanations to the Geological Map of the Börzsöny and Visegrád Mountains*; Hungarian Geological Institute: Budapest, Hungary, 1998; p. 216.
62. Veress, M. A general description of karsts in Hungary. In *Cave and Karst Systems of Hungary*; Veress, M., Leél-Őssy, S., Eds.; Springer: Berlin/Heidelberg, Germany, 2022; pp. 1–21.

diversity

MDPI

Article

Groundwater Amphipods of the Hyporheic Interstitial: A Case Study from Luxembourg and The Greater Region

Dieter Weber [1,2] and Alexander M. Weigand [1,3,*]

1 Musée National d'Histoire Naturelle de Luxembourg, 25 Rue Münster, L-2160 Luxembourg, Luxembourg
2 Senckenberg Deutsches Entomologisches Institut, Eberswalder Straße 90, 15374 Müncheberg, Germany
3 Fondation Faune-Flore, Institut de Recherchesur le Patrimoine Naturel et la Biodiversité, L-2160 Luxembourg, Luxembourg
* Correspondence: alexander.weigand@mnhn.lu

Abstract: Hyporheic interstitials are ecologically dynamic and often neglected aquatic environments. In this study, groundwater amphipods (Niphargidae, Pseudoniphargidae and Crangonyctidae) were sampled in hyporheic interstitials throughout Luxembourg and The Greater Region and specimens were analyzed by DNA barcoding. Sites characterized by gravel or coarse sand and high flow velocities of incoming water were the most species- and specimen-rich. A total of 11 species were detected, of which the *Niphargus aquilex* lineage EF of the *N. aquilex* cryptic species complex and juveniles of *N. schellenbergi* dominated the data set, but an unknown lineage of the *N. aquilex*-complex was also found. Some regions appeared to be devoid of groundwater amphipods. We hypothesize that underlying sandstone formations resulting in low sediment porosity may prevent physical colonization, but also that historical water pollution may have a long-lasting effect, either through the persistence of contaminants in the sediment or low recolonization rates of affected populations. In summary, our approach expanded regional species inventories, confirmed known occurrences, and validated previously questionable or historical morphology-based detections. In addition, the collection of absence data provided valuable insights into local extinctions. Finally, DNA-based distribution data are needed to gather information on the ecological affinities of groundwater amphipods to understudied hyporheic interstitial environments.

Keywords: subterranean biology; hyporheos; DNA barcoding; groundwater; Amphipoda

1. Introduction

Amphipods are a very species-rich group of malacostracan crustaceans (Crustacea: Malacostraca: Amphipoda). They comprise at least 500 known obligate groundwater species, so called stygobionts, of which four families (Niphargidae, Pseudoniphargidae, Bogidiellidae and Crangonyctidae) are present in Central Europe [1,2]. As such, they can be frequently encountered in cave waters, springs and aquifers, where they predominantly but not exclusively act as predators [3]. Exact species numbers are hard to provide, and higher taxonomies are quickly outdated, as both species delimitations and taxonomic placements are the subject of frequent and ongoing scientific discussions [4–7]. However, the integration of molecular genetic methods (e.g., DNA barcoding) in recent years has led to the identification of many cryptic species complexes that are currently being unraveled, and, thus, to a significant increase in the number of known groundwater amphipod species [8–12]. Because of this unclear and dynamic taxonomic situation, many purely morphological records and their associated ecological information are questionable (e.g., [13–16]). This also means that there is a lack of reliable information on the habitat preferences and lifestyles of most groundwater amphipods. There are a few notable exceptions to the ecology of subterranean amphipods in the recent literature, where species identification has been based on molecular data (e.g., [2,17,18]).

The hyporheic interstitial is an often neglected environment, not only for groundwater amphipods, mainly due to sampling difficulties. It is found under and beside flowing water, where groundwater and surface water mix. Due to this mixing, it is usually considered an ecotone and is characterized by sharp gradients of biotic and abiotic factors, such as light, temperature, load of inorganic and organic substances or dissolved oxygen [19,20]. Dissolved oxygen and organic matter are transported into the hyporheic interstitial by downwelling water, whereas upwelling groundwater transports inorganic nutrients into the stream [19]. The type and size of the sediment particles forming the hyporheic interstitial are additional important parameters, as they determine the permeability and accessibility of this environment [19,21], ultimately affecting ecosystem functions. Healthy interstitials increase the resilience of a water body because they can act as a buffer for stressors in the water by filtering, storing and thus removing stressors from the water column [22–24]. Many species use the hyporheic interstitial as a nursery or spawning habitat or as a retreat during unfavorable environmental conditions in the above water column [19]. However, the hyporheic interstitial is also home to a variety of specialized meiofaunal species [19], but the biodiversity of hyporheic communities is poorly understood due to difficulties in the identification of minute individuals, habitat inaccessibility and declining taxonomic expertise [25].

This study aims to increase the knowledge on groundwater amphipods occurring in the hyporheic interstitials of Luxembourg and The Greater Region, the latter including Rhineland-Palatinate and Saarland (Germany), Lorraine (France) and Wallonia (Belgium). In principle, the groundwater amphipod fauna have been already extensively studied in Luxembourg in recent years. However, older taxonomic determinations were based on morphological characters alone [26], or investigations were mainly based on other habitat types, e.g., springs [27] or artificial caverns [28]. At the same time, it has been noted that the identification of groundwater amphipods without DNA-based methods is highly error-prone [28]. This fact was also emphasized by other authors for their regional groundwater amphipod fauna [29–32]. The current setting therefore provided a suitable framework for applying DNA barcoding to assess the diversity of groundwater amphipods in hyporheic interstitials of Luxembourg and The Greater Region. We hypothesized that the integration of molecular data would help identify frequently encountered juveniles and members of cryptic species complexes (e.g., *Niphargus aquilex*-complex and *N. kochianus*-complex), which are known to occur in the study region, and thus will increase the known species diversity of the study area (Hypothesis 1). Furthermore, we hypothesized that small-sized groundwater amphipod species (e.g., some members of *N. aquilex*-complex, *N. kochianus*-complex, *Niphargellus nolli* Schellenberg, 1938 and *Microniphargus leruthi* Schellenberg, 1934) would dominate the hyporheic interstitial due to size limitations of large-sized species (e.g., *Niphargus fontanus* Spence Bate, 1859, *N. schellenbergi* S. Karaman, 1932 and *N. puteanus* (C.L. Koch in Panzer, 1836)) (Hypothesis 2).

2. Materials and Methods

2.1. Sampling Sites and Sampling Method

From June 2017 to July 2021, hyporheic interstitials [33] were investigated in our study area, which included the entire Grand Duchy of Luxembourg, the Province of Luxembourg and some sites in the immediate vicinity of the Province of Liège (both in Belgium), the states of Saarland and Rhineland-Palatinate and the neighboring areas in western Hesse and western Baden-Württemberg (all in Germany), and a single site in the Moselle department (France) (Figure 1, Table S1).

To benefit from low water levels, riverbanks were preferably investigated in the summer at a distance of a couple of m from the stream bank. Whenever possible, gravel banks located on river islands were included in the survey. The Karaman–Chappuis method was applied to collect groundwater amphipods [34,35]. A small pickaxe was used to dig pits in the gravel until they were about 10 cm below the water level. Wherever possible, pits were characterized according to their predominant substrate type ('silt' [<0.063 mm], 'fine and medium sand' [0.063–0.63 mm], 'coarse sand' [0.63–2 mm], 'gravel' [>2 mm]), the inflow direction of the water ('upwelling', 'from side', 'both') and the inflow rate ('<1 L/min', '1–10 L/min', '>10 L/min') (Table S2). The incoming water was investigated for groundwater amphipods for approx. 30 min, but less in the case of high inflow rates. Depending on the speed of the inflow, between 1 and about 50 L of water were examined. The water and a small part of the sediment collected at the bottom of the pit were sieved through a set of sieves (mesh sizes of 5, 1, 0.5 and 0.2 mm). The upper sieve retained coarse gravel and stones. In rare cases, larger amphipods were found in the 1 mm mesh sieve, while most individuals slipped through to the 0.5 mm mesh sieve and rarely to the 0.2 mm sieve. Gammarids (Amphipoda: Gammaridae) were sorted out on the spot. Eyeless white amphipods were transferred individually to 96% undenatured ethanol and later stored at −20 °C.

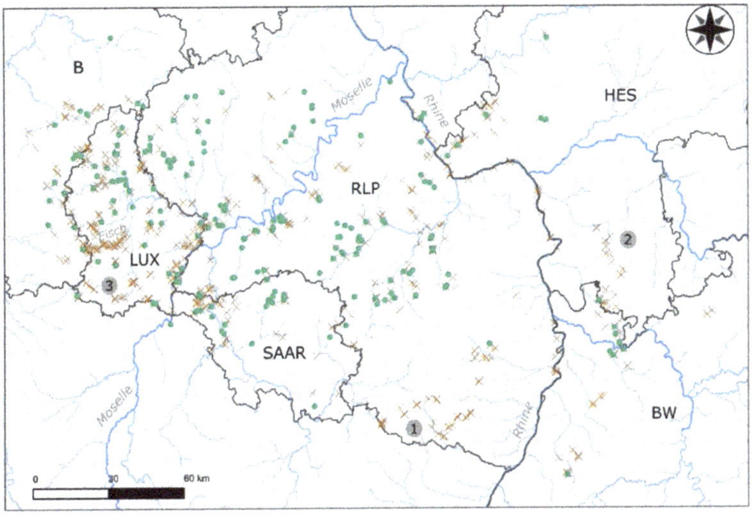

Figure 1. Overview of sampling sites in hyporheic interstitials. Green circles: groundwater amphipods present; orange crosses: no groundwater amphipods found. B = Belgium, BW = Baden-Württemberg (Germany), HES = Hesse (Germany), LUX = Luxembourg, RLP = Rhineland-Palatinate (Germany), SAAR = Saarland (Germany). 1 = Palatinate Forest mountain range (Germany), 2 = western Odenwald mountain range (Germany), 3 = Minett and southern Gutland regions (Luxembourg). The hydrological network and administrative boundaries are indicated, and the map is cropped to show the sampling sites of our core study area.

Pearson's Chi^2 test [36] was used to compare the observed patterns of groundwater amphipod detection with the expectation that the assessed parameters of the collection pits (i.e., sediment type, inflow direction and inflow rate) had no effect on the presence (or the probability of detection) of groundwater amphipods.

2.2. Molecular Identification of Specimens

Specimens were morphologically investigated using a Stereo Zoom Microscope 4002 (Wang BioMedical, Wageningen, The Netherlands). In general, one specimen per site was barcoded, but in the case of different morphotypes, several specimens from the same site were barcoded, even though they may represent different developmental stages of the same species. 273 specimens from 267 sites were DNA barcoded in this study, and enriched by 19 barcoded specimens from 9 interstitial sites in Luxembourg from another source [28]. Specimens were shipped to the laboratory of the Aquatic Ecosystem Research Group (University of Duisburg-Essen, Germany) for DNA barcoding of the 658 bp-long cytochrome *c* oxidase subunit 1 (COI) marker typically used for animals. The obtained COI sequences were trimmed and edited using Geneious R6 version 6.1.8 [37]. The Neighbor Joining algorithm in Mega X [38] was run with p-distance, 1000 bootstrap replicates and the pairwise deletion option to cluster sequences and identify species according to the latest regional taxonomic revisions and annotations [11,28,30,39,40]. Members of the *N. aquilex*-complex were annotated as the respective (sub)lineage.

3. Results

3.1. Characterization of Hyporheic Interstitials and Regional Patterns

A total of 1006 pits were created, of which 267 contained a total of 957 groundwater amphipods (Figure 1). It was not possible to characterize all the investigated hyporheic interstitials according to their predominant sediment type, since in some places, sediment types were highly intermixed, with no predominate sediment type present (Figure 2, Table S2). However, given the respective pool of characterized sampling sites, groundwater amphipods were underrepresented in silty sediments and medium and fine sands, and more frequent than expected in coarse sands [X^2 (3, N = 225) = 22.1698, $p < 0.0001$]. There was also a higher chance of collecting amphipods under upwelling conditions [X^2 (2, N = 172) = 14.3335, $p < 0.001$] and at high inflow rates [X^2 (2, N = 225) = 22.2656, $p < 0.0001$] (Figure 2, Table S2), taking the variable sampling times and investigated inflow volumes into consideration.

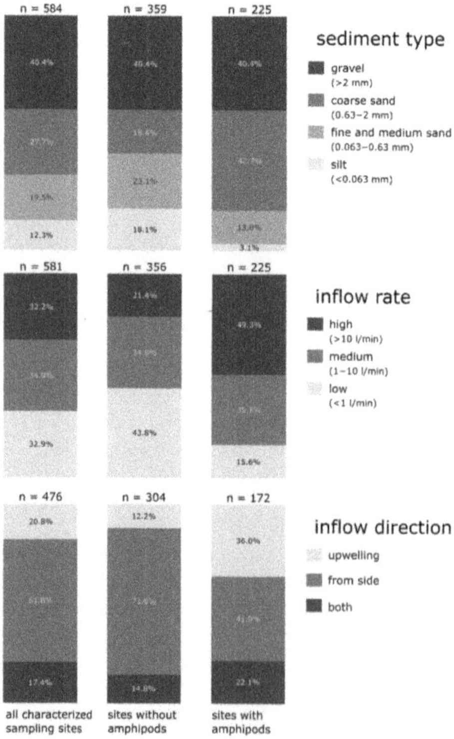

Figure 2. Characterization of collection pits. Indicated are the relative proportions of sediment types (gravel, coarse sand, fine and medium sand, silt), speed of inflow (high, medium, large) and direction of inflow (upwelling, from side, both), respectively, for all characterized collection pits, for sites without groundwater amphipods, as well as for sites with groundwater amphipods. The number of sites characterized (n) is given for each parameter individually, as not all sites could be evaluated for all parameters. More information can be retrieved from Tables S1 and S2.

Groundwater amphipods were collected in the hyporheic interstitials across the whole study area. However, in some areas, there was no accessible hyporheic interstitial due to geological and anthropogenic features, e.g., in deep incisions of straightened streams (Upper Rhine lowlands and West Palatinate moor lowlands, both in Germany) (Figure 1). In other regions, no groundwater amphipods were found in the hyporheic interstitial despite the high sampling effort, e.g., in the Eisch River (Belgium, Luxembourg), in the Rhine River, in the Palatinate Forest and in the western Odenwald mountain range (all three in Germany), as well as in the Minett and southern Gutland regions (both in Luxembourg) (Figure 1).

3.2. *Groundwater Amphipod Species Diversity*

The genetically investigated specimens formed 11 lineages (species richness S = 11), which we were able to taxonomically annotate as *M. leruthi*, five members of the *N. aquilex*-complex *sensu* Schiødte, 1855, *N. puteanus*, *N. fontanus*, *N. schellenbergi*, *N. nolli* and *Crangonyx subterraneus* Spence Bate, 1859 (Table 1, Figure S1). The five entities of the *N. aquilex*-complex were *N. aquilex* A, *N. aquilex* B, *N. aquilex* EF (*sensu* [28,30]), *N. aquilex* G (*sensu* [39]) and a single specimen of a so-far-undescribed lineage, collected in the hyporheic interstitial of the Udelfanger Bach River (the state of Rhineland-Palatinate, Germany). *Niphargus aquilex* EF and *N. aquilex* G contain two separate sublineages (F-1 and F-2, as well as G-1 and G-2, respectively) each, but available data are not sufficient to unambiguously infer the species status of either sublineage. We therefore decided to treat *N. aquilex* EF and *N. aquilex* G as one species each, but to separate our findings of the respective sublineages (Tables 1 and 2, Figure S1). *Niphargus aquilex* EF (sublineage F-1) and *N. schellenbergi* were the most frequently observed taxa in the investigated hyporheic interstitial, occurring at 39.7 (106/267; 114 specimens) and 22.8% (77/267; 101 specimens), respectively, of all sites where groundwater amphipods were detected. The species diversity of groundwater amphipods detected in the interstitials in this study was compared with the known regional groundwater amphipod fauna of the two best-studied areas in our target region, i.e., Luxembourg and Saarland (Table 2). To ensure a fair comparison, only the regionally observed taxa from our study were used for comparison. The findings of *N. aquilex* A and *N. nolli* in Luxembourg are the first records for the country. *Microniphargus leruthi sensu stricto* [11] has been found again in Luxembourg, 60 years after its first finding [41]. Six groundwater amphipod species have been recorded for Saarland despite a decade-long and intensive research in the most diverse biotopes and using DNA-based tools for species identification. While we were not able to detect *N. aquilex* A and I, *N. kochianus* Spence Bate, 1859 (questionable record) and *N. virei* Chevreux, 1896 in the investigated hyporheic interstitials, we can confirm the presence of the species *N. schellenbergi*, *N. aquilex* EF (sublineage F-1) and G (sublineage G-1), as well as *N. nolli* (so-far-questionable record). Furthermore, *N. aquilex* B, which refers to the type species of the cryptic species complex [28], must be added to the regional checklist.

Table 1. Overview of groundwater amphipod species found in the hyporheic interstitial of Luxembourg and The Greater Region. Taxonomic annotations: *Niphargus aquilex* A, B, EF *sensu* [28,30], *aquilex* G *sensu* [39]. The single finding of *Microniphargus leruthi* originates from a cave interstitial. B-Wal = Wallonia (Belgium), D-BW = Baden-Württemberg (Germany), D-Hes = Hesse (Germany), D-RLP = Rhineland-Palatinate (Germany), D-Saar = Saarland (Germany), LUX = Luxembourg, n = number of barcoded specimens, x = species detected in the target region. Included are barcoded specimens from our study and [28].

Species	n	LUX	B-Wal	D-RLP	D-Saar	D-Hes	D-BW
Crangonyx subterraneus Spence Bate, 1859	4		x	x		x	
Microniphargus leruthi Schellenberg, 1934	1	x					
Niphargellus nolli Schellenberg, 1938	6	x		x	x	x	
Niphargus aquilex A	7	x		x			

Table 1. *Cont.*

Species	n	LUX	B-Wal	D-RLP	D-Saar	D-Hes	D-BW
Niphargus aquilex B (*sensu stricto*)	28	x	x	x	x	x	
Niphargus aquilex EF (sublineage F-1)	114	x	x	x	x	x	
Niphargus aquilex EF (sublineage F-2)	11	x	x	x			
Niphargus aquilex G (sublineage G-1)	12			x	x	x	x
Niphargus aquilex G (sublineage G-2)	1						x
Undescribed species of the *N. aquilex*-complex	1			x			
Niphargus fontanus Spence Bate, 1859	1	x					
Niphargus puteanus (C.L. Koch in Panzer, 1836)	5			x			x
Niphargus schellenbergi S. Karaman, 1932	101	x	x	x	x	x	x

Table 2. Comparison of groundwater amphipod species diversity from this study and the literature. Indicated are the species found in the hyporheic interstitials in Luxembourg and Saarland from this study, as well as from published data on groundwater amphipod diversity from Luxembourg [27,28] and Saarland [12]. Taxonomic annotations: *Niphargus aquilex* A, B, EF *sensu* [28,30], *N. aquilex* G, I *sensu* [39]. x = present; (x) questionable morphodetermination.

Species	Luxembourg Interstitial (This Study)	Luxembourg Interstitial [27]	Luxembourg All [28]	Saarland Interstitial (This Study)	Saarland All [12]
Microniphargus leruthi Schellenberg, 1934	x				
Niphargellus nolli Schellenberg, 1938	x			x	(x)
Niphargus aquilex-complex *sensu* Schiødte, 1855		x			
Niphargus aquilex A	x				x
Niphargus aquilex B (*sensu stricto*)	x		x	x	
Niphargus aquilex EF (sublineage F-1)	x		x	x	x
Niphargus aquilex EF (sublineage F-2)	x		x		
Niphargus aquilex G (sublineage G-1)			x	x	x
Niphargus aquilex I					x
Niphargus fontanus Spence Bate, 1859	x		x		
Niphargus kochianus-complex *sensu* Spence Bate, 1859			x		(x)
Niphargus puteanus (C.L. Koch in Panzer, 1836)			x		
Niphargus schellenbergi S. Karaman, 1932	x	x	x	x	x
Niphargus virei Chevreux, 1896					x

4. Discussion

4.1. Overall High Species Diversity in the Hyporheic Interstitial

We hypothesized that the application of DNA barcoding for the groundwater amphipod fauna of the hyporheic interstitial will lead to an increase in species diversity compared to the older literature, as members of cryptic species complexes and juveniles can be readily identified down to the species level (Hypothesis 1) [28,30]. With 11 putative groundwater amphipod species, species richness (*S*) was remarkably high in the interstitials of the study area. In the specific case of Luxembourg (*S* = 7), our study led to the first finding of *N. aquilex* A and *N. nolli* in the country, as well as to the redetection of *M. leruthi sensu stricto* [11] 60 years after its first finding in the country [41]. A recent integrative taxonomic study using DNA barcoding and traditional morphology has already discovered much more species than were previously known for the country [28], although mainly artificial cavities (e.g., mines) and springs were studied. On the contrary, another study systematically investigated the amphipod fauna of 41 springs and 30 hyporheic interstitial sites in Luxembourg, but did not apply DNA barcoding [27]. The detection of two additional new species in the hyporheic interstitials in Luxembourg by our study must thus be related to the combination of a systematic sampling in this specific environment and a DNA-based identification of specimens. We expect that a more systematic investigation of the hyporheic interstitials in Belgium, France and other parts of Germany will likely reveal further unrecognized groundwater amphipod species. However, the finding of *M. leruthi* is not linked to a hyporheic interstitial site, but the species was revealed from a cave interstitial. This may indicate that this environment has not been sufficiently studied in the study area and also warrants further investigation. Most often, caves and artificial cavities are opportunistically studied [42], and samplings are made when 'something' is encountered. This strategy thereby can easily overlook small species/specimens such as the 1.2–2-mm long *M. leruthi* [41,43]. Alternatively, cave lakes or subterranean streams could be sampled by the aid of either bait boxes or plankton nets. In both cases, mainly the fauna of cave lakes and subterranean streams are surveyed, but not those of cave interstitials. Since the transportation of a Bou–Rouch pump [35,44] into caves is not practical, the Karaman–Chappuis method [34] seems to be a suitable and effective alternative. In practice, however, this sampling method may encounter other difficulties, as cave soles are often cemented or, in the case of the presence of calcium^{2+} or iron^{2+}, gravel banks may be heavily sintered.

In addition to the specific case of Luxembourg, we also compared our findings with the recently compiled amphipod checklist of the state of Saarland (Germany) [12]. Our data thus led to a slight increase in the total groundwater amphipod species richness in Saarland (S = 6 to S = 7), in addition to confirming previously questionable records. Furthermore, the finding of *N. aquilex* B highlights the importance of the hyporheic interstitial acting as a habitat of so far poorly known groundwater amphipod species. *Niphargus aquilex* B is considered a rare species in the study area, and was known in the study area only in Luxembourg, where a single specimen was collected by [28] in the Sauer River. However, in the context of our study, we collected 28 specimens of this species throughout the hyporheic interstitials of the whole study area (Luxembourg and The Greater Region). A preference for the hyporheic interstitial can be also revealed for *N. aquilex* EF (sublineage F-1), which was the predominant taxon in our study. However, our still-limited knowledge of the biodiversity of hyporheic interstitials is best illustrated by our discovery of a species new to science within the *N. aquilex*-complex.

4.2. Small Individuals but Not Small Species Dominate in the Hyporheic Interstitial

We speculated that small-sized groundwater amphipod species would dominate in the hyporheic interstitial due to the size benefits when compared to adults of larger species (Hypothesis 2). In the study area, several small species were expected, e.g., *M. leruthi* (adult length 1.2–2 mm) [43,45,46], *C. subterraneus* (1.2–6 mm) [47], *N. nolli* (2.8 mm) [48,49], *N. dimorphopus* Stock & Gledhill, 1977 (3.5–4.5 mm) [50], *N. kieferi* Schellenberg, 1936 (7 mm) [48], *N. kochianus* (5–8 mm) [48–50], *N. laisi* Schellenberg, 1936 (5–7 mm) [48,49] and *N. pachypus* Schellenberg, 1933 (7 mm) [51]. While both *C. subterraneus* and *N. nolli* were indeed found in the hyporheic interstitial in Luxembourg and The Greater Region, albeit in low numbers, the other species were absent. In addition, *M. leruthi* was found at a single site from a cave interstitial. While large(r) adult individuals of groundwater amphipods are physically excluded by the small spaces offered by most hyporheic interstitials, their juveniles are principally able to colonize them. However, those taxonomic findings are usually not recorded, because the morphodetermination of juvenile groundwater amphipods is known to be a very challenging or even impossible task. For example, almost 25 years ago, [27] investigated the hyporheic interstitial of Luxembourg and recorded a total of 154 niphargids, but with several juvenile specimens of *Niphargus* sp. present in the collection (pers. obs. AW). *Niphargus aquilex sensu lato* dominated their dataset (146 specimens, 95%), whereas just eight specimens (5%) were morphologically identified as *N. schellenbergi*. In our study, we genetically identified *N. schellenbergi* as one of the two predominant species in the hyporheic interstitial of Luxembourg. It is considered a large-sized species whose adults reach a length of about 15 mm (according to own investigation, they can be up to 18 mm long) [12,29,52], but apparently juveniles seem to strive well in the hyporheic interstitial. Since a faunal change between the time of the study of [27] and our sampling and in all interstitial habitats in Luxembourg seems unlikely, the more likely explanation is that juveniles of *N. schellenbergi* cannot be identified due to their size or life-period morphological characteristics.

Two other large(r)-sized species, *N. fontanus* and *N. puteanus*, were also revealed, but in low numbers. *Niphargus fontanus* is generally rare, but seems to be slightly more common in Belgium [26]. While the species could not be detected by morphological identification of the material from Luxembourg [27], it was sequenced from one hyporheic interstitial site in the Minett region (Luxembourg). Additionally, *N. puteanus* is generally rare in the study area, with more frequent records in the eastern part [18,28,41]. We detected this species only in hyporheic interstitials located in Rhineland-Palatinate and Baden-Württemberg (both in Germany), which is in accordance with the results of [18].

In summary, the present data indicate that there is no clear pattern of small-sized groundwater amphipod species dominating at least the shallow hyporheic interstitials (<0.5 m depth) that we sampled, but the medium-sized *N. aquilex* EF (sublineage F-1) and juveniles of the large-sized *N. schellenbergi* were predominant. *Niphargus schellenbergi* is by far the most common species in the study area and has so far been found mainly in caves, artificial cavities and springs [12,28,29,52]. Due to the predominance of juveniles in the hyporheic interstitial, it is likely that it has often been overlooked in the past. *Niphargus schellenbergi* must be considered an ecological generalist, colonizing and surviving in different and distinct groundwater-dependent ecosystems and environmental conditions (see [28] and this study). The strong dominance of *N. aquilex* EF among the members of the *N. aquilex*-complex in our data set is striking. Based on our data, *N. aquilex* EF seems to be an interstitial specialist. However, its specialization is not reflected at the morphological level, since other members of the *N. aquilex*-complex are cryptic or at least pseudocryptic [40]. The environmental specializations for hyporheic interstitials might originate from eco-physiological adaptations to better cope with the often fluctuating environmental conditions and steep abiotic and biotic gradients [20,22–24], or the increased numbers of host–parasite interactions when being confronted with the parasites of epigean taxa [39,52]. In this context, we can assume that the hyporheic interstitial represents the main habitat of *N. aquilex* EF (sublineage F-1), which completes its entire lifecycle in this environment. On the contrary, only juveniles of *N. schellenbergi* seem to be physically able to inhabit interstitial sediments, whereas its adults have to migrate into groundwater-dependent environments with larger porosity (e.g., aquifers), connected to hyporheic interstitial sites. This difference in the ecological connection to the hyporheic interstitial also affects the likeliness of recolonization events after environmentally unsuitable conditions, which should be higher for *N. schellenbergi* when compared to interstitial specialists such as *N. aquilex* EF.

4.3. Regional Absence of Groundwater Amphipods in Hyporheic Interstitials

The Karaman-Chappuis method used proved to be effective for collecting groundwater amphipods from different hyporheic interstitial sites. However, only groundwater close to the sediment surface and near the riverbanks can be investigated, whereas a Bou-Rouch pump is closed and reaches down to a depth of 2 m, enabling the investigation of hyporheic interstitials at a certain distance from the river or even of submerged sediments. Acknowledging these sampling limitations and the fact that our results must be interpreted in this context as well, we were able to identify some geographical regions and rivers without any groundwater amphipods in the hyporheic interstitials, although groundwater amphipods are common in other biotopes in these areas (e.g., springs, mines, caves) [12,28,29,41,52]. Their absence in the Palatinate Forest mountain range (Rhineland-Palatinate, Germany) and the western Odenwald mountain range (Baden-Württemberg and Hesse, both in Germany) can be best explained by the geologies of the mountain ranges: Buntsandstein weathers into small sand grains, which results in overall low porosity and thus accessibility of sediments. This situation is enhanced by the weathering clay of the overlying Muschelkalk, which can close the remaining small pores in the hyporheic interstitial and ultimately lead to sediment clogging. As such, a high fraction of silt and fine sands and a low inflow speed characterized the collection pits in those regions, not allowing even small-sized groundwater amphipods to be present. The occurrence of sandstone environments might also be responsible for the apparent absence of groundwater amphipods in interstitial sites in the Minett and southern Gutland regions (both in Luxembourg). In addition, the Minett region is characterized by an intense historical iron industry [53,54], and today is highly populated and urbanized. Both might have led to an accumulation of stressors affecting the regional aquatic fauna in general [55,56].

In addition to the absence of groundwater amphipods in the hyporheic interstitials of some mountain ranges or geographical regions, groundwater amphipods were also absent from a few but well-sampled rivers. In the Eisch River (Belgium and Luxembourg), 24 sites in the main channel and another 9 sites in its tributaries in the close vicinity to the main channel were investigated, but no specimen was found. The Eisch River was contaminated multiple times between 1948 and 1961 by the phenol industry located in Steinfort (upstream from our sampling sites), either directly by the slightly water-soluble phenol with its safety classification H411 (i.e., toxic to aquatic organisms with long-lasting effects [57]), or indirectly by its degradation products, leading to the extinction of the local aquatic fauna [58]. It can be expected that traces of these contaminations are still present in the sediments and aquifers, potentially also influencing the presence of groundwater amphipods or affecting the probability of their recolonization. A similar anthropogenically induced situation could be found in the Rhine River (Germany). Although the river has also been heavily straightened, there are many larger gravel banks along the riverbanks. We were able to examine 31 pits on different gravel banks, all of which remained free of living amphipods. It is unlikely that the interstitials of the Rhine River were never colonized by groundwater amphipods [18]. The most plausible hypothesis is therefore similar to the situation of the Eisch River. Namely, municipal wastewater and frequent pollution by the chemical industry in the 1950s–1970s [58] led to an accumulation of toxic substances in the aquifers and sediments. Although the epigean aquatic fauna have largely recovered [59,60], groundwater amphipods apparently have not returned there yet. Even in the case of a high ecological suitability of the investigated hyporheic interstitials in the Rhine River, it is most likely that a generally slow migration rate prevents groundwater amphipod populations from recolonizing them faster after serious pollution events. Future research should focus on identifying the ecophysiological or physical factors that prevent groundwater amphipod species from being more abundant in hyporheic interstitials. Similarly, one has to generate more knowledge on the actual species diversity in hyporheic interstitials and the ecological preferences of individual groundwater amphipods for this specific environment. Of particular interest could be species from the hyporheic interstitial that are more widely distributed, which could possibly represent complexes of cryptic species, or which can provide new insights into the dispersal capabilities, resiliencies and migration corridors of groundwater amphipods. Likewise, their areas of distribution or regional absence should be investigated in more detail. How deep are the potential chemical pollutants and do they affect the quality of groundwater? If previous pollution still has a significant effect, this could indicate that groundwater (and drinking water) pollution is a long-term process with a slow recovery rate. If the water quality is fine, it might indicate that the recolonization of groundwater and the recovery of ecosystem functions are long-lasting processes. Both aspects are important for groundwater monitoring, especially in relation to drinking water.

Supplementary Materials: The following supporting information can be downloaded at: https://www.mdpi.com/article/10.3390/d15030411/s1, Table S1: Overview of sampling sites and specimen metadata; Table S2: Additional data on the characterization of the collection pits underlying Figure 2; and Figure S1: Neighbor Joining-tree reconstruction based on COI of groundwater amphipod species detected in the hyporheic interstitial of Luxembourg and The Greater Region.

Author Contributions: Conceptualization, D.W. and A.M.W.; methodology, D.W. and A.M.W.; formal analysis, D.W. and A.M.W.; investigation, D.W. and A.M.W.; resources, D.W. and A.M.W.; data curation, D.W. and A.M.W.; writing—original draft preparation, D.W. and A.M.W.; writing—review and editing, D.W. and A.M.W.; visualization, A.M.W.; supervision, A.M.W.; project administration, D.W. All authors have read and agreed to the published version of the manuscript.

Funding: This research received no external funding.

Institutional Review Board Statement: Not applicable.

Data Availability Statement: Metadata of the specimens analyzed in this study can be found in the Supplementary Materials Table S2. DNA barcodes and the associated metadata (i.e., collector, date of collection, locality, coordinates) are additionally deposited in the Barcode of Life Data System project 'MNHNL'.

Acknowledgments: Christian Anders, Verena Schuhmacher and Monika Weber assisted in collection. Julian Wittische is acknowledged for generating the overview map of interstitial sites where groundwater amphipods were collected. The laboratory work was carried out by Dominik Buchner and Martina Weiss at the University of Duisburg–Essen (Germany). We also wish to thank Teo Delić, the remaining four anonymous reviewers and the editor for their critical feedback on our submission.

Conflicts of Interest: The authors declare no conflict of interest.

References

1. Horton, T.; Lowry, J.; De Broyer, C.; Bellan-Santini, D.; Copilaș-Ciocianu, D.; Corbari, L.; Costello, M.J.; Daneliya, M.; Dauvin, J.-C.; Fišer, C.; et al. World Amphipoda Database. Available online: https://www.marinespecies.org/amphipoda/ (accessed on 18 January 2023).
2. Borko, Š.; Trontelj, P.; Seehausen, O.; Moškrič, A.; Fišer, C. A subterranean adaptive radiation of amphipods in Europe. *Nat. Commun.* **2021**, *12*, 3688. [CrossRef] [PubMed]
3. Premate, E.; Borko, Š.; Delić, T.; Malard, F.; Simon, L.; Fišer, C. Cave amphipods reveal co-variation between morphology and trophic niche in a low-productivity environment. *Freshw. Biol.* **2021**, *66*, 1876–1888. [CrossRef]
4. Esmaeili-Rineh, S.; Sari, A.; Delić, T.; Moškrič, A.; Fišer, C. Molecular phylogeny of the subterranean genus *Niphargus* (Crustacea: Amphipoda) in the Middle East: A comparison with European niphargids. *Zool. J. Linn. Soc.* **2015**, *175*, 812–826. [CrossRef]
5. Fišer, C.; Sket, B.; Trontelj, P. A phylogenetic perspective on 160 years of troubled taxonomy of *Niphargus* (Crustacea: Amphipoda). *Zool. Scr.* **2008**, *37*, 665–680. [CrossRef]
6. Fišer, C.; Trontelj, P.; Luštrik, R.; Sket, B. Toward a unified taxonomy of *Niphargus* (Crustacea: Amphipoda): A review of morphological variability. *Zootaxa* **2009**, *2061*, 1–22. [CrossRef]
7. Stoch, F.; Christian, E.; Flot, J.F. Molecular taxonomy, phylogeny and biogeography of the *Niphargus tatrensis* species complex (Amphipoda, Niphargidae) in Austria. *Org. Divers. Evol.* **2020**, *20*, 701–722. [CrossRef]
8. Fišer, C.; Zagmajster, M. Cryptic species from cryptic space: The case of *Niphargus fongi* sp. n. (Amphipoda, Niphargidae). *Crustaceana* **2009**, *82*, 593–614. [CrossRef]
9. Delić, T.; Trontelj, P.; Rendoš, M.; Fišer, C. The importance of naming cryptic species and the conservation of endemic subterranean amphipods. *Sci. Rep.* **2017**, *7*, 3391. [CrossRef]
10. Fišer, C.; Robinson, C.T.; Malard, F. Cryptic species as a window into the paradigm shift of the species concept. *Mol. Ecol.* **2018**, *27*, 613–635. [CrossRef]
11. Weber, D.; Stoch, F.; Knight, L.R.F.D.; Chauveau, C.; Flot, J.-F. The genus *Microniphargus* (Crustacea, Amphipoda): Evidence for three lineages distributed across northwestern Europe and transfer from Niphargidae to Pseudoniphargidae. *Belg. J. Zool.* **2021**, *151*, 169–191. [CrossRef]
12. Weber, D.; Flot, J.-F. Rote Liste und Gesamtartenliste der Grundwasserkrebse (Niphargidae) des Saarlandes. In *Rote Liste Gefährdeter Tiere und Pflanzen des Saarlandes;* Ministerium für Umwelt und Verbraucherschutz und DELATTINIA: Saarbrücken und Landsweiler-Reden, Germany, 2020; pp. 3–9.
13. Rütz, N.K.; Marxsen, J.; Wolters, V. Long-term cultivation of the groundwater amphipod *Niphargus aquilex* (Crustacea). *Hydrobiologia* **2023**, *850*, 269–281. [CrossRef]
14. Husmann, S. Studies on subterranean drift of stygobiont crustaceans (*Niphargus, Crangonyx, Graeteriella*). *Int. J. Speleol.* **1976**, *8*, 81–92. [CrossRef]
15. Hinz, W.; Danneel, I. Vorkommen von *Niphargus a. aquilex* Schiödte (Amphipoda) in Oberflächengewässern bei Düsseldorf. *Nat. Und Heim.* **1974**, *34*, 7–11.
16. Spangenberg, H.-J. Beitrag zur Faunistik von Höhlengewässern im Zechstein des Südharzes und Kyffhäusers. *Hercynia (Neue Folge)* **1973**, *10*, 143–160.

17. Copilaş-Ciocianu, D.; Fišer, C.; Borza, P.; Petrusek, A. Is subterranean lifestyle reversible? Independent and recent large-scale dispersal into surface waters by two species of the groundwater amphipod genus *Niphargus*. *Mol. Phylogenetics Evol.* **2018**, *119*, 37–49. [CrossRef]
18. Weber, D.; Flot, J.-F.; Weigand, H.; Weigand, A.M. Demographic history, range size and habitat preferences of the groundwater amphipod *Niphargus puteanus* (C.L. Koch in Panzer, 1836). *Limnol. Ecol. Manag. Inland Waters* **2020**, *82*, 125765. [CrossRef]
19. Boulton, A.J.; Findlay, S.; Marmonier, P.; Stanley, E.H.; Valett, H.M. The functional significance of the hyporheic zone in streams and rivers. *Annu. Rev. Ecol. Syst.* **1998**, *29*, 59–81. [CrossRef]
20. Gómez-Gener, L.; Siebers, A.R.; Arce, M.I.; Arnon, S.; Bernal, S.; Bolpagni, R.; Datry, T.; Gionchetta, G.; Grossart, H.-P.; Mendoza-Lera, C.; et al. Towards an improved understanding of biogeochemical processes across surface-groundwater interactions in intermittent rivers and ephemeral streams. *Earth-Sci. Rev.* **2021**, *220*, 103724. [CrossRef]
21. Mathers, K.L.; Hill, M.J.; Wood, C.D.; Wood, P.J. The role of fine sediment characteristics and body size on the vertical movement of a freshwater amphipod. *Freshw. Biol.* **2019**, *64*, 152–163. [CrossRef]
22. Lawrence, J.E.; Skold, M.E.; Hussain, F.A.; Silverman, D.R.; Resh, V.H.; Sedlak, D.L.; Luthy, R.G.; McCray, J.E. Hyporheic zone in urban streams: A review and opportunities for enhancing water quality and improving aquatic habitat by active management. *Environ. Eng. Sci.* **2013**, *30*, 480–501. [CrossRef]
23. Stubbington, R.; Dole-Olivier, M.-J.; Galassi, D.M.P.; Hogan, J.-P.; Wood, P.J. Characterization of macroinvertebrate communities in the hyporheic zone of river ecosystems reflects the pump-sampling technique used. *PLoS ONE* **2016**, *11*, e0164372. [CrossRef] [PubMed]
24. Datry, T.; Singer, G.; Sauquet, E.; Jorda-Capdevila, D.; Von Schiller, D.; Stubbington, R.; Magand, C.; Pařil, P.; Miliša, M.; Acuña, V.; et al. Science and management of intermittent rivers and ephemeral streams (SMIRES). *Res. Ideas Outcomes* **2017**, *3*, e21774. [CrossRef]
25. Weigand, A.M.; Macher, J.-N. A DNA metabarcoding protocol for hyporheic freshwater meiofauna: Evaluating highly degenerate COI primers and replication strategy. *Metabarcoding Metagenom.* **2018**, *2*, e26869. [CrossRef]
26. Fišer, C.; Zagmajster, M.; Dethier, M. Overview of Niphargidae (Crustacea: Amphipoda) in Belgium: Distribution, taxonomic notes and conservation issues. *Zootaxa* **2018**, *4387*, 47. [CrossRef]
27. Gerecke, R.; Stoch, F.; Meisch, C.; Schrankel, I. Die Fauna der Quellen und des hyporheischen Interstitials in Luxembourg. Unter Besonderer Berücksichtigung der Milben (Acari), Muschelkrebse (Ostracoda) und Ruderfusskrebse (Copepoda). *Ferrantia* **2005**, *41*, 1–140.
28. Weber, D.; Flot, J.-F.; Frantz, A.; Weigand, A.M. Molecular Analyses of Groundwater Amphipods (Crustacea: Niphargidae) from Luxembourg: New species reveal limitations of morphology-based checklists. *Zootaxa* **2022**, *5222*, 501–533. [CrossRef]
29. Flot, J.-F.; Weber, D. Amphipods from caves of the Grand Duchy of Luxembourg. *Ferrantia* **2013**, *69*, 186–190.
30. McInerney, C.E.; Maurice, L.; Robertson, A.L.; Knight, L.R.F.D.; Arnscheidt, J.; Venditti, C.; Dooley, J.S.G.; Mathers, T.; Matthijs, S.; Erikkson, K.; et al. The ancient Britons: Groundwater fauna survived extreme climate changes over tens of millions of years across NW Europe. *Mol. Ecol.* **2014**, *23*, 1153–1166. [CrossRef]
31. Alther, R.; Bongni, N.; Borko, Š.; Fišer, C.; Altermatt, F. Citizen science approach reveals groundwater fauna in Switzerland and a new species of *Niphargus* (Amphipoda, Niphargidae). *Subterr. Biol.* **2021**, *39*, 1–31. [CrossRef]
32. Rendoš, M.; Delić, T.; Copilaș-Ciocianu, D.; Fišer, C. First insight into cryptic diversity of a Caucasian subterranean amphipod of the genus *Niphargus* (Crustacea: Amphipoda: Niphargidae). *Zool. Anz.* **2021**, *290*, 1–11. [CrossRef]
33. Husmann, S. Eine neue Methode zur Entnahme von Interstitialwasser aus subaquatischen Lockergesteinen. *Arch. Für Hydrobiol.* **1971**, *68*, 519–527.
34. Chappuis, P.-A. Eine neue Methode zur Untersuchung der Grundwasserfauna. Acta Scientiarum Mathematicarum. *Nat. Univ. Fr. Josephinae Kolozsvar* **1942**, *6*, 3–7.
35. Husmann, S. Eine Rammpumpe zur Untersuchung des subaquatischen Stygorheals. *Gewässer Abwässer* **1971**, *50–51*, 115–119.
36. Pearson, K.X. On the criterion that a given system of deviations from the probable in the case of a correlated system of variables is such that it can be reasonably supposed to have arisen from random sampling. *Lond. Edinb. Dublin Philos. Mag. J. Sci.* **1900**, *50*, 157–175. [CrossRef]
37. Kearse, M.; Moir, R.; Wilson, A.; Stones-Havas, S.; Cheung, M.; Sturrock, S.; Buxton, S.; Cooper, A.; Markowitz, S.; Duran, C.; et al. Geneious Basic: An integrated and extendable desktop software platform for the organization and analysis of sequence data. *Bioinformatics* **2012**, *28*, 1647–1649. [CrossRef]
38. Kumar, S.; Stecher, G.; Li, M.; Knyaz, C.; Tamura, K. MEGA X: Molecular Evolutionary Genetics Analysis across computing platforms. *Mol. Biol. Evol.* **2018**, *35*, 1547–1549. [CrossRef]
39. Grabner, D.; Weber, D.; Weigand, A.M. Updates to the sporadic knowledge on microsporidian infections in groundwater amphipods (Crustacea, Amphipoda, Niphargidae). *Subterr. Biol.* **2020**, *33*, 71–85. [CrossRef]
40. Weber, D. *Molecular Phylogeny and Systematics of Central and Western European Niphargids (Crustacea: Amphipoda)*; Université libre de Bruxelles: Bruxelles, Belgium, 2022.
41. Hoffmann, J. Faune des Amphipodes du Grand-Duché de Luxembourg. Crustacea, Malacostraca, Amphipoda. *Arch. de la Sect. Des. Sci. de L'institut Grand-Ducal Nouv. Série* **1963**, *29*, 77–128.
42. Weber, D.; Brad, T.; Stoch, F.; Flot, J.-F. Rediscovery and redescription of *Niphargus enslini* Karaman, 1932 (Amphipoda, Niphargidae) in southern Germany. *Subterr. Biol.* **2021**, *40*, 65–89. [CrossRef]

43. Knight, L.R.F.D.; Gledhill, T. The discovery of *Microniphargus leruthi* Schellenberg, 1934 (Crustacea: Amphipoda: Niphargidae) in Britain and its distribution in the British Isles. *Zootaxa* **2010**, *2655*, 52–56. [CrossRef]
44. Bou, C.; Rouch, R. Un nouveau champ de recherches sur la faune aquatique souterraine. *Comptes Rendus De L'académie Des Sci. Série D* **1967**, *265*, 369–370.
45. Schellenberg, A. Eine neue Amphipoden-Gattung aus einer belgischen Höhle, nebst Bemerkungen über die Gattung *Crangonyx*. *Zool. Anz.* **1934**, *106*, 215–218.
46. Weber, D.; Knight, L.R.F.D.; Salussolia, A. Microniphargus, first results of phylogenetic analysis. In Proceedings of the 18th UIS Congress—Volume I. Karstologia Mémoires, Savoie Technolac, France, 24–31 July 2022; Volume 21, pp. 357–360.
47. Spence Bate, C. On the genus Niphargus (Schiödte). In *Proceedings of the Dublin University Zoological and Botanical Association*; Dublin University: Dublin, Ireland, 1859; Volume 1, pp. 237–240.
48. Schminke, H.K.; Gad, G. *Grundwasserfauna Deutschlands. Ein Bestimmungswerk*; Deutsche Vereinigung für Wasserwirtschaft, Abwasser und Abfall e.V.: Hennef, Germany, 2007.
49. Schellenberg, A. *Krebstiere oder Crustacea. IV: Flohkrebse oder Amphipoda. Die Tierwelt Deutschlands und der angrenzenden Meeresteile nach ihren Merkmalen und nach ihrer Lebensweise*; Fischer: Jena, Germany, 1942.
50. Stock, J.H.; Gledhill, T. The *Niphargus kochianus*-group in north-western Europe. *Crustaceana Suppl.* **1977**, *4*, 212–243.
51. Schellenberg, A. Höhlenflohkrebse des Adelsberger Grottensystems nebst Bemerkung über *Niphargus kochianus*. *Mitt. über Höhlen- und Karstforschung* **1933**, *2*, 32–36.
52. Weigand, A.M.; Kremers, J.; Grabner, D.S. Shared microsporidian profiles between an obligate (*Niphargus*) and facultative subterranean amphipod population (*Gammarus*) at sympatry provide indications for underground transmission pathways. *Limnol. Ecol. Manag. Inland Waters* **2016**, *58*, 7–10. [CrossRef]
53. van de Kerkhof, S. Die Industrialisierung der lothringisch-luxemburgischen Minette-Region. In *Die Industrialisierung Europäischer Montanregionen im 19. Jahrhundert (=Regionale Industrialisierung)*; Pierenkemper, A., Ed.; F. Steiner: Stuttgart, Germany, 2002; pp. 225–276.
54. Konen, J.; Scheffen, J.-L. *Bergwerke: Luxemburgs Vergangene Industrie unter Tage*; Jérôme Konen Productions: Luxembourg, 2015.
55. Weigand, H.; Weiss, M.; Cai, H.; Li, Y.; Yu, L.; Zhang, C.; Leese, F. Fishing in troubled waters: Revealing genomic signatures of local adaptation in response to freshwater pollutants in two macroinvertebrates. *Sci. Total Environ.* **2018**, *633*, 875–891. [CrossRef]
56. Iniguez-Armijos, C.; Tapia-Armijos, M.F.; Wilhelm, F.; Breuer, L. Urbanisation process generates more independently-acting stressors and ecosystem functioning impairment in tropical Andean streams. *J. Environ. Manag.* **2022**, *304*, 114211. [CrossRef]
57. United Nations. *Globally Harmonized System of Classification and Labelling of Chemicals (GHS)*, 9th ed.; United Nations: New York, NY, USA; Geneva, Switzerland, 2021.
58. Neyens, N. Das Steinforter Phenolwerk und die Umweltprobleme an der Eisch. *Nos Cah. Kanton Capellen* **2003**, *3–4*, 75–88.
59. Giger, W. The Rhine red, the fish dead—The 1986 Schweizerhalle disaster, a retrospect and long-term impact assessment. *Environ. Sci. Pollut. Res.* **2009**, *16*, 98–111. [CrossRef]
60. Van Dijk, G.M.; Marteijn, E.C.L.; Schulte-Wülwer-Leidig, A. Ecological rehabilitation of the River Rhine: Plans, progress and perspectives. *Regul. Rivers: Res. Manag.* **1995**, *11*, 377–388. [CrossRef]

diversity

MDPI

Article

Toward an Integrative Overview of Stygobiotic Crustaceans for Aquifer Delimitation in the Yucatan Peninsula, Mexico

Sarahi Jaime [1], Adrián Cervantes-Martínez [1,*], Martha A. Gutiérrez-Aguirre [1], Gerardo Hernández-Flores [1], Roger A. González-Herrera [2], Gabriel Sánchez-Rivera [3], Fernando Enseñat-Soberanis [4] and Víctor H. Delgado-Blas [1]

1 Departamento de Recursos Naturales, Universidad Autónoma del Estado de Quintana Roo, Avenida Andrés Quintana Roo s/n, Cozumel 77600, Mexico; 1518305@uqroo.mx (S.J.); margutierrez@uqroo.edu.mx (M.A.G.-A.); gerardo.hernandez@uqroo.edu.mx (G.H.-F.); blas@uqroo.edu.mx (V.H.D.-B.)
2 Campus de Ciencias Exactas e Ingenierías, Universidad Autónoma de Yucatán, Av. Industrias No Contaminantes por Anillo Periférico Norte, s/n, Mérida 97302, Mexico; roger.gonzalez@correo.uady.mx
3 Facultad de Filosofía y Letras, Universidad Nacional Autónoma de México, Circuito Interior s/n, C.U., Coyoacán, Ciudad de México 04510, Mexico; grivera@uqroo.edu.mx
4 Facultad de Ciencias Antropológicas, Universidad Autónoma de Yucatán, Carretera Mérida-Tizimín, Km. 1, Cholul, Mérida 97305, Mexico; fernando.ensenat@correo.uady.mx
* Correspondence: adcervantes@uqroo.edu.mx

Abstract: The Yucatan Peninsula (YP) presents heterogeneous environments in a karstic landscape that has been formed from permeable sedimentary rocks dating from the Cretaceous period. Its aquifers now face significant pressure from tourism, agriculture, soil use changes and population growth. Aquifer delimitation typically relies on environmental and socioeconomic criteria, overlooking the subterranean fauna. Stygobiotic crustaceans are highly diverse in the YP's subterranean karstic systems, expressing adaptations to extreme environments while often also displaying the primitive morphology of evolutionary relics. With distributions restricted to specific environments, they are potential markers of water reserves. A literature review recovered records of 75 species of crustaceans from 132 subterranean systems in the YP, together with geomorphological, hydrological, hydrogeochemical and historical precipitation data. Fourteen UPGMA clusters were informative for mapping species composition, whereby the "Ring of Cenotes", "Caribbean Cave" and "Cozumel Island" regions were delineated as consolidated aquifers. These aquifers are distinguished by abiotic factors as well: freshwater species dominate the Ring of Cenotes, while marine-affinity species characterize the Caribbean Cave and Cozumel Island aquifers. Stygobiotic crustaceans, being linked to geologically ancient water reserves and having a restricted distribution, offer a complementary tool for aquifer delimitation. Their presence suggests long-term and stable water availability. The use of these unique organisms for integrative aquifer delimitation can provide a way to improve the monitoring networks of regional aquifers.

Keywords: aquifers; Yucatan Peninsula; stygobiotic crustaceans; subterranean systems

1. Introduction

Biodiversity in groundwater environments is comprised of fauna incidentally associated with subterranean water systems (stygoxenes), fauna normally inhabiting both epicontinental and subterranean water systems (stygophiles), and fauna exclusively inhabiting subterranean water environments (stygobionts) [1]. This last group live in extreme,

https://doi.org/10.3390/d17020077

generally hypoxic environments with a total absence of sunlight [1,2] and require specific adaptations such as loss of pigmentation, reduction or loss of eyes, enhanced setation and reduced metabolism, among other unusual features [3]. In general, crustaceans are among the most diverse groups dwelling in subterranean environments, representing approximately 60% of the subterranean species richness worldwide [4,5]. Stygobiotic crustaceans tend to display primitive tagmosis with fused thorax and pleon (or abdomen) and multi-segmented appendages. They also tend to have restricted distributions, with strictly endemic forms being common [1].

In the Yucatan Peninsula (YP), more than 71 stygobiotic species have been recorded in 10 phyla: Porifera, Cnidaria, Gastrotricha, Tardigrada, Nematoda, Annelida, Arthropoda, Mollusca, Echinodermata, and Chordata. Studies by Angyal et al. (2020) [3], Álvarez et al. (2023) [4], Calderón-Gutiérrez et al. (2017) [5], Durán & Álvarez (2021) [6] and Iliffe (1992, 2002) [7,8] have contributed the understanding of the stygobiotic faunal diversity in the region for over the last 30 years. In particular, different researchers have inventoried 29 [7], 47 [5] and 55 [4] species of stygobiotic crustaceans there. The resulting taxonomic catalogs have revealed general distributional patterns, including a notable increase in locally endemic species. The YP, with the world's largest karstic aquifers, is considered one of the most aquatically biodiverse regions of Mexico [9]; nevertheless, an integrated understanding of these aquifers has yet to be achieved.

In this paper, the term "aquifer" is conceptualized in an ecological and socioeconomic context based on the systemic approach, as a porous rock mass that can store, discharge and transfer a significant amount of water, thereby satisfying consumers' needs [10–12]. Until now, few studies of aquifers have attempted to integrate their abiotic components with biotic ones, such as the stygobionts, to establish criteria for delimiting aquifer boundaries and designing monitoring networks for them [6]. In the YP, for example, CONAGUA (2021) [13]—the Spanish acronym for the National Water Commission—has delimited four aquifers based on studies related only to geology [14], edaphology (including geopedomorphology) [15–17], hydrology [18,19] and hydrogeochemistry [20,21], but not the region's stygofauna.

Although there are more than 2000 confirmed sources of natural groundwater in the YP, including those used for different anthropogenic activities, less than 5% of them have been biologically inventoried [3]. Consequently, from the point of view of conservation and sustainable management, only a fragmented and generalized understanding of the local water resources exists. Combined with the standardized and not always locally appropriate environmental legislation that is in force over all the Mexican territory, this could lead to misconceived management practices and water resource use in this region.

The present study is an attempt to integrate information on certain biological aspects of the YP's subterranean waters (presence/absence and specific composition of stygobiotic crustaceans) and various abiotic components including geomorphology, geology, edaphology, hydrology along with historical patterns of precipitation, including the general socioeconomic water uses. Inasmuch as the highly endemic distributional patterns of such crustaceans [1] are per se indicative of differences in water conditions [22]. Inclusion of biotic data may potentially lead to new proposals for delimiting aquifer boundaries. Such an integrated approach could improve the management of water resources in areas with high socioeconomic activity [18,23] and resultant anthropogenic pressure on the aquifers primarily related to tourism, agriculture, changes in landscape use and population growth [15,24,25].

2. Materials and Methods

2.1. Study Area

The YP is composed mainly of heterogeneous karstic landforms formed by the erosion and dissolution of different strata of permeable sedimentary rocks that were laid down during the Cretaceous period (60 ma), late Pliocene and Pleistocene epochs (5.6 to 3.5 ma) and the Holocene period (1.8 ka) [15,24]. The region features many subterranean (or groundwater) habitats such as underground rivers, dry and submerged caves, and water-filled sinkholes, locally called "Cenote" from the Mayan word "Dzonot", meaning "holes with water". Chemical reactions between rainwater or saltwater and carbon dioxide result in the formation of carbonic acid and the dissolution of calcareous rock, a process referred to as karstification. Depending on the nature of the water source, these habitats can be classified as freshwater (rainwater only), marine (intrusive saltwater only) or anchialine (a combination of both) [26].

Because of the karst's permeable rocks, the YP has no important surface rivers [18,20]. The biggest surface water system is the Rio Hondo in the south of Quintana Roo state, where it serves as the border between Mexico and Belize [21]. Consequently, the hydrology of the YP depends mainly on subterranean water systems divided between four aquifers according to CONAGUA [13]: Yucatan Peninsula, Hills and Valleys, Xpujil, and Cozumel Island, which provide several ecosystemic and socioeconomic services for the more than 4,500,000 people who live-in the region [18,23]. These involve water resources for agriculture, industry, households and tourism-related activities in the three states that compose the YP: Quintana Roo, Yucatan and Campeche (Figure 1).

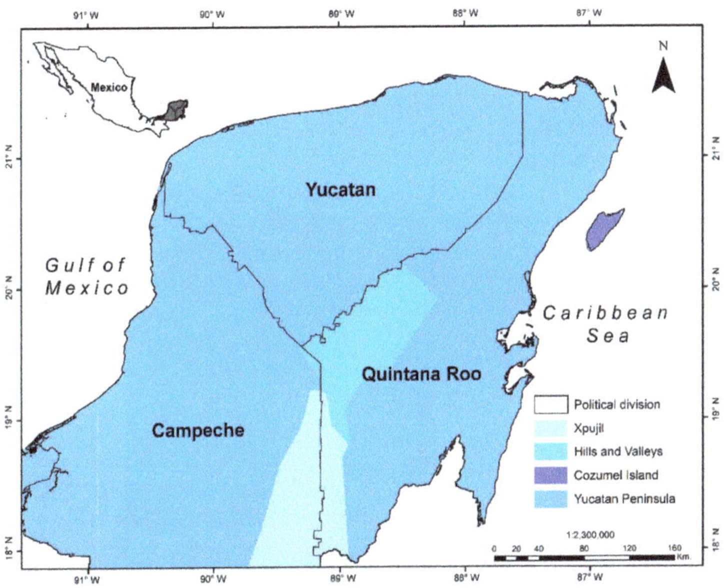

Figure 1. Study area showing the aquifers delimited by CONAGUA and the political divisions of the Yucatan Peninsula, composed of the states of Yucatan, Campeche and Quintana Roo.

2.2. Data Analysis

Based on a literature search for available records of stygobiotic crustaceans in the YP, related to data from relatively well-studied habitats, a database was created. Published literature compiled by Álvarez et al. (2023) [4], Angyal et al. (2020 a, b) [3,27] and Calderón-Gutiérrez et al. (2017) [5] was consulted, as well as specialized books, book chapters and journal articles listed in various online sources, including Web of Science, Google Scholar, Scopus, the World Register of marine Cave Species (WoRCS, https://www.marinespecies.org/worcs/ accessed on 27 November 2024) and the World Register of Marine Species (WoRMS, https://www.marinespecies.org/ accessed on 27 November 2024). New stygobiotic crustaceans recorded after 2022 from anchialine environments in the YP were also included. The finished database (Tables S1 and S3) includes taxonomic information for crustaceans of every rank from class to species as well as the name of each sampling site with its geopositional data and information as to the presence of fresh or salt water there [28]. Additionally, physicochemical variables of the water were recorded when available.

PAST 4.03 software was used to compute similarities in species composition between sites for cluster analysis (useful for identifying groups containing species with distributions limited to a specific area) with the unweighted pair-group method using the arithmetic averages (UPGMA) method and the Dice-Sørensen coefficient (DS). This coefficient, which takes the double occurrence of species in the matrix into account, is generated with the following equation [29–31]:

$$DS = 2a/2a + b + c$$

where a represents the number of species in the first pair-group, b represents the number of species in the second pair-group and c represents the number of shared species present in both groups.

The similarity percentages test (SIMPER) was used to infer which species were present in various groups, versus being restricted to a single group or being an isolated entity. The complete results of this test are shown in Table S2.

Group delimitation was performed using the similarity profile test (SIMPROF) in PRIMER v6 [31], based on an average of 1000 permutations with a significance level of $\alpha = 0.05$. For a better understanding of crustacean-specific group composition, the stygobiotic crustacean groups inferred on the basis of the SIMPROF test were mapped using QGIS version 3.40.1 on the geomorphological model of the YP proposed by Bautista (2023), with the following concepts used in this study [15].

Karst controlled by buried morphostructure is related to the buried crater structures resulting from the Chicxulub impact that was later filled with relatively young rock formations in the Middle Miocene; incipient karst: karstic formations dating from the Pleistocene (1.8 ka) [24]; juvenile karst: karstic formations dating from the Pliocene (3.5 ma); karst with good drainage is related to karts covered mainly with thick shapes of vertisols and luvisols soils; karst with poor drainage is related to karst covered only with leptosols; paludal karst: karstic formations in waterlogged formations; and tectono-karst: kastic formations surrounded by tectonic faults.

2.3. General Characterization of Aquifers

Published information on the abiotic, biotic and anthropogenic components, which in combination constitute the systemic aquifer concept, was consulted. For the abiotic components, features related to geology, geomorphology, edaphology and hydrology were defined according to Bauer-Gottwein et al. (2011) [24], Bautista (2023) [15], Fragoso-Servón et al. (2016) [32], Hernández-Flores et al. (2021) [23], López-Ramos (1975) [14], Perry et al. (2002) [20], Ríos-Ponce et al. (2020) [18] and Sánchez-Sánchez (2015) [21]. For the biotic components, the endemic species inhabiting the area were listed according to the SIMPER test when specialized literature provided corroboration of endemism. The anthropogenic components considered here were the main socioeconomic activities, with water extraction sites and volumes consumed being obtained from Arroyo-Arcos et al. (2021) [33], Bautista (2023) [15], Estrada-Medina (2019) [17], Haro-Zea et al. (2021) [34] and CONAGUA (2021) [13]. Finally, historical precipitation data (averages from 1902–2011) from Ríos-Ponce et al. (2021) [18] were considered, along with the average annual rates of water extraction and use for Cozumel Island from Hernández-Flores et al. (2021) [23].

3. Results

The database created from published literature lists a total of 75 stygobiotic crustacean species—representing five classes, 15 orders, 34 families, and 49 genera—from 132 subterranean aquatic sites in the YP. Most of them have been recorded in the northern part of the peninsula, which is characterized by incipient and young rocks (1.8 ka) as well as older karstic formations (3.5 ma), and in particular from the Ring of Cenotes area of Yucatan state, where the karst is controlled by buried morphostructure. In contrast, the north coast of Quintana Roo and Cozumel Island is characterized by paludal karst. In those places there are records of stygobiotic crustaceans inhabiting karstic environments with good drainage that date from over 5.6 ma.

Cluster analysis of the crustacean fauna of 132 subterranean aquatic sites revealed a total of 14 groups (Figure 2), each of which is internally similar according to the SIMPROF test (Pi = 0.925, $p = 0.01$). In addition, some sites characterized by few species records and no species overlap with other subterranean systems represent isolated entities. SIMPER analysis identified species commonly shared among various groups in the YP (species names marked in bold in Table S2). They include the isopod *Creaseriella anops*, with 7.9% similarity. The amphipod *Tuluweckelia cernua* contributed 5.0%, and the decapods *Creaseria morleyi* and *Typhlatya mitchelli* 5.30% and 4.58%, respectively. The mysids *Antromysis cenotensis* (5.18%) and *Stygiomysis cokei* (3.37%) were also well distributed. Finally, the thermosbaenacean *Tulumella unidens* had a similarity value of 3.37%. Conversely, some species infrequently or never occurred outside a single locality (Tables S2 and S3). Among these are the copepod *Stephos fernandoi* (0.22%) and the remipedes *Xibalbanus cozumelensis* (0.24%) and *X. fuchscockburni* (0.47%).

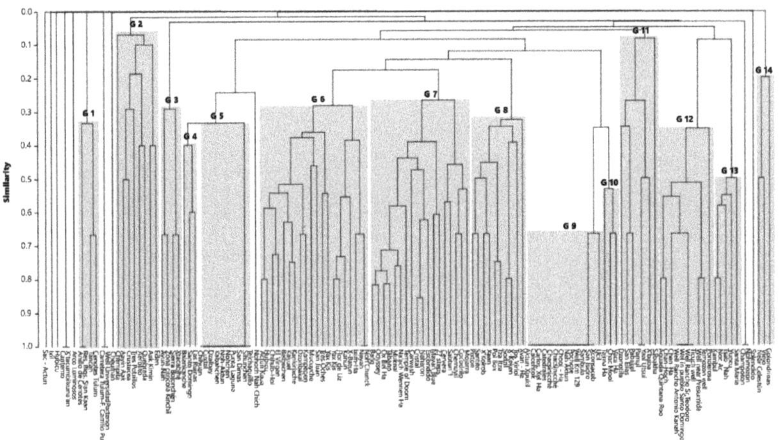

Figure 2. UPGMA clusters inferred with the Dice-Sørensen coefficient for 132 subterranean aquatic systems and 75 stygobiotic crustacean species from the Yucatan Peninsula. Similarity scales represent the similarity between groups, range from 0.0 (no similarity) to 1.0 (complete similarity). Groups were discriminated using the SIMPROF test. G = Group.

In the cluster analysis, Groups 1 and 10 consist of species that are restricted to the northeast coast of the YP, while the species in Group 2 occur only on Cozumel Island. Both regions are characterized by incipient and paludal karst formations and express the highest pluvial precipitation in the entire region (annual mean rainfall: 1200–1500 mm) (Figures 3 and 4). These three groups show marine affinities and include numerous decapods ($n = 9$) and amphipods ($n = 7$). Notable species include the shrimps *Agostocaris zabaletai* and *Triacanthoneus akumalensis*, alongside endemic species such as remipedes of the genus *Xibalbanus* (SIMPER values: 0.24–0.47%). The copepods *Stephos fernandoi* and *Exumella tsonot* also have low SIMPER values (0.01 and 0.03%, respectively). The amphipod *Bahadzia bozanici* stands out with a SIMPER value of 0.9% (Tables S2 and S3). The restricted distributions of species in these groups suggests isolation of the water reserves or aquifers in the northeast and Cozumel from those in the rest of the YP.

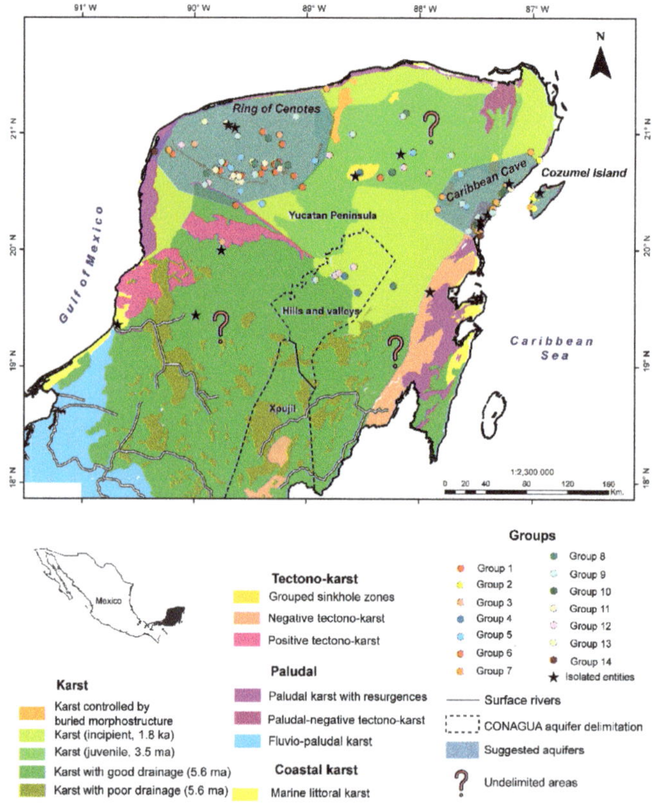

Figure 3. Species composition according to UPGMA cluster analysis using the Dice-Sørensen coefficient. The geomorphological representation is modified from Bautista (2023) [15], and the aquifer delimitation is based on CONAGUA (2021) [13].

Groups 7 and 14 are well differentiated and are distributed mainly alongon the coasts of the eastern region of the YP in the paludal karst (Figure 3), with few records from the western coast. These areas are characterized by the region's lowest pluvial precipitation (mean annual rainfall: 600–800 mm) (Figure 4). The ostracod *Spelaeoecia mayan* and the copepods *Mexicophria cenoticola* and *Urocorycaeus lautus*, which are associated with intrusive sea water, are restricted to these areas.

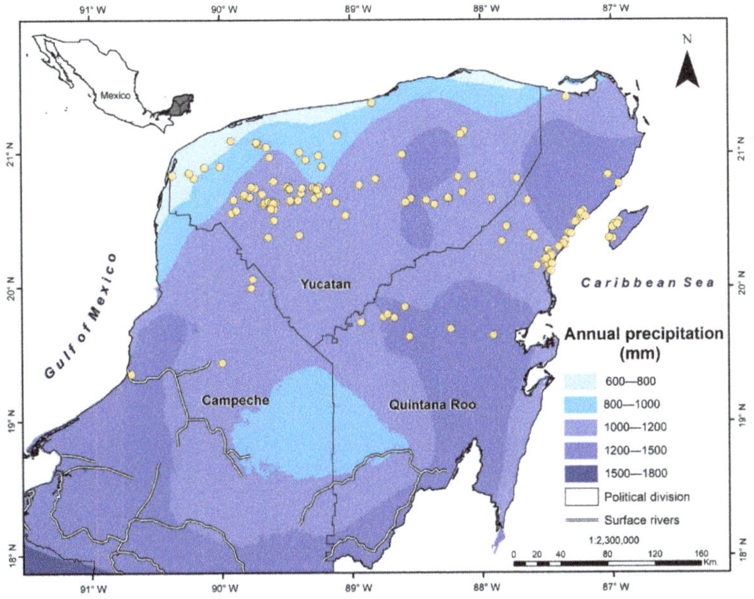

Figure 4. Catalogued subterranean aquatic systems where stygobiotic crustaceans have been recorded in the Yucatan Peninsula. Annual precipitation (mean from 1902–2011) data were modified from Ríos-Ponce et al. (2020) [18].

Groups 6, 11, and 13 are primarily composed of widespread species found across the YP, while some species are restricted to areas associated with karst formations influenced by buried morphostructure and incipient karst (Figure 3). Notable examples include the copepods *Diacyclops chakan*, *Halicyclops cenoticola*, *Mesocyclops chaci*, *Mesocyclops yutsil* and *Prehendocyclops* spp., as well as the isopod *Haptolana bowmani*. These all contribute to a distinct species composition at the relevant sites compared to the eastern regions of the YP.

The remaining Groups (3, 4, 5, 8, 9 and 12) show no clear geographic distribution within the study area, and most of the species comprising them are widely distributed in the YP. Nevertheless, sites from Groups 5 and 9 present the only localities of the isopods *Creaseria morleyi* and *Creaseriella anops*, respectively (Figure 2, Tables S1 and S3), and a few species in the remaining groups are known from either specific sites or single records, e.g., the cladoceran *Streblocerus pygmaeus* and the copepod *Parapseudoleptomesochra subterranea*.

The 12 "isolated entities" discussed above were found scattered around the YP, or with records restricted to a single site featuring species such as the atyid shrimp *Typhlatya campecheae*, catalogued only on areas with good drainage in Campeche state (Figure 3, Table S3).

Species with a restricted distributions within Groups, such as those from Groups 1 and 10 on the east coast of the YP (Caribbean Cave area), Groups 6, 11 and 13 in the Ring of Cenotes area, and Group 2 on Cozumel Island, may be informative for proposing a new and more precise delimitation of the boundaries of aquifers in the YP than CONAGUA's [13] statement (Figure 1). Each of the three proposed aquifers can be mapped (Figure 3: highlighted in light blue) and characterized in terms of its abiotic and biotic features (Table 1). One, named the "Ring of Cenotes", is located within the karstic area controlled by buried morphostructure. Another, named "Caribbean Cave," is located in the northeastern coastal area of the YP. The third one, named "Cozumel Island", underlies that island. These aquifers' names are based on the distributional areas of the stygobiotic fauna, which are coincident with those proposed previously by Álvarez et al. (2023) [4].

Table 1. General characteristics of the proposed aquifers.

Component	Feature	Ring of Cenotes *	Caribbean Cave *	Cozumel Island *
Abiotic	Geology	Formation corresponding to the Late Cretaceous Chicxulub crater [24,35]. Combination of Miocene-Pliocene, Oligocene and Eocene sedimentary rocks [14].	Miocene-Pliocene and patches of Quaternary sedimentary rocks [14].	Developed by block-faulting in the Late Mesozoic and Cenozoic [23]. Miocene-Pliocene sedimentary rocks [14].
	Geomorphology [15]	Incipient karst and karst controlled by morphostructure.	Negative tectono-karst, paludal karst with resurgences, incipient karst, and grouped sinkhole zones.	Paludal and incipient karst.
	Edaphology	Carbonate rocks ($CaCO_3$) from the early Tertiary period, with high permeability [20], mainly composed of leptosols [15].	Soils dominated by pure leptosols, with areas containing a combination of leptosols with luvisols and vertisols [32].	Limestone composed of leptosol with patches of gleysol and arenosol [23].
	Hydrology	Mean annual precipitation 600 to 1200 mm [18]. Water composed of $CaCO_3$ and $CaMg(CO_3)_2$ ions [20]. Halocline may be present in subterranean ecosystems with a depth of more than50 m [3,20].	Mean annual precipitation 1200 to 1800 mm [18]. Heterogeneity of $CaHCO_3$, mixed CaMgCl, $CaSO_4$ and NaCl, contributing to rapid erosion and the development of extensive anchialine environments along the fractures [21].	Mean annual precipitation 1400 to 1500 mm [23] Water composed of $Ca(HCO_3)_2$, characterized by heterogeneous environment with unconfined and highly permeable rocks [23].
Biotic	Endemic stygobiotic crustaceans	Freshwater species *Diacyclops chakan* [36]. *Halicyclops cenoticola* [37]. *Mesocyclops chaci* [36]. *M. yutsil* [36]. *Prehendocyclops abbreviatus* [38]. *P. boxshalli* [38]. *P. monchenkoi* [38]. *Haptolana bowmani* [39].	Mainly species with marine affinities *Bahadzia bozanici* [4]. *Hyalella azteca* [5]. *Exumella tsonot* [40]. *Triacanthoneus akumalensis* [41]. *Xibalbanus fuchscockburni* [42]. *X. tulumensis* [42].	Mainly species with marine affinities *Agostocaris zabaletai* [43]. *Anchialinocaris paulini* [43]. *Bahadzia bozanici* [44]. *Cirolana adriani* [45]. *Procaris mexicana* [46]. *Pseudopolycope helix*. *Speleophria germanyanezi* [47]. *Stenopus hispidus*. *Stephos fernandoi* [48]. *Xibalbanus cozumelensis* [49].

Table 1. *Cont.*

Component	Feature	Ring of Cenotes *	Caribbean Cave *	Cozumel Island *
Anthropogenic	Socioeconomic activities	Agriculture, pig farming, industrial use (beverage bottling), national projects (Tren Maya) [15,17].	Tourist activities (cenote snorkeling-diving), national projects (Tren Maya) [15,33].	Tourist activities (cruise ships, sun and beach, reef diving) [34].
	Water uses and average annual extraction	Mean total extraction: 859,919,519.57 m^3/year. Three main socioeconomic activities requiring high amounts of water extraction: agriculture, urban use, and industrial [13].	Mean total extraction: 907,712,597.72 m^3/year. Three main socioeconomic activities requiring high amounts of water extraction: services, urban use, and aquaculture [13].	Estimated mean extraction: 4,149,684 m^3/year. Three main socioeconomic activities requiring high amounts of water extraction: commercial, hotels and domestic [23].

* Aquifer names proposed by Álvarez et al. (2023) and adopted herein are based on the stygobiotic crustaceans.

4. Discussion

Stygobiotic crustaceans with restricted distributions can be used as complementary evidence for an integrative aquifer delimitation. They have probably been associated with subterranean waters for long periods of geological time and inhabit areas with a long-term stable supply of water. Within the YP, hydrological conditions in different areas are not uniform due to the heterogeneity of environmental conditions. Tectonic faults, different types of sinkholes produced by karstification processes (depending on proximity to or distance from the coast) and diverse rock and soil types will be involved in the differentiated capture, absorption and infiltration of recharge water for the aquifers [20,21,24].

Stygobiotic crustaceans in the YP include both freshwater forms and species of marine origin. Although all stygobiotic crustaceans have been recorded in juvenile and incipient karstic environments (Figure 3), most of the purely freshwater forms are found in the northern area of the YP near the Ring of Cenotes, where the karst is controlled by buried morphostructure. Species with freswater and saltwater affinities are found in paludal karst with resurgences [15], which are areas with anchialine habitats, and a halocline generally found at 18 m depth in the water column [24].

Several hypotheses have been proposed to explain the distributional patterns of stygofauna across the world, frequently invoking Tethyan distributions, marine regressions, deep-sea origins and active and passive migration [50]. For the subterranean crustaceans of the YP, one of the most popular theories relies on marine regressions and transgressions during the Tertiary period [Paleocene (65 ma) to Pliocene (7 ma)], when different areas of the region were sometimes exposed to form barriers that prevented the dispersal of aquatic species [51]. Considering the freshwater crustaceans associated with the Ring of Cenotes, for example, this theory maintains that mysid species such as *Antromysis cenotensis* became confined there after a series of regressions and transgressions. According to this theory, amphipods such as *Mayaweckelia cenoticola*, *M. yucatanensis* and *Tuluweckelia cernua* entered the YP from the sea at the beginning of the Quaternary period and, similarly to the now widespread isopods *Metacirolana mayana* and *Creaseriella anops*, became more associated with freshwater habitats than marine environments [52].

The aquifer subdivisions proposed in this study are the result of integrally considering abiotic, biotic and anthropogenic components (Table 1). The species distributions (freshwater vs. marine affinities) are related to the ionic composition of the subterranean water at the respective sites, as reported by Suarez-Moo et al. (2022) [22]. The majority of inland sinkholes are exclusively freshwater whereas those close to the coast display a marked vertical stratification of surface freshwater and deeper saltwater. In consequence, compared to inland sinkholes, the ichthyofauna and invertebrate diversity of coastal sinkholes are more similar to those of marine environments, a situation that agrees with the present study's findings for crustaceans. It was found that inland aquatic systems in the northern part of the YP harbor stygobiotic crustaceans with freshwater affinities, whereas in near-coastal systems the stygobiotic crustaceans show a clear affiliation with coastal marine environments.

The Ring of Cenotes aquifer mainly contains dissolved $CaCO_3$ [21,23], which is typical of freshwater environments [20]. In contrast, the Caribbean Cave and Cozumel Island aquifers in paludal karst have a different ionic composition, a mixture of $CaHCO_3$, CaCl, MgCl, $CaSO_4$ and NaCl, due to their proximity to the coast [15].

Álvarez et al. (2023) [4] and CONAGUA [13] classified Cozumel as a zone and aquifer respectively independent from the continental part of the YP. Aside from the physical separation of the island, its rocks—calcarenites formed during the Quaternary period (2 ma to 1 ka) from cemented fragments of shells deposited on the shore [23]—are among the youngest in the YP—. Cozumel's insular character might explain why it has several endemic crustacean species of marine origin, including the remipede *Xibalbanus cozumelensis*, the copepods *Stephos fernandoi* and *Speleophria germanyanezi* and the decapods *Agostocaris zabaletai* and *Anchialinocaris paulini* [5]. Similarly, the Caribbean Cave aquifer is inhabited exclusively by the remipedes *Xibalbanus fuchscockburni* and *X. tulumensis*, the amphipods *Bahadzia bozanici* and *Hyalella azteca*, the copepod *Exumella tsonot* and the decapod *Triacanthoneus akumalensis*. Conversely, the Ring of Cenotes (approximately 200 km from the two former aquifers, as shown in Figure 3) species with freshwater affinities such as the copepods *Diacyclops chakan*, *Halicyclops cenoticola*, *Mesocyclops chaci*, *M. yutsil*, *Prehendocyclops abbreviatus*, *P. boxshalli*, and *P. monchenkoi*. All the latter species are only distributed at sites within the same aquifer, which means that they do not spread to other aquifers. This supports the assumption of a limited distribution of crustaceans due to geological isolation.

Although some studies in the YP have made physicochemical measurements of the water, in most cases there has been no standardization of the addressed parameters nor of sample depth. This makes it difficult to gain a general perspective of role of the water in defining the diversity and distribution of the YP's stygofauna. Future research should include a detailed descriptions of water sampling (incorporating as many variables as possible, along with site-specific data such as depth and geolocation) in an effort to standardize methodologies and foster an integrative approach to aquifer management.

The stygobiotic fauna remains unknown in some regions of the YP (question marks in Figure 3). One area is the northern part of Yucatan state, where there are many subterranean water systems [20], juvenile karst is predominant, including some patches of grouped sinkhole zones and the paludal-negative tectono-karst (Figure 3) [15], and water extraction for livestock is important [13]. Similar uncertainties exist in the southeastern and southwestern areas of the YP, with respect to karst with poor and good drainage, including areas classified as positive tectono-karst.

The deficit in studies of the stygobiotic fauna there is mostly due to difficulties in accessing the relevant sites [53]. As for the broader picture, Angyal et al. (2020) [3] stated that less than 5% of the known subterranean ecosystems in the YP have been biologically inventoried. Although this region is considered the most diverse in Mexico [9], with 15 orders containing 34 families of stygobiotic crustaceans, these records could be expanded with other groups that have not yet been recorded or catalogued in the region, such as the malacostracan order Bathynellacea, which commonly inhabit springs, wells, caves, and rivers in Australia and some regions of Europe [54]. However, future investigations will reveal whether they also occur in the Yucatan peninsula caves, as the closest records of the family Parabathynellidae are found in Texas, USA, in karst and non-karst landscapes [55].

The results of this study support the idea of routinely using additional collection techniques to survey stygobiotic microcrustaceans with a body length of less than 1 mm, such as copepods and ostracods, which are usually more species-rich than macrocrustaceans such as malacostracans, but usually go unnoticed [56]. Collecting only crustaceans that are over 2 cm in size, such as amphipods, isopods, mysids, decapods and thermosbaenaceans, is insufficient to provide a complete picture.

On the assumption that stygobiotic species with restricted distributions are likely to be associated with water bodies with ancient geological histories and a degree of stability, we propose that the sites of stygobiotic crustaceans revealed by comprehensive sampling can be used as a complementary criterion for delimiting aquifer boundaries. Adding such data to the environmental criteria that are currently in use will allow more informed management and conservation decisions in areas with high human impact. These impacts include: massive pig and poultry farms, water-intensive agriculture and the use of pesticides in the northern YP [15,17]; high tourist pressure, including cave diving, on cenotes and anchialine systems; and national infrastructure projects such as the completion of the peninsular railroad "Tren Maya", which will attract more visitors demanding water resources but at the same time affect vulnerable fragile environments such as karstic aquifers. It may also be used eventually to assign "isolated entities" to larger aquifers as the true ranges of current "local endemics" become clear in such locations and more species are recorded from there [44] (Figures 2 and 3; Supplementary Material Table S2).

5. Conclusions

This study represents a significant step toward an integrative vision for aquifer delimitation in the YP by considering biotic and abiotic components.

The presence of stygobiotic crustaceans with restricted distributions can be used as complementary evidence for aquifer delimitation, when considering species with restricted distributions, as they are likely to be associated with water from ancient geological periods, indicating areas that could be stable.

The three proposed aquifers (Ring of Cenotes, Caribbean Cave and Cozumel Island) demonstrate distinct biotic and abiotic characteristics that reflect unique hydrological, geological and anthropogenic contexts. The Ring of Cenotes is characterized by freshwater systems with endemic species linked to ancient water stability, nonetheless, disturbed by socioeconomic activities such as agriculture and industry. The Caribbean Cave aquifer hosts a mixure of freshwater and marine species influenced by its proximity to the coast and stratified waters. Cozumel Island is characterized by its marine affinities and a high degree of endemism. It was formed by geological isolation and a younger age of formation than the rest of the YP and is currently dominated by tourism.

The results highlight the heterogeneity of the YP's groundwater systems as revealed by the species composition of their stygobiotic crustaceans. This assessment makes it possible to contribute biotic data to the paradigms for aquifer delimitation. Namely, CONAGUA's current concept of four aquifers can be refined under a new systemic approach involving the diversity and restricted distributional patterns of the groundwater species. This in turn can lead to more informed decision-making regarding water use and management in different areas of the YP.

Many knowledge gaps remain concerning the distribution and composition of stygobiotic species, including crustaceans, in various regions of the YP. More collecting and a greater monitoring effort are essential and should be focused on unexplored areas such as the northern and central YP—characterized by juvenile karst—and zones of grouped sinkholes, along with other areas where future knowledge of the stygobiotic fauna may imply water reserves. Such sampling efforts should not only be limited to macrocrustaceans, but also to microcrustaceans such as copepods and ostracods, which are the most species-rich groups in these habitats.

Supplementary Materials: The following supporting information can be downloaded at: https://www.mdpi.com/article/10.3390/d17020077/s1, Table S1: Database of stygobiotic crustacean records from the Yucatan Peninsula; Table S2: Similarity Percentage (SIMPER) test from stygobiotic crustacean species recorded in the Yucatan Peninsula, including values in each group inferred in the cluster analysis; Table S3: Known records of stygobiotic crustaceans in the Yucatan Peninsula, indicating the Similarity Percentage (SIMPER) of each species and their presence/absence in each group recognized by cluster analysis. FW = Freshwater, MW = Marine water, BW = Brackish water, IE = Isolated entities.

Author Contributions: Conceptualization, S.J., A.C.-M. and M.A.G.-A.; methodology, S.J., A.C.-M. and M.A.G.-A.; software, S.J.; validation, S.J., A.C.-M., M.A.G.-A., R.A.G.-H., G.S.-R., V.H.D.-B., F.E.-S. and G.H.-F.; formal analysis, S.J. and A.C.-M.; investigation, S.J., A.C.-M., M.A.G.-A., R.A.G.-H., G.S.-R., V.H.D.-B., F.E.-S. and G.H.-F.; resources, S.J., A.C.-M. and M.A.G.-A.; data curation, S.J. and M.A.G.-A.; writing—original draft preparation, S.J. and A.C.-M.; writing—review and editing, S.J., A.C.-M., M.A.G.-A., R.A.G.-H., G.S.-R., V.H.D.-B., F.E.-S. and G.H.-F.; visualization, S.J., A.C.-M., M.A.G.-A., R.A.G.-H., G.S.-R., V.H.D.-B., F.E.-S. and G.H.-F.; supervision, A.C.-M.; project administration, S.J., A.C.-M. and M.A.G.-A.; funding acquisition, S.J., A.C.-M., M.A.G.-A., R.A.G.-H., G.S.-R., V.H.D.-B., F.E.-S. and G.H.-F. All authors have read and agreed to the published version of the manuscript.

Funding: The Secretaría de Ciencia, Humanidades, Tecnología e Innovación (SECIHTI), (before-named as CONAHCYT) provided a doctoral scholarship (No. 1012891) to the first author. The group of the vulnerability of continental and coastal water systems (VBSACC), represented by the Universidad Autónoma del Estado de Quintana Roo (UAEQROO), Campus Cozumel, is also thanked for their financial support.

Institutional Review Board Statement: Not applicable.

Data Availability Statement: The data presented in this study are available on request from the corresponding author.

Acknowledgments: The first author thanks the SECIHTI, (before CONAHCYT) for a doctoral scholarship (No. 1012891). The Laboratory of Limnology and Tropical Ecology and the Laboratory of Zooplankton Ecology and Taxonomy in UAEQROO provided facilities to perform this study. We are grateful to Francisco Bautista Zúñiga and Eduardo Cejudo from the Asociación Mexicana de Estudios sobre el Karst (AMEK), who provided valuable data regarding geomorphology and annual precipitation, respectively. Eduardo Suárez Morales improved the early version of this manuscript with his comments. Two anonymous reviewers offered valuable comments for improving an earlier version of this paper.

Conflicts of Interest: The authors declare no conflicts of interest.

References

1. Calderón Gutiérrez, F.; Iliffe, T.M.; Borda, E.; Yáñez Mendoza, G.; Labonté, J. Response and resilience of karst subterranean estuary communities to precipitation impacts. *Ecol. Evol.* **2023**, *13*, e10415. [CrossRef] [PubMed]
2. Mejía-Ortiz, L.M.; Sprouse, P.; Tejeda-Mazariegos, J.C.; Valladarez, J.; Frausto-Martínez, O.; Collantes-Chavez-Costa, A.L.; Ruíz-Cancino, G.; Yáñez, G. Tropical subterranean ecosystems in Mexico, Guatemala and Belize: A review of aquatic biodiversity and their ecological aspects. In *Natural History and Ecology of Mexico and Central America*; Hufnagel, L., Ed.; IntechOpen: London, UK, 2021; p. 18.
3. Angyal, D.; Chávez-Solís, E.M.; Liévano-Beltrán, L.A.; Magaña, B.; Simoes, N.; Mascaró, M. New distribution records of subterranean crustaceans from cenotes in Yucatan (Mexico). *ZooKeys* **2020**, *911*, 21–49. [CrossRef] [PubMed]

4. Álvarez, F.; Durán, B.; Meacham, S. Anchialine fauna of the Yucatan Peninsula: Diversity and conservation challenges. In *Mexican Fauna in the Anthropocene*; Jones, R.W., Ornelas-García, C.A., Pineda-López, R., Álvarez, F., Eds.; Springer: Cham, Switzerland, 2023; pp. 287–298. [CrossRef]
5. Calderón-Gutiérrez, F.; Solís-Marín, F.A.; Gómez, P.; Sánchez, C.; Hernández-Alcántara, P.; Álvarez-Noguera, F.; Yáñez-Mendoza, G. Mexican anchialine fauna—With emphasis in the high biodiversity cave El Aerolito. *Reg. Stud. Mar. Sci.* **2017**, *9*, 43–55. [CrossRef]
6. Durán, B.; Álvarez, F. La cochinilla *Creaseriella anops* (Crustacea: Isopoda: Cirolanidae) de las cuevas anquihalinas de Yucatán, México ¿Existen patrones de variación en la talla? *Rev. Mex. Biodivers.* **2021**, *92*, e923925. [CrossRef]
7. Iliffe, T.M. An annotated list of the troglobiotic anchialine and freshwater fauna of Quintana Roo. In *Diversidad biológica en la Reserva de la Biósfera de Sian Ka'an Quintana Roo, México*; Navarro, D., Suárez-Morales, E., Eds.; CIQROO/SEDESOL: Chetumal, Mexico, 1992; Volume II, pp. 197–215.
8. Iliffe, T. Conservation of anchialine cave biodiversity. In *Hydrogeology and Biology of Post-Paleozoic Carbonate Aquifers*; Martin, J.B., Wicks, C.M., Sasowsky, I.D., Eds.; Karst Water Institute: Gainesville, FL, USA, 2002; pp. 99–102.
9. Mejía-Ortíz, L.M.; Collantes-Chavez-Costa, A.L.; López-Contreras, C.; Frausto-Martínez, O. Subterranean waters in Riviera Maya of the Yucatan Peninsula: Vulnerability and the importance of monitoring. In *Limnology—The Importance of Monitoring and Correlations of Lentic and Lotic Waters*; Massarelli, C., Campanele, C., Eds.; IntechOpen: London, UK, 2023; Volume 11. [CrossRef]
10. Stevanović, Z. Characterization of karst aquifer. In *Karst Aquifers—Characterization and Engineering*; Stevanović, Z., Ed.; Springer: Cham, Switzerland, 2015; pp. 47–125. [CrossRef]
11. Rodríguez-Vaquero, J.E. La configuración de los paisajes hidráulicos. Visión desde el enfoque sistémico. *Nimbus Rev. Climatol. Meteorol. Paisaje* **2006**, *17*, 145–157.
12. Danielopol, D.L.; Griebler, C. Changing paradigms in groundwater ecology—From the "living fossils" tradition to the "new groundwater ecology". *Int. Rev. Hydrobiol.* **2008**, *93*, 565–577. [CrossRef]
13. CONAGUA. *Programa Hídrico Regional 2021–2024. Región Hidrológico-Administrativa XII Península de Yucatán*; Secretaría: Mexico, Mexico, 2021.
14. López-Ramos, E. Geological summary of the Yucatan Peninsula. In *The Gulf of Mexico and the Caribbean*; Nairn, A.E.M., Stehli, F.G., Eds.; Springer: New York, NY, USA, 1975; Volume 3, pp. 257–282.
15. Bautista, F. *El karst de México*; Asociación Mexicana de Estudios Sobre el Karst: Ciudad de México, Mexico, 2023.
16. Bautista, F.; Frausto, O.; Ihl, T.; Aguilar, Y. Actualización del mapa de suelos del estado de Yucatán México: Enfoque geomorfopedológico y WRB. *Ecosistemas Recur. Agropecu.* **2015**, *2*, 303–315.
17. Estrada-Medina, H.; Jiménez-Osornio, J.J.; Álvarez-Rivera, O.; Barrientos-Medina, R.C. El karst de Yucatán: Su origen, morfología y biología. *Acta Univ.* **2019**, *29*, 1–18. [CrossRef]
18. Ríos-Ponce, J.D.; Acosta-González, G.; Cejudo, E. La precipitación histórica y la extracción del agua subterránea en la península de Yucatán: Una reflexión. *Desde El Herb. CICY* **2020**, *12*, 110–118.
19. Cejudo, E.; González, G.A.; Bautista, R.M.L.; Medina, H.E. Water stable isotopes (δ2H and δ18O) in the Yucatan Peninsula, Mexico. *Tecnol. Cienc. Agua* **2022**, *13*, 507–557. [CrossRef]
20. Perry, E.; Velazquez-Oliman, G.; Marin, L. The hydrogeochemistry of the karst aquifer system of the northern Yucatan Peninsula, Mexico. *Int. Geol. Rev.* **2002**, *44*, 191–221. [CrossRef]
21. Sánchez-Sánchez, J.A.; Álvarez-Legorreta, T.; Pacheco-Ávila, J.G.; González-Herrera, R.A.; Carrillo-Bribiezca, L. Caracterización hidrogeoquímica de las aguas subterráneas del sur del Estado de Quintana Roo, México. *Rev. Mex. Cienc. Geol.* **2015**, *32*, 62–76.
22. Suárez-Moo, P.; Remes-Rodríguez, C.A.; Márquez-Velázquez, N.A.; Falcón, L.I.; García-Maldonado, J.Q.; Prieto-Davó, A. Changes in the sediment microbial community structure of coastal and inland sinkholes of a karst ecosystem from the Yucatan Peninsula. *Sci. Rep.* **2022**, *12*, 1110. [CrossRef]
23. Hernández-Flores, G.; Gutiérrez-Aguirre, M.A.; Cervantes-Martínez, A.; Marín-Celestino, A.E. Historical analysis of a karst aquifer: Recharge, water extraction, and consumption dynamics on a tourist island (Cozumel, Mexico). *Ann. Limnol. Int. J. Limnol.* **2021**, *57*, 16. [CrossRef]
24. Bauer-Gottwein, P.; Gondwe, B.R.N.; Charvet, G.; Marín, L.E.; Rebolledo-Vieyra, M.; Merediz-Alonso, G. Review: The Yucatán Peninsula karst aquifer, Mexico. *Hydrogeol. J.* **2011**, *19*, 507–524. [CrossRef]
25. González Herrera, R.A.; Albornoz Euán, B.S.I.; Sánchez Y Pinto, I.A.; Osorio Rodríguez, J.H. El acuífero yucateco. Análisis del riesgo de contaminación con apoyo de un sistema de información geográfica. *Rev. Int. Contam. Ambient.* **2018**, *34*, 667–683. [CrossRef]
26. van Hengstum, P.J.; Cresswell, J.N.; Milne, G.A.; Iliffe, T.M. Development of anchialine cave habitats and karst subterranean estuaries since the last ice age. *Sci. Rep.* **2019**, *9*, 11907. [CrossRef]
27. Angyal, D.; Simões, N.; Mascaró, M. Uptaded checklist, historical overview and illustrated guide to the stygobiont Malacostraca (Arthropoda: Crustacea) species of Yucatan (Mexico). *Subterr. Biol.* **2020**, *36*, 83–108. [CrossRef]

28. Cervantes-Martínez, A.; Durán-Ramírez, C.; Elías-Gutiérrez, M.; García-Morales, A.E.; Gutiérrez-Aguirre, M.; Jaime, S.; Macek, M.; Maeda-Martínez, A.M.; Martínez-Jerónimo, F.; Mayén-Estrada, R.; et al. Freshwater diversity of zooplankton from Mexico: Historical review of some of the main groups. *Water* **2023**, *15*, 858. [CrossRef]
29. Alvarez, F.; Iliffe, T.M.; Benitez, S.; Brankovits, D.; Villalobos, J.L. New records of anchialine fauna from the Yucatan Peninsula, Mexico. *Check List* **2015**, *11*, 1505. [CrossRef]
30. Palacio, F.X.; Apodaca, M.J.; Crisci, J.V. *Análisis multivariado para datos biológicos teoría y su aplicación utilizando en lenguaje R, Primera ed.*; Fundación de Historia Natural Félix de Azara: Ciudad Autónoma de Buenos Aires, Argentina, 2020.
31. Ríos-Jara, E.; Galván-Villa, C.M.; Rodríguez-Zaragoza, F.A.; López-Uriarte, E.; Bastida-Izaguirre, D.; Solís-Marín, F.A. Los equinodermos (Echinodermata) de Bahía Chamela, Jalisco, México. *Rev. Mex. Biodivers.* **2013**, *84*, 263–279. [CrossRef]
32. Fragoso-Servón, P.; Bautista, F.; Pereira, A.; Frausto, O. Distribución de suelos en ambientes tectonokársticos en la porción este de la Península de Yucatán, México. *GEOS* **2016**, *2*, 265–273.
33. Arroyo-Arcos, L.; Bolan-Sorchini, S.; Hernández-Aguilar, M.L. Potencial y aprovechamiento turístico del sistema de cenotes de la Riviera Maya. In *Los territorios kársticos de la Península de Yucatán: Caracterización, manejo y riesgos*; Bautista, F., Ed.; Asociación Mexicana de Estudios Sobre el Karst: Ciudad de México, Mexico, 2021; pp. 133–145.
34. Haro-Zea, K.L.; Alcázar-Zamacona, J.P.; González-Damián, A. Competitividad de la industria turística en Cozumel, Quintana Roo. *Rev. Acad. ECO* **2019**, *20*, 81–100.
35. González-Herrera, R.; Sánchez-y-Pinto, I.; Gamboa-Vargas, J. Groundwater-flow modeling in the Yucatán karstic aquifer, Mexico. *Hydrogeol. J.* **2002**, *10*, 539–552. [CrossRef]
36. Fiers, F.; Reid, J.W.; Iliffe, T.M.; Suárez-Morales, E. New hypogean cyclopoid copepods (Crustacea) from the Yucatan Peninsula, Mexico. *Contrib. Zool.* **1996**, *66*, 65–102. [CrossRef]
37. Rocha, C.E.F.; Iliffe, T.M.; Reid, J.W.; Suárez-Morales, E. A new species of *Halicyclops* (Copepoda, Cyclopoida, Cyclopidae) from cenotes of the Yucatan Peninsula, Mexico, with an identification key for the species of the genus from the Caribbean region and adjacent areas. *Sarsia* **1998**, *83*, 387–399. [CrossRef]
38. Rocha, C.E.F.; Iliffe, T.M.; Reid, J.W.; Suárez-Morales, E. *Prehendocyclops*, a new genus of the subfamily Halicyclopinae (Copepoda, Cyclopoida, Cyclopidae) from cenotes of the Yucatan Peninsula, Mexico. *Sarsia* **2000**, *85*, 119–140. [CrossRef]
39. Botosaneanu, L.; Iliffe, T.M. Four new stygobitic cirolanids (Crustacea: Isopoda) from the Caribbean—With remarks on intergeneric limits in some Cirolanids. *Bull. L'institut R. Sci. Nat. Belg. Biol.* **1997**, *67*, 77–121.
40. Suárez-Morales, E.; Iliffe, T.M. A new *Exumella* (Crustacea: Copepoda: Ridgewayiidae) from anchialine waters of the western Caribbean, with comments on regional biogeography. *Bull. Mar. Sci.* **2005**, *77*, 409–423.
41. Alvarez, F.; Iliffe, T.M.; Gonzalez, B.; Villalobos, J.L. *Triacanthoneus akumalensis*, a new species of alpheid shrimp (Crustacea: Caridea: Alpheidae) from an anchialine cave in Quintana Roo, Mexico. *Zootaxa* **2012**, *3154*, 61–68. [CrossRef]
42. Neiber, M.T.; Hansen, F.C.; Iliffe, T.M.; Gonzalez, B.C.; Koenemann, S. Molecular taxonomy of *Speleonectes fuchscockburni*, a new pseudocryptic species of Remipedia (Crustacea) from an anchialine cave system on the Yucatán Peninsula, Quintana Roo, Mexico. *Zootaxa* **2012**, *3190*, 31–46. [CrossRef]
43. Mejía-Ortíz, L.M.; Yañez, G.; López-Mejía, M. Alvarez. Anchialocarididae, a new family of anchialine decapods and a new species of the genus *Agostocaris* from Cozumel Island, Mexico. *Crustaceana* **2017**, *90*, 381–398. [CrossRef]
44. Mercado-Salas, N.F.; Morales-Vela, B.; Suárez-Morales, E.; Iliffe, T.M. Conservation status of the inland aquatic crustaceans in the Yucatan Peninsula, Mexico: Shortcomings of a protection strategy. *Aquat. Conserv.* **2013**, *23*, 939–951. [CrossRef]
45. Ortiz, M.; Cházaro-Olvera, S. A new species of cirolanid isopod (Peracarida, Isopoda) collected from cenote Aerolito, Cozumel Island, northwestern Caribbean. *Crustaceana* **2015**, *88*, 152–163. [CrossRef]
46. Mejía-Ortiz, L.M.; López-Mejía, M.; Pakes, J.; Hartnoll, R.G.; Zarza-González, E. Morphological adaptations to anchialine environments in species of five shrimp families (*Barbouria yanezi, Agostocaris bozanici, Procaris mexicana, Calliasmata nohochi* and *Typhlatya pearsei*). *Crustaceana* **2013**, *86*, 578–593. [CrossRef]
47. Suárez-Morales, E.; Cervantes-Martínez, A.; Gutiérrez-Aguirre, M.A.; Iliffe, T.M. A new *Speleophria* (Copepoda, Misophrioida) from an anchialine cave of the Yucatán Peninsula with comments on the biogeography of the genus. *Bull. Mar. Sci.* **2017**, *93*, 863–878. [CrossRef]
48. Suárez-Morales, E.; Gutiérrez-Aguirre, M.A.; Cervantes-Martínez, A.; Iliffe, T.M. A new anchialine *Stephos* Scott from the Yucatan Peninsula with notes on the biogeography and diversity of the genus (Copepoda, Calanoida, Stephidae). *ZooKeys* **2017**, *671*, 1–17. [CrossRef]
49. Olesen, J.; Meland, K.; Glenner, H.; Van Hengstum, P.J.; Iliffe, T.M. *Xibalbanus cozumelensis*, a new species of Remipedia (Crustacea) from Cozumel, Mexico, and a molecular phylogeny of *Xibalbanus* on the Yucatán Peninsula. *Eur. J. Taxon.* **2017**, *316*, 1–27. [CrossRef]
50. Neiber, M.T.; Hartke, T.R.; Stemme, T.; Bergmann, A.; Rust, J.; Iliffe, T.M.; Koenemann, S. Global Biodiversity and phylogenetic evaluation of Remipedia (Crustacea). *PLoS ONE* **2011**, *6*, e19627. [CrossRef]

51. Suárez-Morales, E. Historical biogeography and distribution of the freshwater calanoid copepods (Crustacea: Copepoda) of the Yucatan Peninsula, Mexico. *J. Biogeogr.* **2003**, *31*, 1051–1063. [CrossRef]
52. Suárez-Morales, E.; Rivera-Arriaga, E. The aquatic fauna of karstic environments in the Yucatan Peninsula, Mexico: An updated overview. In *Aquatic Ecosystems of Mexico: Status and Scope*; Munawar, M., Lawrence, S.G., Munawar, I.F., Malley, D.F., Eds.; Backhuys Publishers: Kerkwerve, The Netherlands, 2000; pp. 151–164.
53. Benítez, S.; Iliffe, T.M.; Quiroz-Martínez, B.; Alvarez, F. How is the anchialine fauna distributed within a cave? A study of the Ox Bel Ha system, Yucatan Peninsula, Mexico. *Subterr. Biol.* **2019**, *31*, 15–28. [CrossRef]
54. Camacho, A.I. An overview of the distribution of the Parabathynellidae (Crustacea, Syncarida, Bathynellacea) on the Iberian Peninsula and Balearic Islands. *Graellsia* **2003**, *59*, 63–78. [CrossRef]
55. Camacho, A.I.; Mas-Peinado, P.; Hutchins, B.T.; Schwarts, B.F.; Dorda, B.A.; Casado, A.; Rey, I. New stygofauna from Texas, USA: Three new species of Parabathynellidae (Crustacea: Bathynellacea). *J. Nat. Hist.* **2021**, *55*, 979–1007. [CrossRef]
56. Galassi, D.M.P.; Huys, R.; Reid, J.W. Diversity, ecology and evolution of groundwater copepods. *Freshw. Biol.* **2009**, *54*, 691–708. [CrossRef]

diversity

MDPI

Article

Deep Troglomorphy: New Arrhopalitidae (Collembola: Symphypleona) of Different Life Forms from the Snezhnaya Cave System in the Caucasus

Robert S. Vargovitsh

Schmalhausen Institute of Zoology, National Academy of Sciences of Ukraine, Bohdan Khmelnitsky Str. 15, 01030 Kyiv, Ukraine; arete@izan.kiev.ua

Abstract: The Snezhnaya (=Snow) Cave System (depth −1760 m; total length 40.8 km), located in the West Caucasus, is inhabited by distinctly troglomorphic collembolan species from several families. Two new species of the family Arrhopalitidae occur in the deep parts of the system. Both are highly morphologically specialized; however, they evolved into different troglobiont life forms. *Arrhopalites profundus* sp. nov. with extremely long claws, lamellate mucro, and an enlarged sensory organ of third antennal segment lives exclusively in hygropetric and epineustonic habitats, whereas *Pygmarrhopalites rystsovi* sp. nov., with greatly elongated and multi-subsegmentated antennae and long legs, prefers open terrestrial spaces. The troglomorphy, as well as ecological life forms of the obligate cave-dwelling Arrhopalitidae, in general, are discussed and classified as neustonic-hygropetric, atmobiont, intrasubstrate, and intermediate troglomorphs.

Keywords: *Arrhopalites*; *Pygmarrhopalites*; springtails; taxonomy; biospeleology; deep caves; life form; troglobiont; West Caucasus

1. Introduction

The West Caucasus could be considered an important hotspot of the subterranean biodiversity on the southern border of Europe and Asia, which holds numerous terrestrial and aquatic obligate cave-dwelling invertebrates—troglobionts and stygobionts (e.g., [1–3])—but it is still insufficiently investigated.

Our biospeleological samplings in this region show that Collembola, presumably, is the most diverse group among the terrestrial arthropod cave dwellers in the Caucasus, with an average of about ten species per well-sampled cave and reaching up to thirty species per cave at maximum limits (unpublished data). However, this fauna remains poorly known. Especially enigmatic are the fauna of the two high-mountainous limestone massifs in Abkhazia (Arabika and Bzyb massifs) with classic plateaus karst of Alpine type and hundreds of caves, including nine abysses exceeding 1 km in depth and two of the world's deepest caves over 2 km in depth. Hitherto, six collembolan species have been known from these massifs. Four species were described from the currently second world deepest Krubera Cave in the Arabika Massif: *Anurida stereoodorata* (Jordana and Baquero, 2012), *Deuteraphorura kruberaensis* (Jordana and Baquero, 2012), *Schaefferia profundissima* (Jordana and Baquero, 2012), and *Plutomurus ortobalaganensis* (Jordana and Baquero, 2012) [4]. In the outstanding Snezhnaya Cave System (Bzyb Massif), two genuine troglomorphic species have been described: *Typhlogastrura morozovi* (Babenko, 1987) [5] and the yet monotypic *Troglaphorura gladiator* (Vargovitsh, 2019) [6]; several other troglobionts, among seventeen sampled in this cave morpho-species (unpublished), are awaiting taxonomic investigation.

Up to date, five obligate cave-dwelling species of the family Arrhopalitidae, belonging to three genera, are known from the Caucasus: *Arrhopalites abchasicus* (Vargovitsh, 2013), *A. macronyx* (Vargovitsh, 2012), *Pygmarrhopalites dbari* (Vargovitsh, 2017), *P. kovali* (Vargovitsh, 2017), and the monotypic *Troglopalites stygios* (Vargovitsh, 2012) [7–9].

Besides, troglophilous *A. caecus* (Tullberg, 1871), *P. principalis* (Stach, 1945), and *P. pygmaeus* (Wankel, 1860) were listed from Georgian caves [10].

Here, I describe two more highly specialized species of the genera *Arrhopalites* (Börner, 1906) and *Pygmarrhopalites* (Vargovitsh, 2009), with both inhabiting the same deep parts of Snezhnaya Cave but representing different life forms correlated with delicate microhabitat preferences.

In the final part of this work, the troglomorphic features of the family Arrhopalitidae are listed and discussed, and ecological life forms of troglobionts are defined and classified.

2. Materials and Methods

2.1. Locality

Snezhnaya Cave and Souvenir Cave: the Khipsta karst massif on the Bzyb Mountain Range, Gudauta District, Abkhazia (Figure 1A).

The Bzyb (or Bzybsky) Massif is a fragment of a belt of limestone karst relief, stretching along the edge of the southern macroslope of the West Caucasus, with over 400 caves explored [11]. Late Quaternary glaciation existed here, and together with adjacent Arabika Massif, it can be considered classical mid and high-mountain karst [12]. The Khipsta karst massif (156 km^2) (Figure 1A,B) with the three highest mountains, around 2500 m a.s.l., is located on the southern slope of the Bzyb Mountain Range, between the rivers Khipsta and Aapsta. At altitudes of 400–500 m a.s.l. in the south, the massif borders on the Colchis Lowland. Up to an altitude of 1700–1800 m, it comprises a belt of mountain-forest vegetation and a belt of mountain-meadow vegetation at higher altitudes. The Khipsta Massif is composed of Jurassic, Cretaceous, and Paleocene rocks, and it comprises karst of naked, covered, and green (within the forest zone) types [12,13].

The Snezhnaya Cave System is located in the southern slope of the Khipsta Massif and consists of six interconnected caves with entrances exposed at different altitudes in the meadow (2389 m, 2015 m, 1970 m) and forest (1505 m, 1329 m, 1318 m) zones (Figure 1A). It is a branched, tree-like river system fed by numerous sub-vertical channels. A total spatial development of the system is 40,840 m, and a total vertical amplitude is 1760 m [14], which ranges it as the fourth deepest in the world and the longest in the Caucasus. The system is developed in a large anticlinal fold in limestones, dolomitized limestones, and dolomites of Upper Jurassic and Lower Cretaceous carbonate rocks. Its upper parts are sub-vertical, whereas lower parts are mainly sub-horizontal with running waters. It contains an underground river with the average discharge of about 500 L/s, but it drastically increases during seasonal floods. The air and water temperature in the cave system gradually increases, with depth, from 0–2 °C to 6.5 °C, and relative air humidity along the system is close to 100% [15].

Souvenir Cave: With 1464 m length and −408 m depth, it is also located in the Khipsta Massif, with an entrance exposed at 1850 m a.s.l. (Figure 1C), south to the Khipsta Mountain, at a distance 0.4 km west to the huge entrance of Snezhnaya Cave (Figure 1D). Flowing streams and three siphons start from about −300 m, descending to the bottom part. Hydrologically, Souvenir Cave obviously belongs to the Snezhnaya Cave System [13], but terrestrial connection is yet unknown. Air and water temperature along the cave, in August 2018 and 2019, ranged from 2.8 °C to 5.7 °C.

Figure 1. Type locality of *Arrhopalites profundus* sp. nov. and *Pygmarrhopalites rystsovi* sp. nov. (**A**)—Map with locations of the six entrances of the Snezhnaya Cave System (blue marks with black spot) and Souvenir Cave; (**B**)—Landscape in the Khipsta Massif of the Bzyb Mt Range; (**C**)—Entrance to Souvenir Cave; (**D**)—Entrance to Snezhnaya Cave; (**E**)—Fragment of Crystal Meander (−1100 m) in Snezhnaya Cave—a type habitat of *A. profundus* sp. nov. and *P. rystsovi* sp. nov.

2.2. Sampling

Springtails were sampled in two ways: (a) using the mouth aspirator on the water surface of small pools, usually scattered on the walls along the cave river and on rocks, near water (Figure 1E); (b) using Barber's traps [16] installed at a distance, from several to tens of meters, from water. Fixation and preservation: in 96% alcohol.

2.3. Microscopic Study

Specimens were mounted on permanent slides using Hoyer's medium [17] (p. 38), and they were studied under the light microscope Zeiss Axio Imager M1 (Center of collective using of scientific equipment "Animalia" at Schmalhausen Institute of Zoology). Figures were designed and adjusted with the help of Helicon Focus Pro v. 7.6.6 (Kharkiv, Ukraine), CorelDRAW v. 23 (Ottawa, Canada) and Adobe Photoshop v. 21 (San Jose, CA, USA) software.

2.4. Measurements

Mounted body and its parts were measured from microscope (Zeiss Axio Imager M1, Jena, Germany) photographs using AxioVision SE64 Rel. 4.9.1 software. Claws were measured from the most basal proximal point of the outer margin to the tip. Measurements of adult specimens are given in Table 1, and the length ratios between selected structures are in Table 2. Body/claw ratio is considered an index of the claw's troglomorphy [6].

Table 1. Lengths (in µm) and angles for body parts of adult types of *Arrhopalites profundus* sp. nov. and *Pygmarrhopalites rystsovi* sp. nov.

Body Part	*Arrhopalites profundus*			*Pygmarrhopalites rystsovi*		
	Females, *n* = 2 + 1?		**Male**	**Females, *n* = 6**		**Male**
	Holotype	***n* = 2**	***n* = 1**	**Min–Max**	**Mean**	***n* = 1**
Total (without appendages)	~1000	875; ?	920	1000–1150	1098	~950
Body	~700	620; ?	640	730–830	792	~700
Head	387	250; ?	305	340–385	353	302
Head dorsum longest seta	25	26; ?	23	21–24	23	18
Eye diameter	5.5	?	5.8	11.6–13.8	12.7	13
Antenna	755	?	660	1160–1270	1197	1130
Ant I	53	?	49	51–56	54	53
Ant II	110	?	102	146–160	151	138
Ant III	182	?	140	252–288	264	249
Ant IV	440	?	395	705–812	748	714
Ant III organ rods	15	15.2; ?	14.8	10–11.6	11	11.5
Tibiotarsus I	257	240; ?	218	333–368	348	323
Tibiotarsus II	250	240; ?	213	326–364	340	323
Tibiotarsus III	302	305; ?	255	405–456	422	398
Claw I	135	133; 148	107	80–88	83	81
Claw II	135	133; 146	106	79–87	82	77
Claw III	118	114; 131	93	67–80	73	70
Empodium I	88	80; 95	65	39–46	42	40
Empodium II	75	84; 93	60	41–45	43	39
Empodium III	72	65; 90	62	41–48	44	42
Th II seta **m1**	30	31; ?	30	18–23	21	18
Th II sensilla **a**	?	?	12,6	4.3–12	7.0	?
Th III sensilla **a**	13	?	14,7	16–17	16.4	16
Trichobothria **AB** distance	87	?	72	41–56	50	50
Trichobothria **BC** distance	60	?	54	57–82	72	72
Trichobothria **ABC** angle	97°	?	96°	154–168°	160°	166°
Abd **c1** seta	11.5	9; ?	10.2	31–41	34	35
Abd IV longest (**dI-dIII)-1** seta	85	88; ?	86	127–136	130	127

Table 1. *Cont.*

Body Part	*Arrhopalites profundus*			*Pygmarrhopalites rystsovi*		
	Females, n = 2 + 1?		Male	Females, n = 6		Male
	Holotype	n = 2	n = 1	Min–Max	Mean	n = 1
Abd VI longest circumanal seta	56	47; ?	52	43–58	52	46
Appendices anales, length	37	34; ?	-	28–31	30	-
App. anales, basal shaft width	2.8	2.3; ?	-	2.2–2.4	2.3	-
App. anales, subapical max. width	2.8	2.1; ?	-	2.3–5	3.6	-
Manubrium	231	224; 207	186	225–240	230	221
Dens	265	240; 260	223	307–350	320	302
Mucro	140	120; 112	108	155–168	160	157

Table 2. Proportions (mean in females) for some body parts of adult types of *Arrhopalites profundus* sp. nov. and *Pygmarrhopalites rystsovi* sp. nov.

Ratio	*Arrhopalites profundus*		*Pygmarrhopalites rystsovi*	
	Females	Male	Females	Male
Antenna/head	1.95	2.16	3.40	3.74
Ant II/I	2.08	2.08	2.78	2.60
Ant III/I	3.43	2.86	4.87	4.70
Ant IV/I	8.30	8.06	13.81	13.47
Head/tibiotarsus I	1.27	1.20	1.02	0.93
Tibiotarsus II/I	0.99	0.84	0.98	1.00
Tibiotarsus III/I	1.22	1.17	1.21	1.23
Tibiotarsus I/claw I	1.85	2.38	4.18	3.99
Tibiotarsus II/claw II	1.83	2.01	4.16	4.19
Tibiotarsus III/claw III	2.62	2.74	5.83	5.69
Claw I/empodium I	1.60	1.65	1.99	2.03
Claw II/empodium II	1.69	1.77	1.91	1.97
Claw III/empodium III	1.70	1.50	1.65	1.67
Dens/mucro	1.95	2.06	2.00	1.95
Trichobothria **AB**/**BC** distance	1.45	1.33	0.69	0.69
Abd IV **dI-1**/Th2 **m1** seta	2.84	2.87	6.44	7.06
Abd IV **dI-1**/claw III	0.75	0.92	1.80	1.81
Longest Abd IV/circumanal seta	1.70	1.65	2.52	2.76
Body total/tibiotarsus III	3.31	3.61	2.60	2.39
Body total/furca	1.54	1.78	1.55	1.40
Body total/**dI-1** seta	10.85	10.70	8.44	7.48
Body total/claw I	6.99	8.60	13.45	11.73

2.5. Chaetotaxy Nomenclature

Chaetotaxy nomenclatures are used following Betsch and Waller [18] for the head, Fjellberg [19] for the mouthparts, Vargovitsh [9,20] for the great abdomen, Bretfeld [21] for the fifth abdominal segment and partly for great abdomen, Betsch [22] for the sixth abdominal segment, and Nayrolles [23–25] for appendages.

Based on the Nayrolles [23] system, I use, here, a simple tibiotarsal formula, which looks visually convenient for taxonomical purposes in Arrhopalitidae with number of setae in the whorl **I**, **II**, **III**, **IV**, **V**, region **F** (e.g., 9, 8, 8, 8, 7, 4).

2.6. Abbreviations

Ant—antennal segment, Th—thoracic segment, Abd—abdominal segment, a.s.l.—above sea level. The setae, setal rows and whorls are marked in bold in the text.

3. Results

Taxonomy

Family Arrhopalitidae Stach, 1956.

3.1. Genus Arrhopalites Börner, 1906

Arrhopalites profundus sp. nov.

Figures 2–6, Tables 1 and 2.

urn:lsid:zoobank.org:act:FA9937AE-1E30-4F47-9716-C8A4E2C7111F

Diagnosis

Body unpigmented, about 1 mm long. Eyes 1 + 1. Setae of head dorsum are slender. Antenna 2× of head; Ant III with a large sensory organ about 0.8× of segment width; Ant IV subdivided into seven, more or less distinct, subsegments often separated from each other by several annuli, with 14 whorls of setae. Trochanter I, II, and III with 4, 4, 5 setae. Femur I, II, III with 11, 12, 13 setae. Tibiotarsus I, II, III with 42, 41, 43 setae. Claws are slender and extremely elongated, without tunica, I and II with tiny inner tooth, III without inner tooth. Empodia are short, I and II without and III with a distinct corner tooth. Manubrium with 5 + 5 setae. Dens without spines, posterior side with 14 and anterior side with 3, 2, 1, 1, 1 setae. Mucro with a small spoon-like apex, edges of posterior lamellae without separated teeth. Trichobothria **ABC** form an angle close to a right. Posterior setae of great abdomen about 3× longer than anterior. Sixth abdominal segment with ordinary smooth unbroadened circumanal setae; dorsal valve with 10 + 2 + 10 setae; appendices anales are rod-like with apical denticulation. Belongs to *Arrhopalites caecus* group of species s. str. [7].

Type material

Holotype: female, dissected and mounted on two slides (head with antennae and forelegs on one slide, body with midlegs and hindlegs on another slide): W Caucasus, Abkhazia, Bzyb Mountain Range, Khipsta Massif, Snezhnaya Cave, N 43°15′53.3″; E 40°43′05.6″; 1970 m elevation, Crystal Meander, −1100 m, water surface, 27.viii.2018, R.S. Vargovitsh leg.

Paratypes: There is 1 female, 1 specimen of unknown sex (fragment of adult), and 1 juvenile, with the same data as for holotype; 1 male (Figure 5A): 22.viii.2017, same locality and collector.

Holotype (female on slides C-1136-9-1 and C-1136-9-2) and 4 paratypes (male on slide C-1059a-1; damaged fragment of juvenile on slide C-1059a-2; damaged fragment of adult specimen on slide C-1059d-1; female on slide C-1136-8) are kept in the collection of the Schmalhausen Institute of Zoology, National Academy of Sciences of Ukraine, Kyiv.

Description

Female. Body length about 1 mm, unpigmented.

Head (Figure 2A). Eyes 1 + 1, 5.5 μm in diameter, unpigmented, sometimes hardly noticeable. Clypeal area includes six rows, **a** to **f**, and median region **M** with three setae. Interantennal area: row α: 2 + 2, row β: 1 + axial + 1. Dorsal area (rows **A** to **D**): axial seta present in rows **A**, **B**, and **C**; all setae of head dorsum are slender (Figures 2E and 6A). Chaetotaxy of the head posterior side, as in Figure 2A (dashed circles).

Chaetotaxy of mouth region. Labrum (Figure 2A): labral/prelabral chaetotaxy: 4 5 5/6.

Labium: submentum with 4 setae, mentum with 5 setae; labial palp (Figures 2C and 6C) with 5 proximal setae, 5 papillae (**A**, **B**, **C**, **D**, **E**) with deeply inserted setae, with 13 associated guard setae (**a1**, **b1–4**, **d1–4**, **e1–4** of which **a1** is blunt and strongly curved), and 3 hypostomal setae (**H**, **h1**, **h2**). Maxilla as in Figure 6C. Maxillary outer lobe as in Figure 2B: apical seta of the maxillary outer lobe, with short and thin, hardly noticeable branch at the base, sublobal plate with three sublobal hairs. Oral fold with two setae.

Antenna (Figure 2D) about 2× of head. Length ratio of antennal segments I/II/III/IV = 1/2.1/3.4/8.3. Ant I with 6 setae, **p** as microseta (6 μm). Ant II with 14 setae, one of which is about 2× longer than others. Ant III without swelling, with the following chaetotaxy: 15 simple setae of which **Api** and **Ape** are especially thin, two large sense rods (15 μm) in shallow pits and **Aai** as small (4 μm) bent and blunt sensillum (Figures 2D and 5B–D). Ant IV subdivided into 7 subsegments or pseudosubsegments (almost indistinct in female and more distinct in male), often separated from each other by 0–4 weakly developed and hardly visible annuli. (Pseudo)subsegmental formula: 1 + 5 + 1. Ant IV bears 14 whorls of setae: 4 on apical subsegment (**AI–AIII** + **M1**), 5 on median subsegments (**M2–M6**), and 5 on basal subsegment (**BA** + **BM1–BM3** + **BB**).

Legs. Foreleg (Figure 3A): precoxa 1 and 2 and coxa with 1, 0, 1 seta, respectively (Figure 4A). Trochanter with 3 anterior and 1 posterior setae. Femur with 11 setae, **a4** turned perpendicularly to the longitudinal axis of the segment. Tibiotarsus with 42 setae, tibiotarsal formula: 9, 8, 8, 8, 6, 3; seta **Ja** of distal whorl **I**, curved and distinctly spine-like; whorl **V** without **ai** and **pi** setae; region **F** with primary setae **e**, **ae**, **pe**. Pretarsus with 1 anterior and 1 posterior setulae. Foot complex (Figure 5E,H). Claw: slender and much elongated (up to 148 μm), without tunica, with tiny inner tooth and a pair of long subapical lateral teeth; about 2× shorter than tibiotarsus. Empodium: thin, without corner tooth; 1.6× shorter than claw; tip of empodial filament not reaching tip of claw.

Midleg (Figure 3B): precoxa 1 and 2 with 1, 1 seta, respectively, precoxal process present, coxa with 3 setae and microsensillum (Figure 4A). Trochanter with 4 setae: anterior trochanteral organ, 2 anterior and 1 posterior simple setae. Femur with 12 setae. Tibiotarsus with 41 setae (9, 8, 8, 8, 5, 3); **Ja**, curved and distinctly spine-like, whorl **V** without **ai** and **pi** setae; region **F** with primary setae **e**, **ae**, **pe**. Pretarsus with 1 anterior and 1 posterior setulae. Foot complex (Figure 5F). Claw: slender and much elongated (up to 146 μm), without tunica, with tiny inner tooth (sometimes unnoticed) and a pair of long subapical lateral teeth; about 2× shorter than tibiotarsus. Empodium: thin, without a corner tooth; 1.7× shorter than claw; tip of empodial filament not reaching tip of claw.

Hindleg (Figure 3C): precoxa 1 and 2 with 1, 1 seta, respectively, process on precoxa 1 present, coxa with 3 setae and microsensillum (Figure 4A). Trochanter with 5 setae: anterior trochanteral organ, 3 anterior and 1 posterior simple setae. Femur with 13 setae, 2 posterior ones as microsetae. Tibiotarsus with 43 setae (9, 8, 8, 8, 6, 4); **Ja** of whorl **I** as simple seta, whorl **V** without **ai** seta, region **F** with primary setae **e**, **ae**, **pe** and secondary seta **FSa**. Pretarsus with 1 anterior and 1 posterior setulae. Foot complex (Figure 5G). Claw: slender and elongated (up to 131 μm) but shorter than claw I and II, without tunica, without inner tooth, but with pair of long subapical lateral teeth; about 2.7× shorter than tibiotarsus. Empodium: broad, with a distinct corner tooth; 1.7× shorter than claw; tip of empodial filament not reaching tip of claw.

Lengths ratio of tibiotarsi I/II/III = 1/1/1.2. Tibiotarsus I about 1.3× shorter than head.

Ventral tube with 1 + 1 subapical microsetae.

Tenaculum (Figure 4C): each ramus with three teeth and basal process; anterior lobe with 1 apical seta.

Furca (Figure 4B). Manubrium with 5 + 5 posterior setae. Dens with 22–23 setae (8 anterior and 14–15 posterior). There are 3, 2, 1, 1, 1 setae on the anterior side, **Ia–IVa**, **Ba**, and **IIae** are thick, but **Ia** is not spine-like (Figure 2K). The posterior side is without spine-like setae; however, **Ie**, **Ipe**, and **Ii** are somewhat thickened (Figures 2J and 6E). Mucro (Figure 6E): posterior lamellae with teeth usually accreted together, so margins look rather wavy; anterior lamella smooth and well-developed; tip rounded or spoon-like but not broadened. Dens about 2× as long as mucro.

Great abdomen (Figure 4A). Segments Th II and III bearing single short (13–15 μm) and bent sensillum in row **a** and three setae in row **m** with **m1** (Figure 2L) about three times shorter than most posterior setae (**dI–dIII)-1** of posterior dorsal complex (Figure 2M). Abd I bears single row with 5 setae. Trichobothrial complex (Figure 6B): **ABC** form an angle close to right (96–97°), and **AB** is 1.3–1.5× longer than **BC**; seta **p** is located above the level of trichobothrium **B**; seta **b1** lies on line **BC**, a little closer to **C**; microseta **c1** (11 μm) lies in front of trichobothrium **C** and seta **c2**—below **C**. Posterior lateral complex with five setae in two rows (2 + 3) and furca base complex with eight setae in two rows, neosminthuroid seta is absent. Central dorsal complex with the three usual setae. Posterior dorsal complex, with about 14 long setae arranged in 3 longitudinal rows, the longest of which (the most posterior setae of rows **dI–dIII**) = 85–88 μm but shorter than hind claw (Table 1). Ventral complex with 1–2 setae.

Small abdomen. The fifth abdominal segment (Figure 4A), with two setae and trichobothrium **D** in row **a**, as well as two setae in row **p**. Genital field with 3 + 3 microsetae along the anterior margin of the genital opening.

The sixth abdominal segment (Figures 4D and 6D) without any cuticular spines. Dorsal valve with 10 + 2 axial + 10 setae. Each of the lateral valves bears 15 setae. Setae of the circumanal row are slender and not modified, with the longest = 47–56 μm, which is 1.7× shorter than posterior setae of the great abdomen (Figure 2F,G). Appendices anales (Figure 2H,I) (34–37 μm) rod-like, laterally smooth, apically not broadened, and denticulated; sitting on globular basal papilla.

Male. Body 0.9 mm long (Figure 5A), a little smaller than females. Subsegmentation of Ant IV is clearer than in female, with more distinct annulation (Figure 2D). Apex of Ant IV with 11 small setae. Dens without seta **IIpe** (in Figure 4B, marked as variable). Dorsal anal valve of Abd VI with 7 + 2 axial +7 setae, genital opening surrounded by about 10 short setae per side (Figure 4E). Otherwise, male chaetotaxy corresponds to that of females.

Variability

Dens setae **IIpe** (in male) and **IVpi** (in female) missing. Apical whorl of Ant IV with 12 small setae (in female) or 11 (in male). In male seta of left coxa II forked (Figure 4A). Inner tooth of claw I and II very small, sometimes looks absent.

Bionomy and distribution

Specimens of *Arrhopalites profundus* sp. nov. were found exclusively on the water surface in small pools along the cave river at the depth of 1100 m in Snezhnaya Cave (Figure 1E). The air temperature in the sampling locality was 5.8 °C in August 2017 and 6.2 °C in August 2018, and water temperature 5.2 °C (in 2017) and 4.8 °C (in 2018). They co-occur with highly troglomorphic *T. gladiator*.

Etymology

The species name is a Latin adjective 'profundus' in meaning 'deep, obscure', referring to considerable cave depth at which *Arrhopalites* species was recorded for the first time.

Taxonomic remarks

The new species belongs to *A. caecus* species group s. str. with 3, 2, 1, 1, 1 setae on the anterior side of dens. It resembles two highly troglomorphic species with extremely elongated claws—*A. macronyx* from the West Caucasian Anukhvinskaya Cave [9] and *A. gul* Yosii, 1966 from three Japanese caves [26]. Apart elongated claws all these species share the absence of distinctly spine-like setae on dens, short empodia not reaching the tip of corresponding claw, long antennae with 7 subsegments on Ant IV with several annuli between neighboring subsegments, long posterior setae of great abdomen (3× longer than anterior setae).

From *A. gul* the new species differs by shorter antennae (2× of head in *A. profundus* sp. nov. vs. 3× in *A. gul*), enlarged sense rods of antennal III organ (1.5× shorter than segment width vs. 3× shorter in *A. gul*), thin setae on head dorsum (stout in *A. gul*), thin and smooth setae on female Abd VI (broadened and barbered near the basis in *A. gul*), presence of corner tooth on empodium III (absent in *A. gul*), absence of teeth on posterior lamellae of mucro (present in *A. gul*), unbroadened tip of mucro (distinctly broadened in *A. gul*).

Very close to the new species is *A. macronyx* with which, apart from those mentioned above, it shares important similarities. It also shares enlarged antennal III organ, the same antenna/head ratio, absence of spine-like setae on head dorsum, mucro with large anterior lamella and with unbroadened tip, etc. *A. profundus* sp. nov. differs from it by: slender and smooth circumanal setae (broadened and basally serrated in *A. macronyx*), appendices anales with apical denticulation (smooth and acuminated tip in *A. macronyx*), posterior lamellae of mucro without teeth (teeth partly but not completely accreted in *A. macronyx*), and the presence of very small inner tooth on claw I and II (absent in *A. macronyx*). Some details of chaetotaxy are also different: Ant III with 18 setae, including 3 sensilla (vs. 20 in *A. macronyx*); femur III with 13 setae (vs. 12 in *A. macronyx*); setae **Vai** and **Vpi** of tibiotarsi I and II absent (vs. variable in *A. macronyx*); seta **3** of central dorsal complex of great abdomen is of normal length (vs. microseta in *A. macronyx*); lateral valves of Abd VI with 15 setae each (vs. 17 in *A. macronyx*).

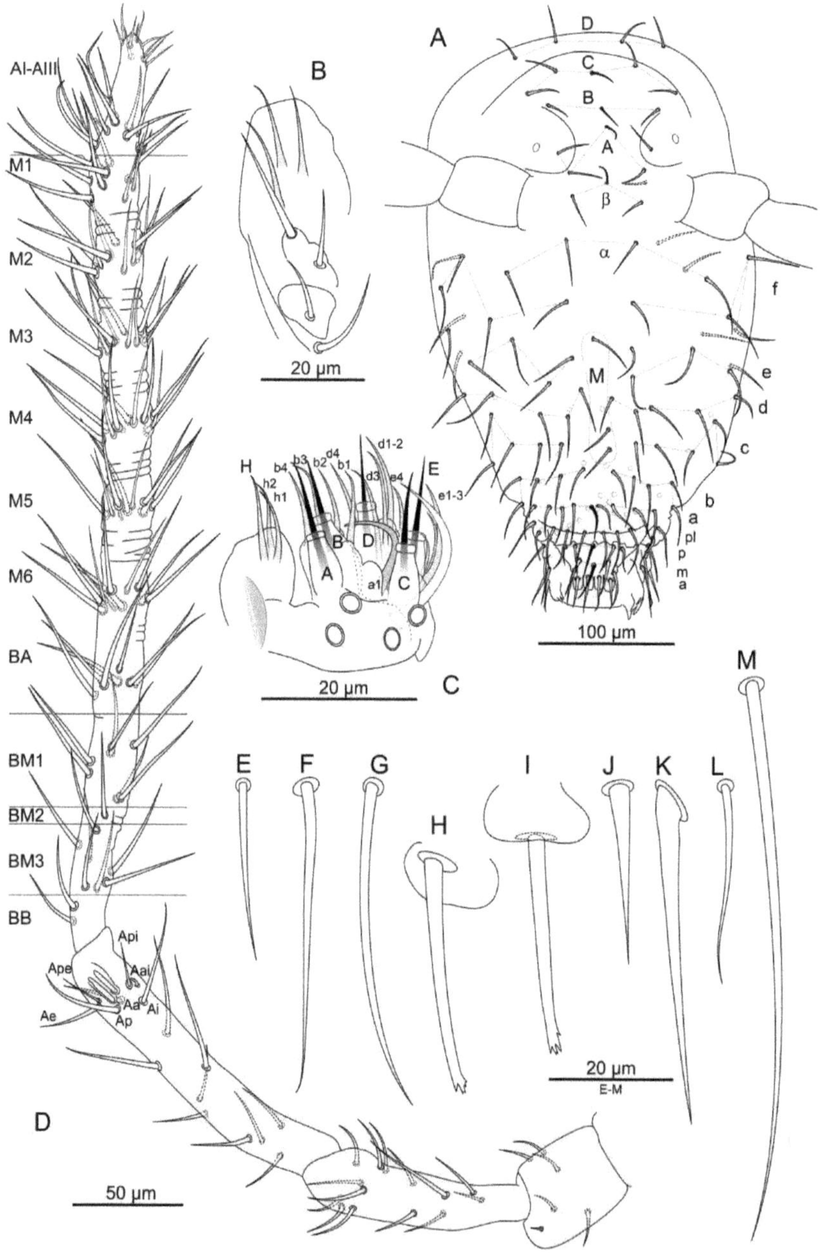

Figure 2. *Arrhopalites profundus* sp. nov. (**A**)—Chaetotaxy of head; (**B**)—Maxillary outer lobe; (**C**)—Chaetotaxy of labial palp; (**D**)—Chaetotaxy of antenna; (**E**)—Seta of head vertex; (**F**)—seta **ms1** of Abd VI; (**G**)—Seta **mps2** of Abd VI; (**H**,**I**)—Appendices anales; (**J**)—Seta **Ie** of dens; (**K**)—Seta **Ia** of dens; (**L**)—Seta **m1** of mesothorax; (**M**)—Seta **dIII-1** of great abdomen. Nomenclatures of setae in (**A**): after [18], (**C**): [19], (**D**): [25], (**F**,**G**): [22], (**J**,**K**): [24], (**L**,**M**): [20].

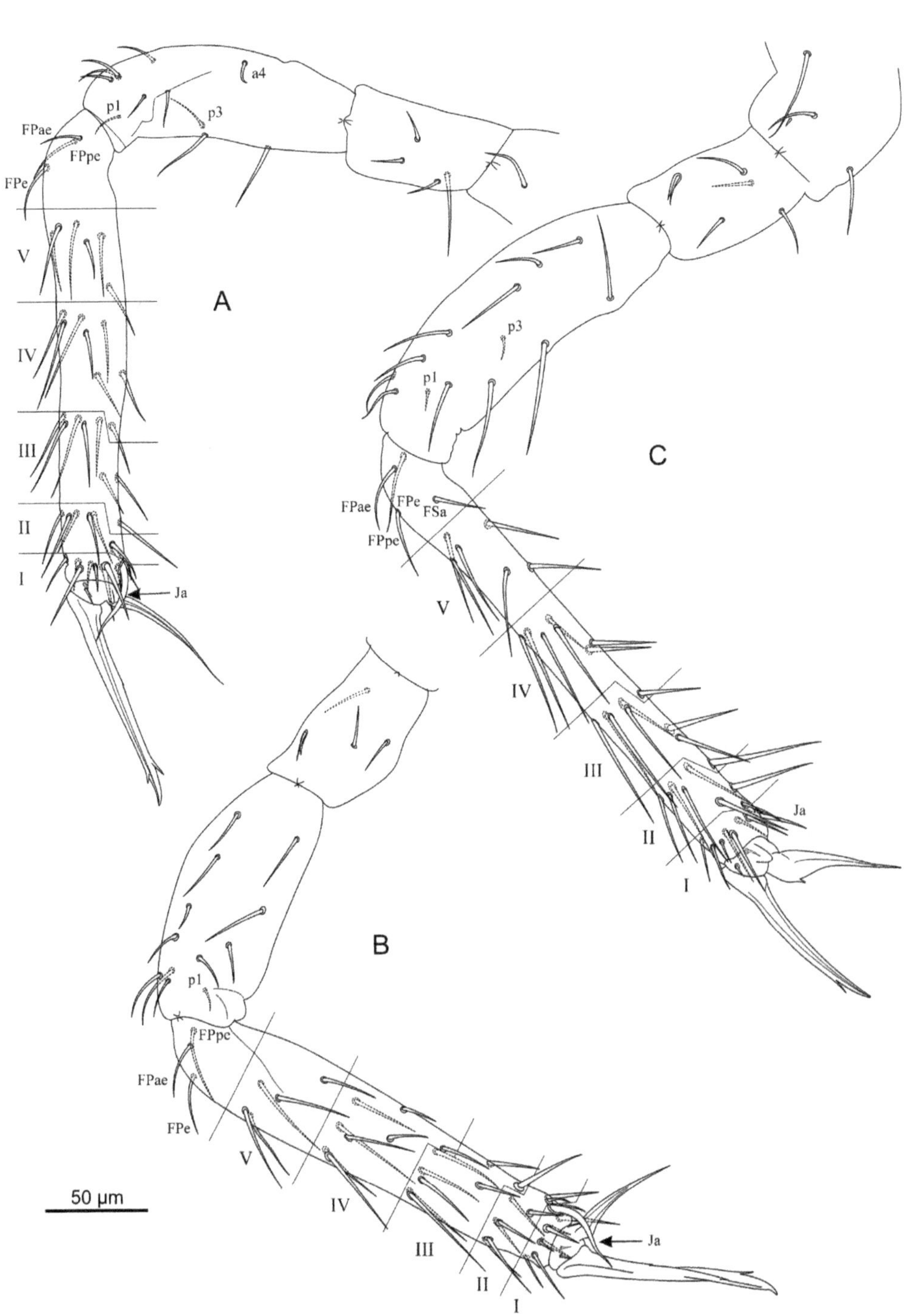

Figure 3. *Arrhopalites profundus* sp. nov. (**A**)—Chaetotaxy of foreleg; (**B**)—Chaetotaxy of midleg; (**C**)—Chaetotaxy of hindleg. Nomenclature of setae in (**A–C**): after [23].

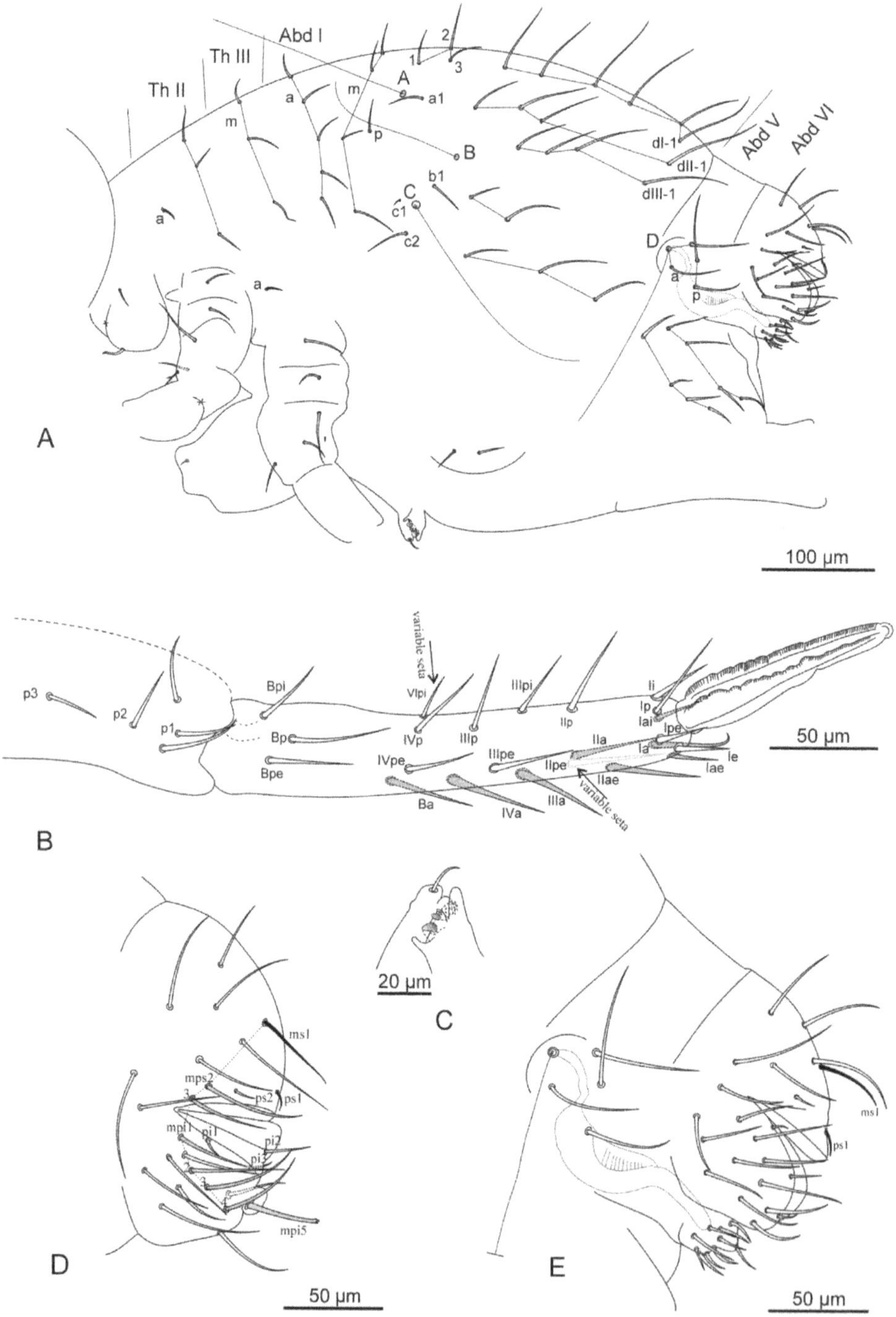

Figure 4. *Arrhopalites profundus* sp. nov. (**A**)—Chaetotaxy of great abdomen; (**B**)—Chaetotaxy of furca; (**C**)—Tenaculum; (**D**)—Chaetotaxy of female Abd VI; (**E**)—Chaetotaxy of male Abd V–VI. Nomenclatures of setae in (**A**): after [9,20], (**B**): [24], (**D**,**E**): [22].

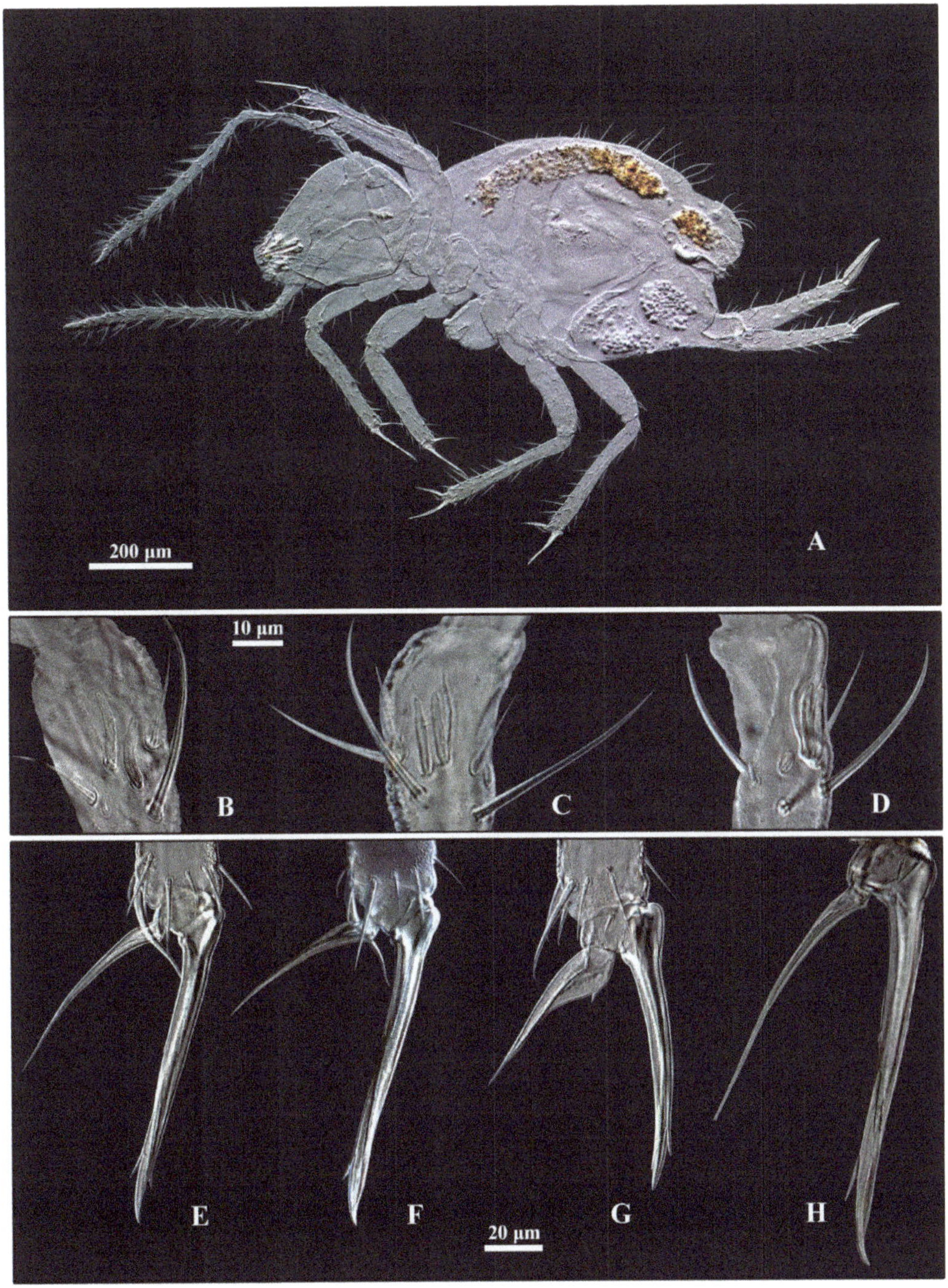

Figure 5. *Arrhopalites profundus* sp. nov. (**A**)—Habitus of mounted male; (**B–D**)—Antennal III organ: (**B**)—Female; (**C**)—Male, right antenna; (**D**)—Male, left antenna; (**E–H**)—Foot complex: (**E**)—Foreleg, male; (**F**)—Midleg, male; (**G**)—Hindleg, male; (**H**)—Foreleg, female.

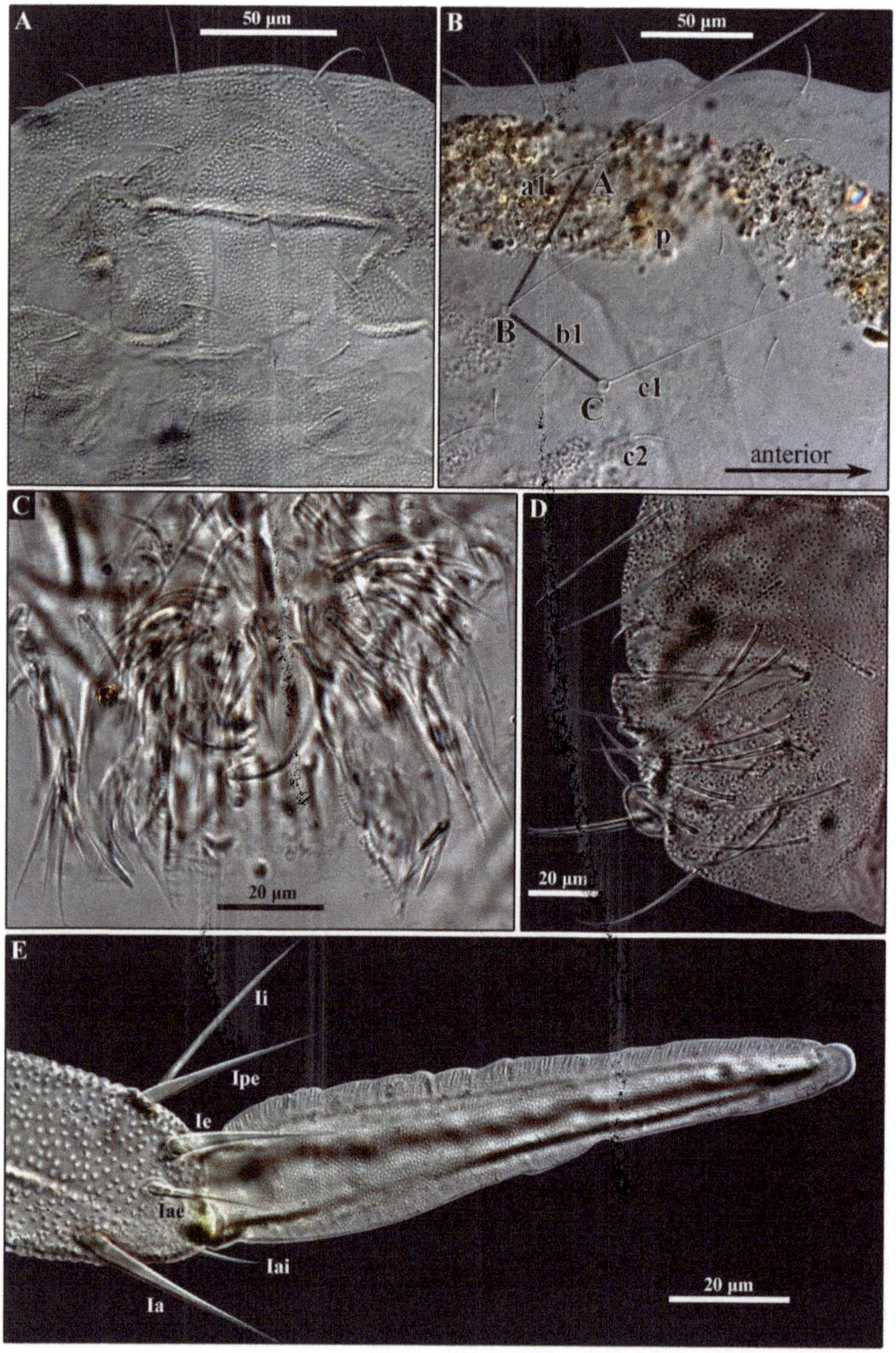

Figure 6. *Arrhopalites profundus* sp. nov. (**A**)—Head dorsum; (**B**)—Chaetotaxy of trichobothrial complex; (**C**)—Mouthparts; (**D**)—Female Abd VI, lateral view; (**E**)—Mucro and distal part of dens. Nomenclatures of setae in (**B**): after [20], (**E**): [24].

3.2. Genus Pygmarrhopalites Vargovitsh, 2009

Pygmarrhopalites rystsovi sp. nov.

Figures 7–11 and Tables 1 and 2.

urn:lsid:zoobank.org:act:5282117D-D944-4F1F-AD6F-D40F8039E3FF

Diagnosis

Body unpigmented, about 1.1 mm long, possessing 1 + 1 eyes without pigmentation. Setae of head dorsum slender. Antenna 3.5× of head; Ant III with subbasal papilla; Ant IV subdivided into 9–10 subsegments with several annulations between them and with 15 whorls of setae. Trochanter I, II, III with 4, 5, 5 setae. Femur I, II, III with 12, 13/14, 12 setae. Each tibiotarsus with 44 setae (9, 8, 8, 8, 7, 4). Claws slender and elongated, without tunica, claw I with tiny (or unnoticeable) and II–III with distinct small inner tooth. Empodia are short, with tips not reaching corresponding claw's tip, and without corner tooth. Manubrium with 7 + 7 setae. Dens posterior side with 16 setae, of which **Ie** as spine and **IIIpe** and **Ii** weakly spine-like near basis; anterior side with 3, 2, 1, 1 setae. Mucro, with broadened tip and about 30 teeth on inner, and partly accreted teeth on outer posterior lamella. Trichobothria **ABC** form an angle about 160° and **AB** < **BC**. Sixth abdominal segment with unbroadened circumanal setae; dorsal valve with 10 + 2 + 10 setae (**ms5** absent); appendices anales apically tapered with lateral subapical fringes or weakly palmated. Intermediate between *Pygmarrhopalites principalis* and *pygmaeus* group of species [27] (p. 66).

Type material

Holotype: female on slide: W Caucasus, Abkhazia, Bzyb Mt. Range, massif of Khipsta Mt., Snezhnaya Cave (N 43°15′53.3″; E 40°43′05.6″; 1970 m elevation), Crystal Meander, −1100 m, 27.viii.2018, R.S. Vargovitsh leg.

Paratypes: There are 4 females and 1 juvenile, with the same data as for holotype; 1 male, W Caucasus, Abkhazia, Bzyb Mt. Range, massif of Khipsta Mt., Snezhnaya Cave, −800 m, 7th Choke (Ozhidaniya Hall), traps, 27.ii.-9.iii.2011, V.V. Rystsov leg.; 1 female, W Caucasus, Abkhazia, Bzyb Mt. Range, massif of Khipsta Mt., Souvenir Cave, −60 m, on trap surface, 24.viii.2019, R.S. Vargovitsh leg.

Holotype (female on slide: C-1136-1) and 7 paratypes (male on 4 slides: C-766-1, C-766-2, C-766-3, C-766-4; female on 3 slides: C-1136-2-1, C-1136-2-2, C-1136-2-3; female on slide: C-1136-3; female on 2 slides: C-1136-4-1, C-1136-4-2; female on slide: C-1136-5; juvenile on slide: C-1136-6; female on slide: C-1216) are kept in the collection of the Schmalhausen Institute of Zoology, National Academy of Sciences of Ukraine, Kyiv.

Description

Female. Length about 1.1 mm, unpigmented.

Head (Figure 7A). Eyes 1 + 1, cornea well-marked, about 13 μm in diameter, unpigmented. Clypeal area includes six rows **a** to **f**: row **a**: 4 + axial + 4 setae, rows **b**, **c**, and **e** with 5 + 5 setae each, rows **d** and **f** with 6 + 6 setae each; median region **M** with 1 seta.

Interantennal area: row **α**: 2 + 2, row **β**: 1 + axial + 1. Dorsal area (rows **A** to **D**): axial seta is present in rows **A**, **B**, and **C**; all setae of head dorsum are slender and relatively short (Figures 7F and 10E). Chaetotaxy of the head posterior side, as in Figure 7A (dashed circles).

Chaetotaxy of mouth region. Labrum (Figure 7A): labral/prelabral chaetotaxy: 4 5 5/6.

Labium: submentum with 4 setae, mentum with 5 setae; labial palp (Figure 7C) with 6 proximal setae, 5 papillae (**A**, **B**, **C**, **D**, **E**) with deeply inserted setae, and with 13 associated guard setae (**a1**, **b1–4**, **d1–4**, **e1–4** of which **a1** is strongly curved) and 3 hypostomal setae (**H**, **h1**, **h2**). Maxillary outer lobe, as in Figure 7B: apical seta with a short and thin subparallel branch at the base, sublobal plate with three sublobal hairs. Oral fold with two setae.

Antenna (Figure 7D,E) about 3.5× of head, subequal or longer than body. Length ratio of antennal segments I/II/III/IV = 1/2.8/4.9/13.8. Ant I with seven setae, **p** as microseta (6 μm). Ant II with 15 setae, 2 of which are over 2× longer than others. Ant III with subbasal papilla (Figure 10B), with following chaetotaxy: 17 simple setae of which **Api** and **Ape** especially thin, as well as 2 relatively large (11 μm) sense rods with 2 longitudinal ribs each in common pit, and **Aai** as small (5 μm) blunt sensillum (Figures 7D and 10C,D).

Ant IV subdivided into 9–10 subsegments, separated from each other by small intercalar pseudosubsegment, without setae or/and by 2–4 annuli (Figure 7E). Subsegmental formula: 1 + 7(8) + 1. Ant IV bears 15 whorls of setae: 4 on apical subsegment (**AI–AIII** + **M1**), 7 on median subsegments (**M2–M8**), and 4 on basal subsegment (**BA** + **BM1–BM2** + **BB**).

Legs (Figure 8). Foreleg (Figure 8A): precoxa 1 and 2 and coxa with 1, 0, 1 seta, respectively (Figure 9A). Trochanter with three anterior and one posterior setae, two of which in the basal part of the segment are longer and stronger than those in subapical. Femur with 12 (in one case in male, asymmetrically 13 setae: variable seta marked in Figure 8A), **a4** thin, not modified and usually not turned perpendicularly to the longitudinal axis of the segment. Tibiotarsus with 44 setae, tibiotarsal formula: 9, 8, 8, 8, 7, 4. Seta **Ja** in distal whorl **I**, weakly curved and slightly thickened but not spine-like; anterior secondary seta **Sa** in region **F** present. Pretarsus with one anterior and one posterior setulae. Foot complex (Figure 11C). Claw: slender and elongated, without tunica, with (sometimes without) tiny inner tooth and a pair of well-marked subapical lateral teeth; about 4.2× shorter than tibiotarsus. Empodium: thin, without corner tooth; smoothly shaped in the broadest part; 2× shorter than claw; empodial tip not reaching the tip of the claw.

Midleg (Figure 8B): precoxa 1 and 2 with 1, 1 setae, respectively, precoxal process present, coxa with 3 setae and microsensillum (Figure 9A). Trochanter with five setae: anterior trochanteral organ, three anterior, and one posterior simple setae. Femur with 13 (usually) or 14 setae (asymmetrically in 2 specimens: position of variable seta marked in Figure 8B). Tibiotarsus with 44 setae (9, 8, 8, 8, 7, 4) as in foreleg. Seta **Ja** in distal whorl **I** is not modified. Pretarsus with one anterior and one posterior setulae. Foot complex (Figure 11D). Claw: slender and elongated, without tunica, with small inner tooth and a pair of well-marked subapical lateral teeth; about 4.2× shorter than tibiotarsus. Empodium: without a corner tooth, but the widest part is angular shaped; 1.9× shorter than claw; tip of empodium not reaching tip of the claw.

Hindleg (Figure 8C): precoxa 1 and 2, with 1, 1 setae, respectively, process on precoxa 1 present, coxa with 3 setae and microsensillum (Figure 9A). Trochanter with five setae: anterior trochanteral organ, three anterior, and one posterior simple setae. Femur with 12 setae, 2 posterior ones as microsetae. Tibiotarsus with 44 setae (9, 8, 8, 8, 7, 4) and **Ja** of whorl **I** as simple seta. Pretarsus with one anterior and one posterior setulae. Foot complex (Figure 11E). Claw: slender and elongated, without tunica, with a small inner tooth and a pair of well-marked subapical lateral teeth; about 5.8× shorter than tibiotarsus. Empodium: broad, untoothed, and the widest part smoothly shaped; 1.7× shorter than claw; with very short apical filament, not reaching the tip of the claw.

Lengths ratio of tibiotarsi I/II/III = 1/1/1.2. Tibiotarsus I as long as the head.

Ventral tube (Figure 9A) with 1 + 1 subapical microsetae.

Tenaculum (Figure 9C): each ramus with three teeth and a basal process; anterior lobe with two apical setae.

Furca (Figure 9B). Manubrium with 7 + 7 posterior setae. Dens with 23 setae (7 anterior and 16 posterior). There are 3, 2, 1, 1 setae on the anterior side. Posterior side with **Ie** as relatively small spine (Figure 7J), **IIpe** and **Ii** are thickened in the basal part (Figure 7I). Mucro (Figure 11F): posterior lamellae with free or somewhat accreted teeth: inner lamella with about 30–35 teeth, outer with less teeth; anterior lamella smooth and moderately developed; tip spoon-like. Dens about 2× as long as mucro.

Great abdomen (Figure 9A). Segments Th II and III bearing single sensillum in row **a** (7 µm in Th II and 16 µm in Th III) and 3 setae in row **m**, with **m1** (Figure 7G) not modified and about 6–7 times shorter than most posterior setae of posterior dorsal complex (Figure 7H). Abd I bears a single row with five setae. Trichobothrial complex (Figure 11B): **ABC** form a very obtuse angle (about 160°), and **AB** is about 1.4× shorter than **BC**; seta **p** is located below the level of trichobothrium **B**; seta **b1** lies on line **BC**, a little closer to **C**; **c1** as normal seta (35 µm, not microseta) lies in front of trichobothrium **C** and seta **c2**—below **C**. Trichobothrium **B** shorter than others. Posterior lateral complex, with six setae in two rows (3 + 3), and furca base complex, with 9 setae in two rows (5 + 4), neosminthuroid seta is absent. Central dorsal complex, with seta **3** distinctly longer than **1** and **2** (Figure 9A). Posterior dorsal complex with about 18 long setae, arranged in 3 longitudinal rows (**dI**: 7 setae, **dII**: 6, **dIII**: 5), the longest of which (the most posterior setae of rows **dI–dIII**) = 127–136 μm, which is about 1.8× longer than hind claw (Figure 7H and Table 2). Ventral complex is usually with two setae.

Small abdomen. Fifth abdominal segment (Figure 9A), with two setae and trichobothrium **D** in row **a** and two setae in row **p**. Genital field with 5 + 5 (also 4 + 4, 4 + 5 or 4 + 6) microsetae along the anterior margin of the genital opening.

Sixth abdominal segment (Figure 9E–G and Figure 11A). Dorsal valve with 10 + 2 axial + 10 setae, **ms5** absent. Each lateral valves bears 18 setae. Setae of circumanal row (**ms1**, **mps1–3** and **mpi1–3**) slightly stronger than others but not broadened (thinner than posterior setae of great abdomen) and relatively short (the longest = 43–58 μm, which is about 2.5× shorter than longest posterior seta of great abdomen) (Figure 7K,L). Appendices anales (Figure 7M–O) (30 μm): rod-like apically tapering shaft, which in the distal half, slightly palmate or bears rough lateral fringes, either subequal with shaft diameter or longer; sitting on subglobular somewhat elongated basal papilla.

Male. Body 0.95 mm long, a little smaller than that of females. Chaetotaxy of head, body, and appendages are the same as in female, apart from the small abdomen (Figure 9D): Abd V, with about 10 + 10 short setae associated with the genital opening (vs. about 5 + 5 in female); dorsal anal valve of Abd VI, with 7 + 2 axial + 7 setae (vs. 10 + 2 + 10 in female) and each of lateral valves with 14 setae (vs. 18 in female).

Variability

Femur I with 12 (usually) or 13 (in one case asymmetrically) setae (Figure 8A). Femur II with 13 or 14 setae (Figure 8B). Claw I with very small inner tooth or without tooth. Female genital plate with 5 + 5 or 4 + 4 or 4 + 5 or 4 + 6 microsetae.

Bionomy and distribution

Specimens of *Pygmarrhopalites rystsovi* sp. nov. were found in the terrestrial habitat on rocky surfaces, at the depth of −1100 m, in Snezhnaya Cave. Additionally, some specimens were captured in pitfall traps at −800 m and −1100 m in Snezhnaya Cave, as well as at −60 m in Souvenir Cave. No specimens were observed on the water surface yet. The air temperature at the sampling localities was: in Snezhnaya Cave: 5.0 °C at −800 m [28] and 5.8–6.2 °C at −1100 m (own data); in Souvenir Cave: 3.0 °C at −60 m (own data).

Etymology

The species name is dedicated to Valentin Rystsov—the head of St. Petersburg Club of Speleologists—who has explored the Snezhnaya Cave System during 11 expeditions since 1986; in 2011, he collected the first specimen (the only male in the type series) of the new species.

Taxonomic remarks

According to Bretfeld's [27] divisions of species groups, and due to the structure of appendices anales (fringes subequal to shaft diameter or somewhat longer), the new species occupies an intermediate position between *P. pygmaeus* and *principalis* species group. Regardless of this, *P. rystsovi* sp. nov. could be compared with congeners having such distinguishable characters as presence of the only spine (**Ie**) on dens and/or presence of subbasal papilla on the third antennal segment. Of the genus *Pygmarrhopalites*, 19 species possess the only spine (**Ie**) on dens, and 12 species have subbasal papilla on Ant III. Combination of these two characters is very rare, and within large genus *Pygmarrhopalites*, with about 100 described species, it occurs in only two troglobiont troglomorphic species: *P. aggtelekiensis* (Stach, 1930) and *P. commorus* (Christiansen and Bellinger, 1996) [29,30]. The impression of similarity with these species enforces the trolomorphic shape of claws (slender, elongated, without tunica, and with short empodia), unmodified (not broadened) circumanal setae, simple (not spine-like) setae on head dorsum, and long antenna, especially in *P. aggtelekiensis*, which also has the same number of subsegments (9–10) on the fourth antennal segment as in the new species.

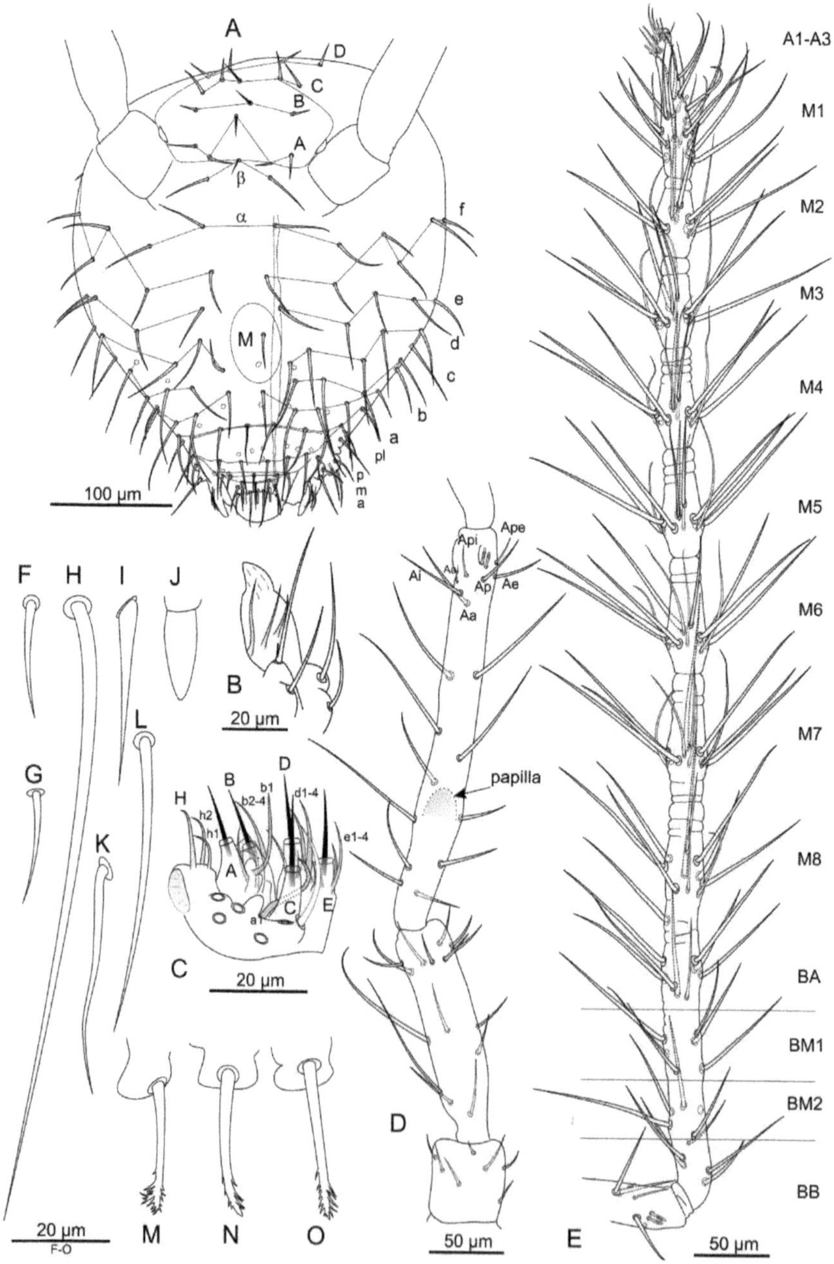

Figure 7. *Pygmarrhopalites rystsovi* sp. nov. (**A**)—Chaetotaxy of head; (**B**)—Maxillary outer lobe; (**C**)—Chaetotaxy of labial palp; (**D**)—Chaetotaxy of antennal segments I–III; (**E**)—Chaetotaxy of antennal segment IV; (**F–O**)—Shape of setae: (**F**)—Seta of head vertex; (**G**)—**m1** of mesothorax; (**H**)—**dIII-1** of great abdomen; (**I**)—**Ii** of dens; (**J**)—**Ie** of dens; (**K**)—**ms1** of Abd VI; (**L**)—**mps2** of abd VI; (**M–O**)—Appendices anales. Nomenclatures of setae in (**A**): after [18], (**C**): [19], (**D**,**E**): [25], (**G**,**H**): [20], (**I**,**J**): [24], (**K**,**L**): [22].

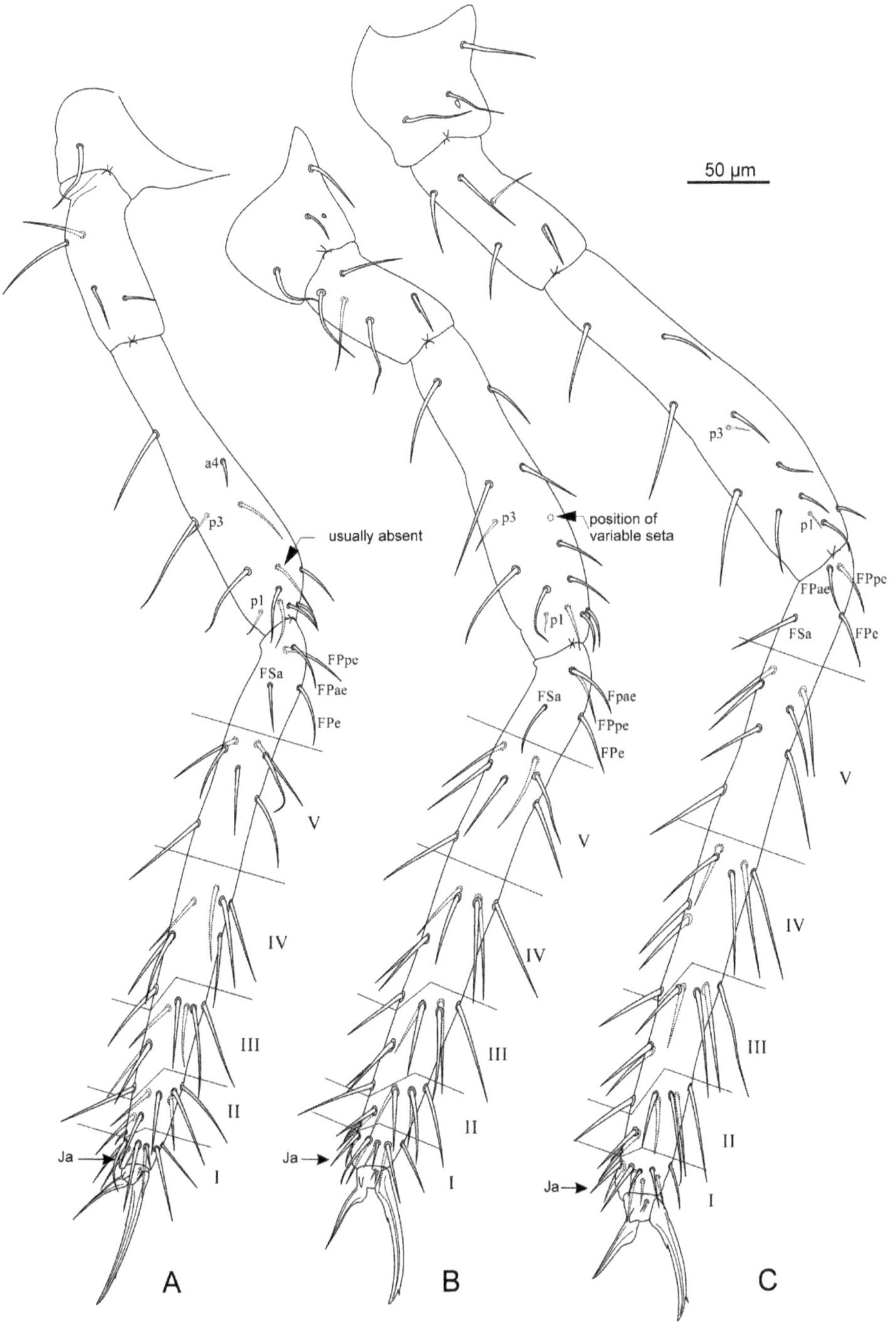

Figure 8. *Pygmarrhopalites rystsovi* sp. nov. (**A**)—Chaetotaxy of foreleg; (**B**)—Chaetotaxy of midleg; (**C**)—Chaetotaxy of hindleg. Nomenclature of setae in (**A–C**): after [23].

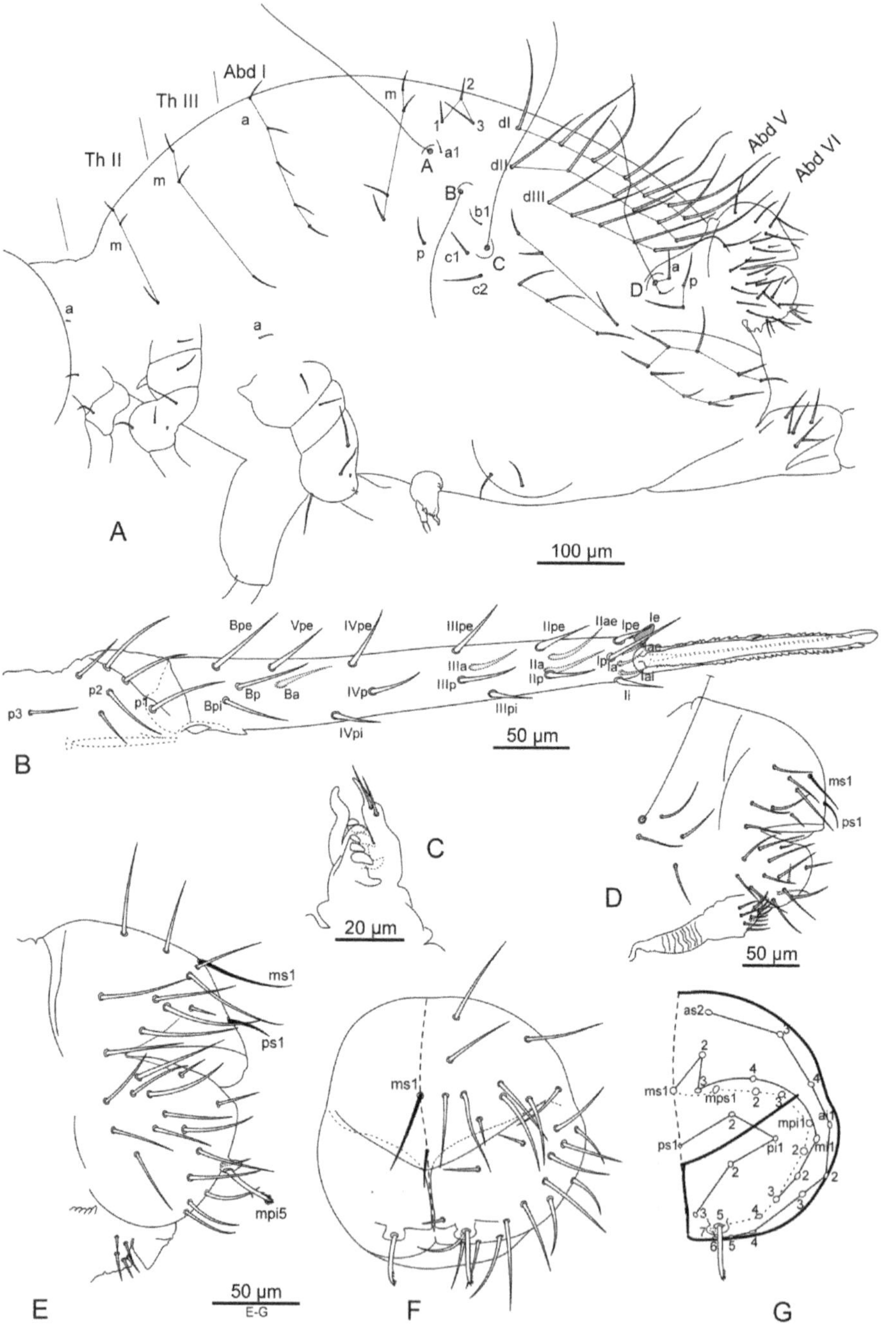

Figure 9. *Pygmarrhopalites rystsovi* sp. nov. (**A**)—Chaetotaxy of great abdomen; (**B**)—Chaetotaxy of furca; (**C**)—Tenaculum; (**D**)—Chaetotaxy of male Abd V–VI; (**E**–**G**)—Chaetotaxy of female Abd VI: (**E**)—Lateral view; (**F**)—Posterior view; (**G**)—The same as (F) but schematic with setae labeling. Nomenclatures of setae in (**A**): after [9,20], (**B**): [24], (**D**–**G**): [22].

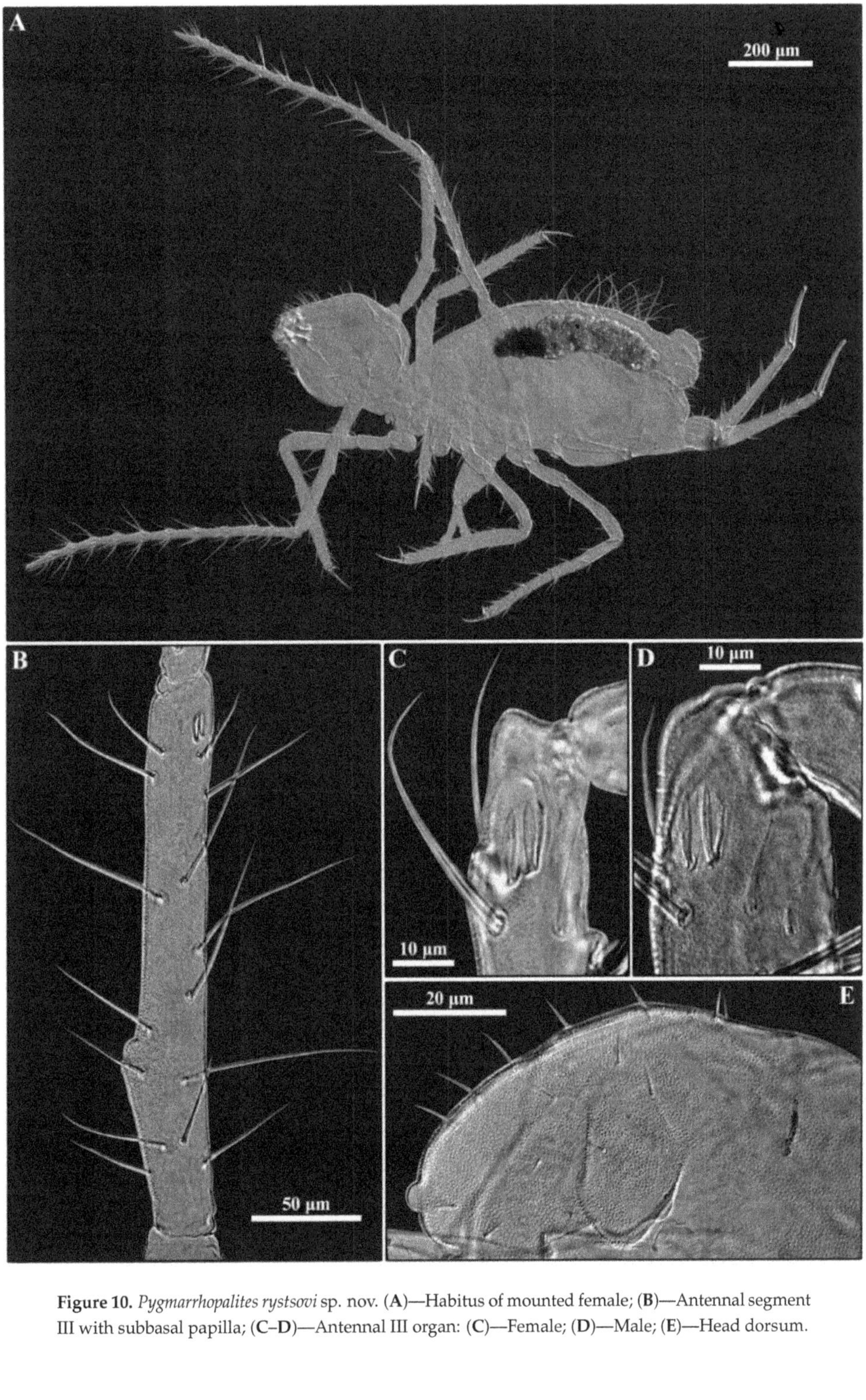

Figure 10. *Pygmarrhopalites rystsovi* sp. nov. (**A**)—Habitus of mounted female; (**B**)—Antennal segment III with subbasal papilla; (**C–D**)—Antennal III organ: (**C**)—Female; (**D**)—Male; (**E**)—Head dorsum.

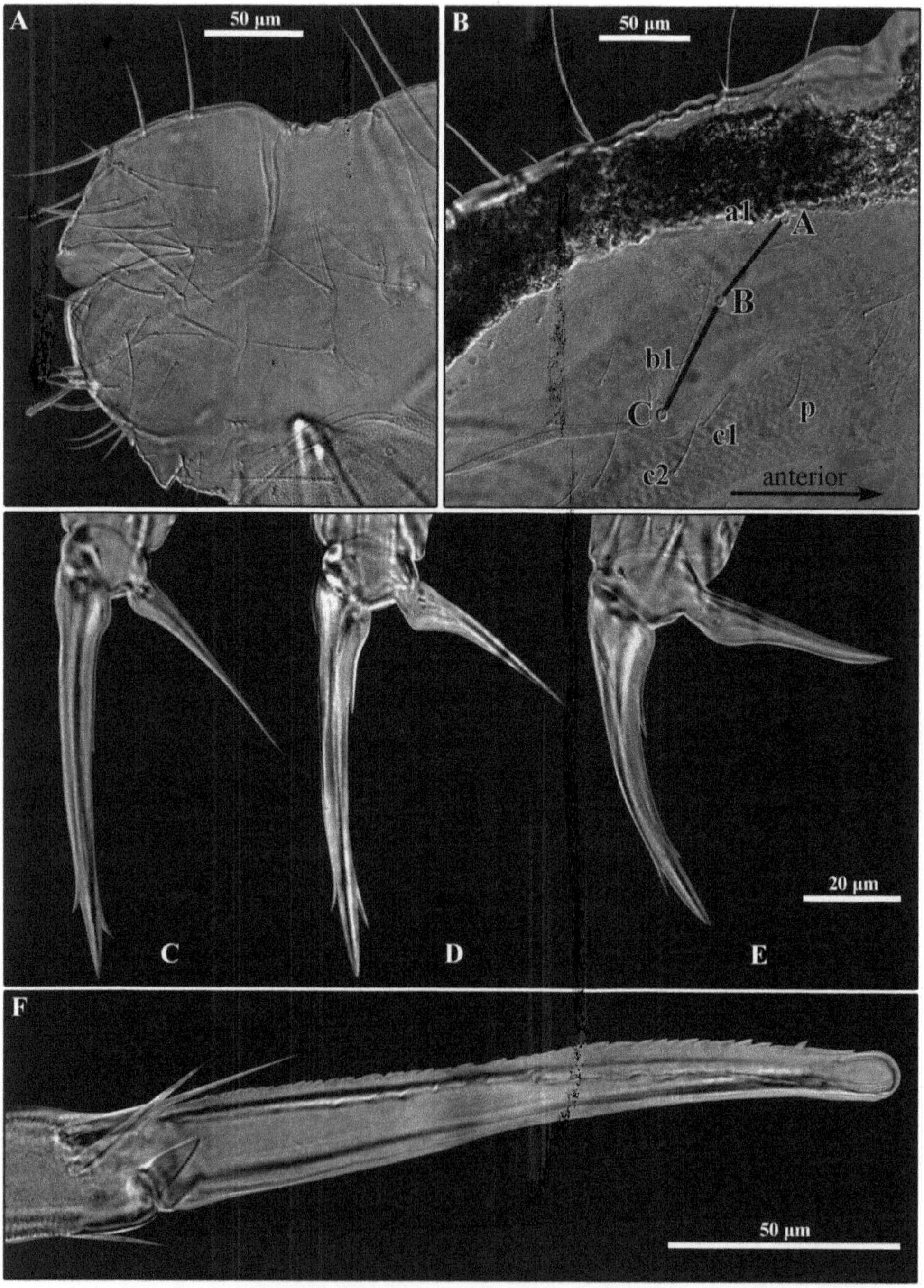

Figure 11. *Pygmarrhopalites rystsovi* sp. nov. (**A**)—Female small abdomen; (**B**)—Chaetotaxy of trichobothrial complex; (**C–E**)—Foot complex: (**C**)—Foreleg; (**D**)—Midleg; (**E**)—Hindleg; (**F**)—Mucro and distal part of dens. Nomenclature of setae in (**B**): after [20].

From the European *P. aggtelekiensis* (N Hungary and S Slovakia), the new species differs by: antenna 3.5× longer than the head (vs. 2.5× in *P. aggtelekiensis*); presence of several annuli (sutures) between the subsegments of Ant IV (vs. only one suture between subsegments); circumanal axial seta **ms1** simple (vs. bifurcated); all empodia without a corner tooth (vs. empodium I and II with tooth); appendices anales, with fringes subequal to shaft diameter or longer (vs. very short).

From the Nearctic *P. commorus* (USA, Virginia), the new species differs by: shorter body (~1.1 mm vs. up to 1.5 mm in *P. commorus*); longer antenna (3.5× of head) with 9–10 subsegments in Ant IV (vs. 1.7–2.1× of head and Ant IV with 6–7 subsegments); presence of several annuli (sutures) between subsegments of Ant IV (vs. only 1 suture); empodium I without a corner tooth (vs. with clear tooth); manubrium with 7 + 7 setae (vs. 4 + 4 setae); appendices anales with distinct lateral and apical fringes (vs. subcylindrical with finely ciliated apex).

4. Discussion

Although inhabiting the same underground realm at considerable depth and sharing basic subterranean abiotic factors, each of the species described above has its own micro-environmental preferences, which obviously led to the evolution into two different troglobiont life forms with subsequently common and different troglomorphic traits.

Below, I consider some aspects of troglomorphy, as well as life forms for the troglobionts of the family Arrhopalitidae, and particularly for the new species described above.

4.1. *Reasons of Different Troglomorphy Level in Arrhopalitidae*

Among 142 species of the family Arrhopalitidae (including the two described above) [31], 88 (or 62%) are considered troglobionts occurring exclusively in caves: 20 of 40 in genus *Arrhopalites* (50%), 67 of 101 in *Pygmarrhopalites* (66%), and 1 in the monotypic genus *Troglopalites* (Vargovitsh, 2012) (100%).

The level of troglomorphy in these troglobionts ranges from weakly marked to extremely high [9,26,29]. Besides, troglomorphic manifestations in highly specialized forms are different. Presumably, it depends on several reasons (drivers) that are listed below.

4.1.1. Ecological Evolutionary Trajectory of Certain Lineages

In general, epigean Arrhopalitidae inhabit dark and moist places, have a strongly reduced visual apparatus (1 + 1 ommatidia, rarely 2 + 2 or absent), and more or less reduced pigmentation. That presumes they are already preadapted to the subterranean life but differently. Particularly, the ancestors of cave-dwelling Arrhopalitidae lived either in litter, moss, etc. (as many *Pygmarrhopalites* species), where weak portions of light occasionally still might penetrate and space volume is not very tight, or in soil and lower litter layers (typical for most *Arrhopalites*), where light is completely absent and space volume between particles is tight. Correspondingly, the species of hemiedaphic ecological lineages possessed body and eye pigmentation and appendages of moderate length, whereas the species of euedaphic lineages lacked pigmentation and had shortened antennae and legs. Consequently, colonization of caves led both lineages to finer cave-depending morphological adaptations or troglomorphies. These adaptations involve parallel or convergent evolution of different lineages [32]. However, morphological signs of past epigean life remain traceable in cave dwellers to a certain extent. This is manifested in the remnants of pigmentation and the relative length of the appendages. Thus, troglobiont species of the genus *Arrhopalites* are usually (but not always) unpigmented and have shorter antennae and appendages than *Pygmarrhopalites* species. Diversification within each of these genera is also observed.

4.1.2. Evolutionary Age of Troglobiont Species

The level of troglomorphy correlates with the evolutionary age of the species, as well as time when colonization of subterranean (cave) habitats occurred. Thus, in Holarctic, the 'old' troglobionts—Tertiary thermophile relicts—are considered highly troglomorphic, whereas the 'young' troglobionts—Quaternary glacial and hydrophilic relicts—are weakly to moderately troglomorphic [33,34]. Normally, in addition to regressive evolutionary features like pigment and eye reduction, the 'old' troglobionts possess well-developed progressive troglomorphisms, such as appendage and claw elongation and the modification of sensory structures (e.g., Japanese *A. gul*, Carpathian *P. aggtelekiensis*, several Caucasian species).

4.1.3. Habitat Preferences of Species in Subterranean Realm

Cave-dwelling Arrhopalitidae species occupy several ecological niches regarding preferable substrates. The latter includes: (a) open terrestrial habitats such as the surface of cave walls, stones, floor, clay, and speleothems; (b) water surface of underground pools, rimstones, lakes, or siphons, as well as rocky surfaces with a thin water layer; (c) closed places such as under stones, in piles of bat guano, or inside other substrates. Using of these niches may be particularly overlapped by the same species, but the most specialized forms, apparently, are strictly determined (e.g., as observed in the Caucasian *A. macronyx*, *A. profundus* sp. nov. and *T. stygios*) [9].

4.2. *Troglomorphic Features in Arrhopalitidae*

Like in other cave animals, both regressive (reduction or loss of morphological structures) and progressive (adaptive development of morphological structures) evolutionary features associated with cave life [32] are present in Arrhopalitidae.

Regressive features. Given that, in Arrhopalitidae, the eye reduction or microphthalmy is not a cave-dependent feature because it was already present in epigean ancestors, the reduction/loss of pigmentation in certain species probably remains the only feature of regressive evolution in troglobionts of this family. Several patterns of this reduction, with combinations of different body and eye pigmentation, could be mentioned: (a) black or dark eyes in combination with light grayish (e.g., *Pygmarrhopalites kristiani* (Vargovich, 2005)) [35], brownish (*P. pseudoprincipalis* (Vargovitsh, 2009)) [20], or unpigmented (*P. ruseki* (Nosek, 1975)) [36] body in 'young' troglobionts; (b) unpigmented eyes with reddish (*P. dbari*) [8] or unpigmented (*P. tauricus* (Vargovitsh, 2009)) [20] body in 'old' troglobionts, as well as in preadapted 'young' troglobionts (*Arrhopalites loczyi* (Loksa, 1960) [37], *A. karabiensis* (Vargovitsh, 2009) [20])—the descendants of soil-dwelling lineages. It is interesting that reddish diffuse pigmentation, as well as complete depigmentation may occur even in highly troglomorphic species (e.g., *P. aggtelekiensis*, *T. stygios*). Unusual combinations may also happen: black eyes and light grayish body together, with the highly troglomorphic slender and distinctly elongated claws in *P. kristiani* [35].

Progressive features. Progressive or constructive troglomorphy is clearly affected by the cave environment [32].

In Arrhopalitidae, the troglomorphic (cave-dependent) features are: elongation of appendages (antennae, legs, furca), neustonic-hygropetric morphological adaptations (slendering and elongation of claws, modification of mucronal lamellae), and development of sensorial structures (enlarged antennal III organ, elongated and thinned setae, etc.) [8,9,20]. They are listed below.

4.2.1. Elongated Antennae with Multi-Subdivided Fourth Antennal Segment

In non-cave-dwelling species of the genus *Arrhopalites* (living in the soil, lower-litter layer, under the bark of trees) antennae are short, usually 1.3–1.5 times longer than the head; besides, Ant IV of these species is not subdivided or it is unclearly subdivided, e.g., [38–40]. The similar pattern is also known in non or weakly troglomorphic cave-dwelling *Arrhopalites* species (e.g., in *A. karabiensis*, *A. glabrofasciatus* (Zeppelini, Brito, and Lima, 2018)) [20,41]. In moderately or strictly cave-dependent forms, the antennae are usually distinctly longer, about 2× the size of the head (e.g., in the Caucasian *A. abchasicus*, *A. macronyx* and *A. profundus* sp. nov.), or reaching even about 2.9× the size of the head in the Japanese *A. gul* [7,9,26]. Subdivision of Ant IV becomes more developed and reaches up to 6–8 subsegments, usually with several annuli between them.

In most epigean species of the genus *Pygmarrhopalites*, antennae are usually about 1.6–1.8× the size of the head, with 5–6 subsegments on Ant IV. In weakly troglomorphic or non-troglomorphic cave dwellers the pattern is similar (e.g., *P. pseudoprincipalis*, *P. ruseki*, *P. slovacicus* (Nosek, 1975)) [20,36,42]. In more specialized species, antenna/head ratio reaches about 2.5× (*P. aggtelekiensis*, *P. kaprusi* (Vargovitsh, 2009), *P. tauricus*) or even 3–4× in most troglomorphic forms (*P. dbari*: 3×; *P. rystsovi* sp. nov.: 3.5×; *P. uenoi* (Yosii, 1956): 4×). Subdivision of Ant IV reaches 9–10 (*P. aggtelekiensis*, *P. delamarei* (Nosek and Paoletti, 1984), *P. altus* (Christiansen, 1966), *P. tauricus*) or even 15 subsegments (*P. uenoi*) [8,20,29,43,44].

In the monotypic *T. stygios*, this feature is also highly troglomorphic: antenna/head ratio is 3×, and Ant IV bears 13 subsegments [9].

4.2.2. Elongated Legs

Relative length of legs in cave-dwelling Arrhopalitidae varies from not elongated, as in epigean relatives, to distinctly elongated in highly specialized species. However, sometimes, leg segments are relatively short in highly troglomorphic—but strictly neustonic-hygropetric—forms. Particularly, this could be seen from comparison of body/tibiotarus III length ratio (2.6× in *P. rystsovi* sp. nov.—distinctly elongated; 3.3× in *T. stygios* and *A. profundus* sp. nov., and 4.3× in *A. macronyx*—not elongated) [9]. For comparison, in soil-dwelling *Arrhopalites* species, the body/tibiotarsus III ratio is about 4.5–5×, e.g., [39,40].

4.2.3. Elongated and Slender Claws

This character is a classical manifestation of troglomorphy, which convergently evolved in cave dwellers of several collembolan families [6,45–49], and the family Arrhopalitidae is among them. Only a few species of the genus *Arrhopalites* possess distinctly elongated claws, but in the two Caucasian species, the character is spectacular. The index of claws troglomorphy (body/claw I length ratio) in *A. macronyx* and *A. profundus* sp. nov. is about 7, representing the highest level of claw troglomorphy among Symphypleona. For comparison, in soil-dwelling and epigean species (e.g., *A. persicus* (Vargovitsh and Kahrarian, 2020), *A. potapovi* (Vargovitsh, 2015), *A. prutensis* (Vargovitsh and Busmachiu, 2015)), the body/claw I length ratio is about 25–30×, which is prominently different from highly troglomorphic ones [38–40]. The slender elongated claws in Arrhopalitidae are usually accompanied by relatively short empodia, far from reaching the tip of the corresponding claw.

4.2.4. Elongated Furca and Modified Lamellae of Mucro

In troglomorphic Arrhopalitidae, the mean body/furca length ratio is about 1.9× (in *A. macronyx*), 1.7× (*T. stygios*), 1.6× (*A. profundus* sp. nov.), and 1.5× (*P. rystsovi* sp. nov.), whereas in soil-dwelling and epigean *Arrhopalites* species, this ratio is about 2.3–2.7×, which means the furca in troglomorphic species is relatively longer. Besides, accreted teeth on lateral lamellae, as well as well-developed ventral lamella on mucro, are exclusive features of the highly troglomorphic water-dependent species (*A. macronyx*, *A. profundus* sp. nov., *T. stygios*) [9].

4.2.5. Enlarged Sensory Organ of the Third Antennal Segment

Antennal III organ in Arrhopalitidae consists of two sense rods, accompanying setae, and a short sensillum. It is especially well-developed in highly troglomorphic species, which are strongly associated with the surface of water [9]. Particularly, in *A. profundus* sp. nov., the length of each sense rod is about 15 μm, which is about three times larger than in soil-dwelling/epigean congeners of the same body size.

4.2.6. Thinning of the Spine-like Setae on Head and Dens

The shape of spiny setae, in weakly or moderately troglomorphic Arrhopalitidae, is similar to their homologues in epigean relatives. In highly troglomorphic species, a tendency towards attenuation and thinning of the spine-like setae is evident, especially on the head dorsum and the dens of the furca. Thus, epigean species of *Pygmarrhopalites* basically possess two strong external and three internal spine-like setae on the posterior surface of dens. Unlike them, highly troglomorphic species bear only one weakened external spine and thinned spine-like setae, or even ordinary setae, instead of other spiny ones (e.g., *P. kaprusi, P. tauricus, P. aggtelekiensis, P. boneti* (Stach, 1945), *P. uenoi, P. hungaricus* (Loksa, 1967), etc.) [20,29,44,50]. Simultaneously, these *Pygmarrhopalites* species possess thin setae on the head dorsum. In most troglomorphic water-depending *Arrhopalites* species (*A. macronyx, A. profundus* sp. nov.) and *T. stygios*, all spines on the dens and head dorsum are replaced with ordinary setae [9]. Probably, this replacement could mean switching the protective or supporting function (spines) to the sensitive function (setae), which is possibly preferred in cave habitats.

4.2.7. Elongated Setae of Posterior Dorsal Complex

In soil-dwelling and epigean *Arrhopalites* species (e.g., *A. potapovi, A. persicus, A. prutensis*), the body/posterior seta **dI-1** of great abdomen length ratio is about 24–31× [38–40]. In highly troglomorphic *A. macronyx*, this ratio is about 15× [9], in *A. profundus* sp. nov. about 11×, and in *P. rystsovi* sp. nov. even about 8×. This comparison indicates a distinct elongation of the posterior abdominal setae in specialized cave dwellers.

4.3. *Life Forms in Cave-Dwelling Arrhopalitidae*

The cave environment [51], in addition to darkness, high humidity, and relatively constant temperature, provides cave-dwelling Collembola with a large open space in combination with often wet, rocky, and clay surfaces, water film, as well as spaces with small diameters within substrates.

In his ecological classification of collembolan life forms, Christiansen [52,53] defined category troglomorphs for cave dwellers with features: "elongate antennae and body form of the atmobios plan, the eyeless, pigmentless characteristics of the euedaphon, and the elongate or modified ungues of the aquatic form".

Depending on the confinement to certain habitats inside caves, and taking for a basis the Christiansen's [52,53] definition of troglomorph life form, the following specific life forms (or sub-forms) of troglobiont Arrhopalitidae can be distinguished.

4.3.1. Neustonic-Hygropetric Troglomorphs (Cave Water & Wet-Stone Walkers)

Preferable habitats: surface of cave water bodies (pools, lakes, rimstones, streams) and cave hygropetric (walls with thin water films moving down in mostly laminar flow [54]. Specific traits in Arrhopalitidae are: (a) slender and significantly elongated claws with a reduced/absent inner tooth—the main indicative adaptation—while other segments of the legs are not elongated; (b) enlarged antennal III organ; (c) spine-like setae on dens are replaced with attenuated setae; (d) lateral lamellae of mucro are with partly or completely accreted teeth, and ventral mucronal lamella is well-developed. Examples: *A. macronyx, A. profundus* sp. nov., *T. stygios* [9]. These species are strongly associated with water-surface habitats, have not been found remote from water, and failed to be caught by terrestrial Barber's pitfall traps.

4.3.2. Atmobiont Troglomorphs (Cave Terrestrial Walkers)

Preferable habitats: open terrestrial substrates—walls, rocks, clay, speleothems, etc., but they may also occur on water surfaces. Specific traits in Arrhopalitidae: (a) elongated antennae and legs, while claws might not be elongated or only slightly elongated; (b) lateral lamellae of mucro are with normal teeth, and ventral lamella is not enlarged; (c) antennal III organ is not hypertrophied; (d) spine-like setae on dens are not much thinned or are moderately thinned. Examples: *P. kaprusi*, *P. skelicus* (Vargovitsh, 2009), *P. carpathicus* (Vargovich, 1999) [20,55]. Normally, these species can be easily caught using Barber's traps.

4.3.3. Intermediate Troglomorphs (Cave Terrestrial & Water Walkers)

The pure neustonic-hygropetric and atmobiont troglomorphs are rather rare. Most of troglobiont Arrhopalitidae species regularly occur in search of food, both in open terrestrial and water-surface habitats. Consequently, their morphological adaptations show combined or intermediate states. Traits from both life forms, determined by shared abiotic factors and developed to varying degrees each, are: elongated antennae with an enlarged number of Ant IV subsegments, elongated legs, slender and moderately elongated claws, elongated abdominal setae, etc. Examples: *A. abchasicus*, *A. peculiaris* (Vargovitsh, 2009), *P. aggtelekiensis*, *P. dbari*, *P. tauricus*, *P. rystsovi* sp. nov. [7,8,20,29] It is probably the most diverse life (sub)form among the cave Arrhopalitidae. Unlike strict water walkers, these species can be caught using terrestrial pitfall traps.

4.3.4. Intrasubstrate Troglomorphs (Cave-Dwelling within-Substrate Inhabitants)

Preferable habitats: substrates with small space diameter, such as inside bat guano piles or other organic substrates, under stones, etc. Hypothetically, such a life form should exist among the Arrhopalitidae, and its appearance should be similar to that of an edaphic life form (relatively small body with short antennae, legs, furca, and unmodified foot complex). Probably, several Neotropic cave-dwelling *Arrhopalites* species described by Zeppelini (e.g., [56]) belong to this group.

5. Conclusions

The deep caves of the Bzyb Massif in the West Caucasus are inhabited by a highly specialized relictual and strictly endemic speleofauna, including collembolans of the family Arrhopalitidae.

To date, the findings of new species described from Snezhnaya Cave, *A. profundus* sp. nov. and *P. rystsovi* sp. nov., are the deepest (up to −1100 m) records of cave-dwelling Arrhopalitidae. Besides, both new species are among the most troglomorphic members of the family, but they represent different life forms.

Progressive troglomorphic (cave-dependent) features, differently exposed in Arrhopalitidae species, depending on habitat, include: (a) elongation of antennae with additional subdivision of the fourth antennal segment; (b) elongation of legs and furca; (c) thinning and elongation of claws with reduction in its inner tooth; (d) lammelization of mucro; (e) enlargement of sensory organ of third antennal segment; (f) attenuation of spiny setae on dens and head dorsum; (g) elongation of abdominal dorsal posterior setae.

Adaptation to the certain habitats inside caves led to the development of specific morphological features, indicating the presence of several life forms in cave-dwelling Arrhopalitidae: (a) neustonic-hygropetric troglomorphs preferably inhabiting the surface of cave water bodies and wet walls; (b) atmobiont troglomorphs preferably inhabiting open terrestrial substrates; (c) intermediate troglomorphs equally sharing both neustonic-hygropetric and open terrestrial habitats; (d) intrasubstrate troglomorphs preferably inhabiting substrates with small space diameter.

Funding: This research received no external funding.

Institutional Review Board Statement: Not applicable.

Informed Consent Statement: Not applicable.

Data Availability Statement: Not applicable.

Acknowledgments: I am obliged to Alexander Koval (St. Petersburg), Illya Turbanov (Sevastopol), Eugeny Romanov (Sevastopol), Igor Pelkin† (Sochi), Sergey Yashin (Sochi) and Oleg Merkurev (Sochi) for invaluable help during fieldworks in the Snezhnaya Cave System and Souvenir Cave in 2017–2019, and to Roman Dbar (Sukhum) for the administrative support during expeditions in Abkhazia.

Conflicts of Interest: The author declares no conflict of interest.

References

1. Antić, D.Ž.; Makarov, S.E. The Caucasus as a major hotspot of biodiversity: Evidence from the millipede family Anthroleucosomatidae (Diplopoda, Chordeumatida). *Zootaxa* **2016**, *4211*, 1–205. [CrossRef] [PubMed]
2. Fiera, C.; Arbea, J.I.; Vargovitsh, R.S.; Barjadze, S. A synthesis on troglobitic springtails in Europe. *J. Zool. Syst. Evol. Res.* **2021**, *59*, 1874–1890. [CrossRef]
3. Turbanov, I.S.; Palatov, D.M.; Golovatch, S.I. The state of the art of biospeleology in Russia and other countries of the former Soviet Union: A review of the cave (endogean) invertebrate fauna. 2. Arachnida—Acknowledgments. *Entomol. Rev.* **2016**, *96*, 1297–1333. [CrossRef]
4. Jordana, R.; Baquero, E.; Reboleira, S.; Sendra, A. Reviews of the genera *Schaefferia* Absolon, 1900, *Deuteraphorura* Absolon, 1901, *Plutomurus* Yosii, 1956 and the *Anurida* Laboulbène, 1865 species group without eyes, with the description of four new species of cave springtails (Collembola) from Krubera-Voronya cave, Arabika Massif, Abkhazia. *Terr. Arthropod Rev.* **2012**, *5*, 35–85. [CrossRef]
5. Babenko, A.B. New species of springtails of the genus *Typhlogastrura* (Collembola, Hypogastruridae) from caves of the USSR. *Zool. Zh.* **1987**, *66*, 463–472. (In Russian)
6. Vargovitsh, R.S. Cave water walker: An extremely troglomorphic *Troglaphorura gladiator* gen. et sp. nov. (Collembola, Onychiuridae) from Snezhnaya Cave in the Caucasus. *Zootaxa* **2019**, *4619*, 267–284. [CrossRef]
7. Vargovitsh, R.S. Cavernicolous *Arrhopalites abchasicus* sp. nov. (Collembola: Symphypleona: Arrhopalitidae) from the West Caucasus with a key to the World species of the genus. *Zootaxa* **2013**, *3666*, 16–30. [CrossRef]
8. Vargovitsh, R.S. Two new troglobiont *Pygmarrhopalites* species of the *principalis* group (Collembola: Arrhopalitidae) from the West Caucasus. *Zootaxa* **2017**, *4250*, 23–42. [CrossRef]
9. Vargovitsh, R.S. New troglomorphic Arrhopalitidae (Collembola: Symphypleona) from the Western Caucasus. *Zootaxa* **2012**, *3174*, 1–21. [CrossRef]
10. Barjadze, S.; Djanashvili, R. Checklist of the springtails (Collembola) of Georgia. *Caucasian Entomol. Bull.* **2008**, *4*, 187–193. [CrossRef]
11. Klimchouk, A. Caucasus, Georgia. In *Encyclopedia of Caves and Karst Science*; Gunn, J., Ed.; Fitzroy Dearborn: Chicago, IL, USA, 2004; pp. 412–417.
12. Vakhrushev, B.A.; Dublyansky, V.N.; Amelichev, G.N. *Karst of the Bzybsky Ridge. The Western Caucasus*; Peoples' Friendship University of Russia: Moscow, Russia, 2001; pp. 1–165. (In Russian)
13. Gusev, A.S. Hydrology of underground water of Khipsta Massif (Abkhazia). *Probl. Geogr.* **2018**, *147*, 107–133. (In Russian)
14. Mavlyudov, B.R.; Shelepin, A.L. Systema Snezhnaya. In *Atlas of Caves of Russia*; Shelepin, A.L., Ed.; Russian Geographical Society: Moscow, Russia, 2019; pp. 751–762. (In Russian)
15. Mavlyudov, B.R. The Snezhnaya-Mezhennogo-Illyuziya cave system in the western Caucasus. *Bol. Geol. Min.* **2016**, *127*, 219–235.
16. Barber, H.S. Traps for cave-inhabiting insects. *J. Elisha Mitchell Sci. Soc.* **1931**, *46*, 259–267.
17. Mercet, R.G. Los enemigos de los parásitos de las plantas. *Trab. Mus. Nac. Ci. Nat.* **1912**, *6*, 1–306.
18. Betsch, J.-M.; Waller, A. Chaetotaxic nomenclature of the head, thorax and abdomen in Symphypleona (Insecta, Collembola). *Acta Zool. Fenn.* **1994**, *195*, 5–12.
19. Fjellberg, A. The labial palp in Collembola. *Zool. Anz.* **1999**, *237*, 309–330.
20. Vargovitsh, R.S. New cave Arrhopalitidae (Collembola: Symphypleona) from the Crimea (Ukraine). *Zootaxa* **2009**, *2047*, 1–47. [CrossRef]
21. Bretfeld, G. The chaetotaxy of the small abdomen of the Symphypleona (Insecta, Collembola) and its phylogenetic interpretation. *Acta Zool. Fenn.* **1994**, *195*, 13–17.
22. Betsch, J.-M. An ontogenetically focused chaetotaxical scheme in Symphypleona (Collembola): The 6th abdominal segment. *Pedobiologia* **1997**, *41*, 13–18.
23. Nayrolles, P. Chétotaxie tibiotarsale des Collemboles Symphypléones. *Trav. Lab. Ecobiol. Arthrop. Edaph. Toulouse* **1988**, *5*, 1–19.
24. Nayrolles, P. Chétotaxie furcale des Collemboles Symphypléones. *Trav. Lab. Ecobiol. Arthrop. Edaph. Toulouse* **1990**, *6*, 27–50.
25. Nayrolles, P. La chétotaxie antennaire des Collemboles Symphypléones. *Trav. Lab. Ecobiol. Arthrop. Edaph. Toulouse* **1991**, *6*, 1–94.
26. Yosii, R. Results of the speleological survey in South Korea 1966. IV. Cave Collembola of South Korea. *Bull. Natl. Sci. Mus. Tokyo* **1966**, *9*, 541–561.
27. Bretfeld, G. *Synopses on Palaearctic Collembola: Symphypleona*; Staatliches Museum für Naturkunde Görlitz: Görlitz, Germany, 1999; Volume 2, pp. 1–318.
28. Mazina, S.E.; Bazarova, E.P.; Bashirov, A.M.; Gabbasova, E.R.; Gusev, A.S.; Kopachevskiy, Y.Y.; Kochetkov, S.N.; Savinov, V.P.; Shadrin, V.O. Abiotic Components of an Ecosystem of a Snezhnaya (Illyuzia-Mezhonnogo-Snezhnaya) Cave: A Condition and Problems of Research. In *Mineralogy of Technogenesis–2011*; Potapov, S.S., Ed.; IMin UrO RAS: Miass, Russia, 2011; pp. 219–238. (In Russian)
29. Stach, J. The species of the genus *Arrhopalites* occurring in European caves. *Act. Mus. Hist. Nat. Acad. Polon. Litt. Sci.* **1945**, *1*, 1–47.
30. Christiansen, K.; Bellinger, P. Cave *Arrhopalites* new to science. *J. Cave Karst Stud.* **1996**, *58*, 168–180.

31. Bellinger, P.F.; Christiansen, K.A.; Janssens, F. Checklist of the Collembola of the World (1996–2022). Available online: http://www.collembola.org (accessed on 1 August 2022).
32. Christiansen, K. Morphological Adaptations. In *Encyclopedia of Caves*; Culver, D.C., White, W., Eds.; Elsevier Academic Press: Amsterdam, The Netherlands; Boston, MA, USA, 2005; pp. 386–397.
33. Vandel, A. *Biospeleology. The Biology of Cavernicolous Animals*; Pergamon Press: Oxford, UK, 1965; pp. 1–524.
34. Kováč, L'.; Parimuchová, A.; Miklisová, D. Distributional patterns of cave Collembola (Hexapoda) in association with habitat conditions, geography and subterranean refugia in the Western Carpathians. *Biol. J. Linn. Soc.* **2016**, *119*, 571–592. [CrossRef]
35. Vargovich, R.S. *Arrhopalites kristiani* sp. n. (Collembola: Symphypleona: Arrhopalitidae) from a cave in Eastern Carpathians. *Vestn. Zool.* **2005**, *39*, 67–73.
36. Nosek, J. A new springtail from the caves of the Moravský kras (Moravian Karst) *Arrhopalites ruseki* sp. n. *Speleol. Věstn.* **1975**, *6*, 45–48.
37. Loksa, I. Faunistisch-systematische und ökologische Untersuchungen in der Lóczy-Höhle bei Balatonfüred (Biospeologica Hungarica, XI). *Ann. Univ. Sci. Bp. Rolando Eötvös Nomin. Sect. Biol.* **1960**, *3*, 253–266.
38. Vargovitsh, R.S. *Arrhopalites potapovi* sp. nov. (Collembola, Symphypleona) from Russia. *Zootaxa* **2015**, *3955*, 101–112. [CrossRef]
39. Vargovitsh, R.S.; Buşmachiu, G. A new species and new records of Arrhopalitidae (Collembola: Symphypleona) from the Republic of Moldova. *Zootaxa* **2015**, *3973*, 539–552. [CrossRef] [PubMed]
40. Vargovitsh, R.S.; Kahrarian, M. A new species of *Arrhopalites* Börner, 1906 (Collembola, Symphypleona, Arrhopalitidae) from Iran with an updated key to *A. diversus* group of species. *Zootaxa* **2020**, *4759*, 338–350. [CrossRef] [PubMed]
41. Zeppelini, D.; Brito, R.A.; Lima, E.C.A. Three new species of Collembola (Arthropoda: Hexapoda) from Central Brazilian shallow caves: Side effects of long term application of environmental law on conservation. *Zootaxa* **2018**, *4500*, 59–81. [CrossRef] [PubMed]
42. Nosek, J. *Arrhopalites slovacicus* a new species of Collembola from the Domica Cave. *Rev. Suisse Zool.* **1975**, *82*, 599–602. [CrossRef]
43. Christiansen, K. The genus *Arrhopalites* in the United States and Canada. *Int. J. Speleol.* **1966**, *2*, 43–73. [CrossRef]
44. Yosii, R. *Monographie zur Höhlencollembolen Japans*; Contributions from the Biological Laboratory Kyoto University: Kyoto, Japan, 1956; Volume 3, pp. 1–109.
45. Christiansen, K. Convergence and parallelism in cave Entomobryinae. *Evolution* **1961**, *15*, 288–301. [CrossRef]
46. Thibaud, J.-M.; Massoud, Z. Un nouveau genre d'insectes collemboles Hypogastruridae cavernicole du Pays Basque. *Mém. Biospéol.* **1983**, *10*, 317–319.
47. Deharveng, L. Collemboles cavernicoles VIII. Contribution a l'etude des Oncopoduridae. *Bull. Soc. Entomol. Fr.* **1988**, *92*, 133–147. [CrossRef]
48. Deharveng, L.; Thibaud, J.-M. *Bessoniella procera* n. g., n. sp., nouvel Orchesellidae cavernicole relictuel des Pyrénées (Insecta, Collembola). *Bull. Mus. Natl. Hist. Nat. Sér. 4* **1989**, *11*, 397–405.
49. Lukić, M.; Houssin, C.; Deharveng, L. A new relictual and highly troglomorphic species of Tomoceridae (Collembola) from a deep Croatian cave. *ZooKeys* **2010**, *69*, 1–16. [CrossRef]
50. Loksa, I. Vier neue Höhlencollembolen aus Ungarn (Biospeologica Hungarica, XXIII). *Opusc. Zool. Bp.* **1967**, *6*, 289–296.
51. Howarth, F.G.; Moldovan, O.T. Where Cave Animals Live. In *Cave Ecology*; Ecological Studies; Moldovan, O., Kováč, L'., Halse, S., Eds.; Springer: Cham, Switzerland, 2018; Volume 235, pp. 23–37. [CrossRef]
52. Christiansen, K. Proposition pour la classification des animaux cavernicoles. *Spelunca* **1962**, *2*, 76–78.
53. Christiansen, K. Bionomics of Collembola. *Annu. Rev. Entomol.* **1964**, *9*, 147–178. [CrossRef]
54. Sket, B. The cave hygropetric—A little known habitat and its inhabitants. *Arch. Hydrobiol.* **2004**, *160*, 413–425. [CrossRef]
55. Vargovich, R.S. A new species of Collembola of the genus *Arrhopalites* (Entognatha, Hexapoda) from a cave in Eastern Carpathians. *Vestn. Zool.* **1999**, *33*, 89–92.
56. Zeppelini, D. The genus *Arrhopalites* Börner, 1906 (Collembola, Appendiciphora, Arrhopalitidae) in the Neotropical Region, with description of four new cave species from Brazil. *Zootaxa* **2006**, *1124*, 1–40. [CrossRef]

diversity

MDPI

Article

The Diversity of Subterranean Terrestrial Arthropods in Resava Cave (Eastern Serbia)

Nikola Vesović [1,*], Christo Deltshev [2], Plamen Mitov [3], Dragan Antić [1], Dalibor Z. Stojanović [1], Dejan V. Stojanović [4], Katarina Stojanović [1], Milenka Božanić [1], Aleksandra Ignjatović-Ćupina [5] and Srećko Ćurčić [1]

1 Institute of Zoology, Faculty of Biology, University of Belgrade, Studentski Trg 16, 11000 Belgrade, Serbia; dragan.antic@bio.bg.ac.rs (D.A.); dstojanovic@bio.bg.ac.rs (D.Z.S.); k.bjelanovic@bio.bg.ac.rs (K.S.); mika.zunic@bio.bg.ac.rs (M.B.); srecko@bio.bg.ac.rs (S.Ć.)

2 National Museum of Natural History, Bulgarian Academy of Sciences, 1 Tsar Osvoboditel Blvd., 1000 Sofia, Bulgaria; deltshev@gmail.com

3 Department of Zoology and Anthropology, Faculty of Biology, University of Sofia, 8 Dragan Zankov Blvd., 1421 Sofia, Bulgaria; pl_mitov@yahoo.com

4 Institute of Lowland Forestry and Environment, University of Novi Sad, Antona Čehova 13d, 21000 Novi Sad, Serbia; dejanstojanovic021@yahoo.co.uk

5 Department of Plant and Environmental Protection, Faculty of Agriculture, University of Novi Sad, Trg Dositeja Obradovića 8, 21000 Novi Sad, Serbia; cupinas@polj.uns.ac.rs

* Correspondence: nikola.vesovic@bio.bg.ac.rs; Tel.: +381-11-2638-284

Abstract: The Balkan region is rich in limestone deposits, which have created one of the largest hotspots of subterranean biodiversity. This paper gives an overview of the diversity of subterranean terrestrial arthropods in Resava Cave in eastern Serbia. This cave is protected and has the status of a natural monument. At the same time, it is one of the most visited caves in Serbia and its surroundings. Our study comprises the results of three years of biospeleological investigations of the famous Serbian cave in combination with data from the few available literature sources on the arthropod fauna of the cave. The arthropod samples were collected both manually and with pitfall traps. A total of 107 arthropod species from the four major subphyla were registered in the cave: 66 species of Hexapoda, 27 species of Chelicerata, 11 species of Myriapoda and three species of Crustacea. For four troglobitic, 16 troglophilic and 87 trogloxenic species recorded in the cave, descriptions of their microhabitats and information on their distribution in the cave are given. Considering the medium size of Resava Cave and the lack of permanent water flow in the two main levels on the one hand, and the large number of arthropod species recorded on the other, the cave is relatively rich in hypogean terrestrial arthropod fauna compared to other caves in Serbia that have been biospeleologically studied so far.

Keywords: Arthropoda; subterranean fauna; troglobite; troglophile; trogloxene; karst; limestone; biospeleology; Balkans

1. Introduction

The subterranean terrestrial arthropods in Europe are generally the descendants of a tropical epigean fauna that lived in the area at the beginning of the Tertiary and later disappeared [1]. The process of karstification created a wide range of hypogean ecological niches, resulting in the formation of a large refugial zone for the original epigean fauna. For this reason, part of the fauna survived the dramatic climatic changes that followed the Tertiary in Europe [1]. The Serbian karst is inhabited by a large number of endemic cavernicolous animals belonging to the Palaeo-Mediterranean, Laurasian, Palaeo-Aegean and South- or North-Aegean phyletic series [2].

The main reasons for the extraordinary diversity of the Balkan troglobitic fauna include: (i) the diverse epigean fauna that populated the region in the distant past; (ii) the continuity of continental phases in different areas of the region; (iii) the presence of thick limestone layers and subsequent karstification; (iv) climatic conditions favoring the colonization of subterranean habitats; and (v) the divergent differentiation of various lower and higher taxa in numerous isolated hypogean niches [3]. Recent works on the unprecedented subterranean diversity in the Balkan region in Slovenia [4,5], Croatia [6] and Bosnia and Herzegovina [7] support these claims.

Similar phenomena of karstification and the development of a rich subterranean fauna also occurred on other continents, as recent studies on subterranean biodiversity in North America [8] and South America [9] confirm. Finally, the Asian limestone masses are the most impressive, with only the vast karst area in southern China, covering some 550,000 km^2 [10], having the greatest potential to be the world's largest hotspot of subterranean diversity with many specialized hypogean taxa [11,12]. Recently, numerous taxa from the karst of China have been described as new to science [13,14].

In Serbia, caves and pits can be found in carbonate rocks. These are mostly Mesozoic limestones that belong to many different geotectonic units [15]. Most of the caves and pits are located in the eastern (i.e., the Carpatho-Balkanides) and western parts of the country (Jadar Block and the Inner Dinarides) (Figure 1). The mountains in eastern Serbia represent a direct extension of the Southern Carpathians, which stretch from Romania, but they do not directly connect to the Balkanides [15].

The distribution of caves in the Serbian Carpatho-Balkanides coincides with the presence of two carbonate platforms in this area (Figure 1), both dating from the Upper Jurassic and Lower Cretaceous [15]. The larger one is known as the Kučaj-Tupižnica carbonate platform [16], while the much smaller is the Miroč carbonate platform [17]. The karst areas of the Carpatho-Balkanides in eastern Serbia are characterized by an extremely complex and variable relief (Figure 2). Cretaceous limestones are more common there, while Triassic limestones are rarer. However, Cenozoic formations are also widespread (mainly from the Oligocene and Neogene) [15]. The areas extending from the Beljanica, Kučajske Planine, Rtanj, Devica, Svrljiške Planine and Belava Mts. to Mt. Vlaška Planina belong to the central part of the Kučaj-Tupižnica carbonate platform [16], where Resava Cave is located (Figures 1 and 3).

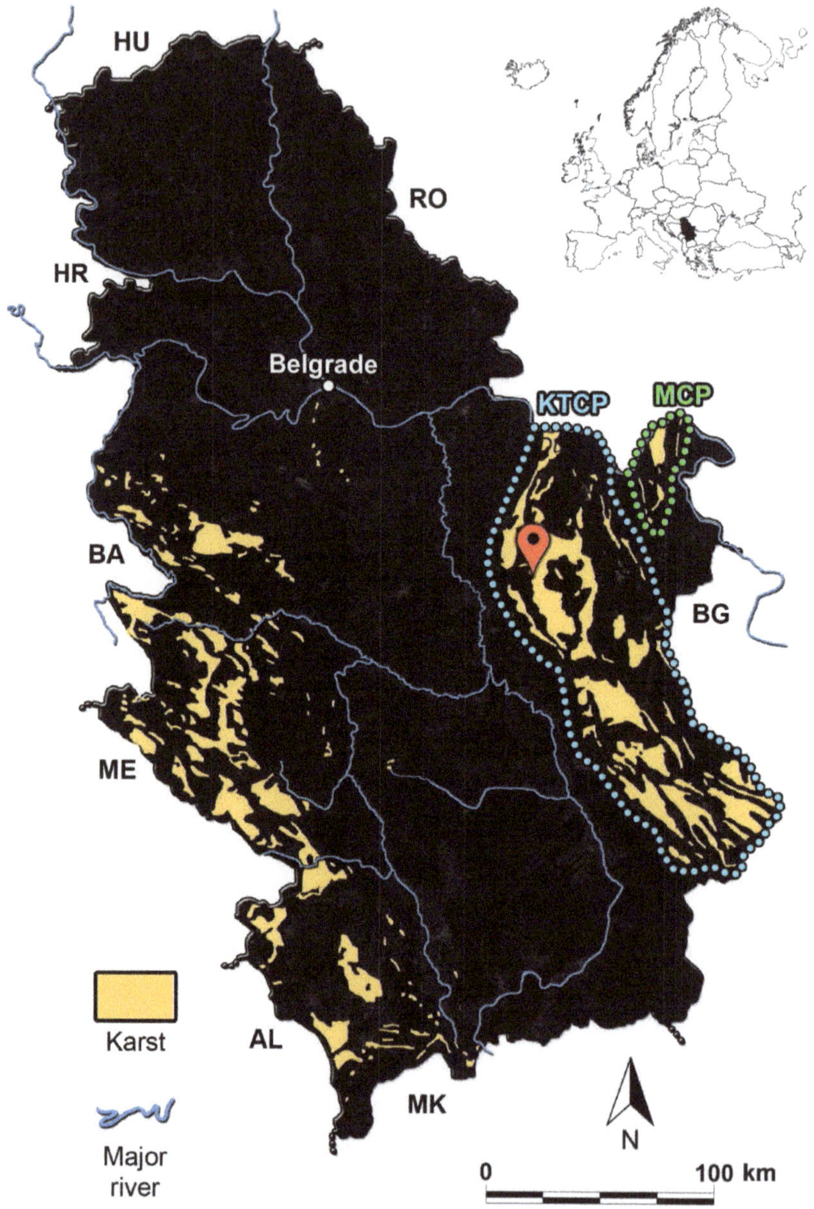

Figure 1. A map of Serbia with karst areas and the location of Resava Cave (red pin) (modified after [18]). The carbonate platforms of the Carpatho-Balkanides in eastern and southeastern Serbia are framed by the dotted lines. AL—Albania; BA—Bosnia and Herzegovina; BG—Bulgaria; HR—Croatia; HU—Hungary; KTCP—Kučaj-Tupižnica carbonate platform; MCP—Miroč carbonate platform; ME—Montenegro; MK—North Macedonia; RO—Romania.

Figure 2. Karst landscapes of the Carpatho-Balkanides in eastern Serbia. (**A**) Lazar's Canyon, Kučajske Planine Mts.; (**B**) Suva Prerast, Vratna Gates, Mt. Miroč; (**C**) one of the peaks of Mt. Stol; (**D**) the ridge of Mt. Veliki Krš. Photos: N. Vesović.

Figure 3. Resava Cave in eastern Serbia. (**A**) The cave entrance seen from the inside; (**B**) flowstone formation in the Menza Gallery; (**C**) formation of a new stalactite in the presence of trickling water; (**D**) corallite speleothems, a feature of the Coral Canals; (**E**) manual collection of hypogean terrestrial arthropods. Photos: N. Vesović and S. Ćurčić.

In contrast to a large number of comprehensive and detailed speleological studies, very few biospeleological studies have been conducted on the cave's living world, as evidenced by the small number of articles published to date on the fauna of Resava Cave [15,19–24].

Although the literature on the terrestrial arthropod fauna of Resava Cave is sparse, reporting only seven species [15,19–24], the existing references are of great importance as they contain descriptions of four new arthropod species to science from Resava Cave (a spider, a millipede, a dipluran and a ground beetle). All four species are blind, depigmented and have other morphological features typical of true cave dwellers (troglobites).

The spider *Centromerus serbicus* Deltshev, 2002 is endemic to several caves in the Kučajske Planine Mts., including Resava Cave [20,21]. It is a small, blind spider of yellow-reddish color and with long legs. It was previously reported that Resava Cave is inhabited by another spider, *Porrhomma convexum* (Westring, 1851) (with Holarctic distribution), as well as the pseudoscorpion *Neobisium* cf. *carpaticum* Beier, 1935 (inhabits a wider Carpathian area in Europe), in addition to *C. serbicus* [15].

The millipede described from Resava Cave, *Serbosoma kucajense* (B. Ćurčić & Makarov, 1998), is endemic to Resava Cave (its type locality) and the nearby Vrtačelje (=Ledena Pećina) Pit [19,25]. It is completely blind and depigmented and has an elongated cylindrical body. This is one of the two diplopod species known from the cave according to the available literature data [19,24]. The report of another millipede from the cave (*Trachysphaera* sp.) was recently published [24]. *Trachysphaera* sp. is a very small, white millipede with a characteristic ribbed sculpture on the tergites. In case of danger it can curl up into a ball. It feeds on decaying organic matter of plant origin.

As far as hexapods are concerned, two species have been recorded so far [22,23]. The dipluran *Plusiocampa christiani* Condé & Bareth, 1996 is endemic to the subterranean habitats of eastern Serbia. In addition to Resava Cave, it also occurs in several other caves on the Kučajske Planine Mts. and Mt. Beljanica [23]. It is blind, completely depigmented and has long antennae and caudal filaments. The other hexapod species, the ground beetle *Duvalius petrovici* S. Ćurčić, Vrbica, Antić & B. Ćurčić, 2014, named in honor of Prof. Jovan Petrović, the first explorer of Resava Cave, is a stenoendemic of the cave [22]. It is a completely blind and almost depigmented beetle, with a large number of sensory hairs on the body.

The aim of our research in Resava Cave during the three-year period (2020–2022) was to determine the diversity of terrestrial arthropods in Resava Cave, which is the first detailed multi-year attempt of this kind for a cave on the territory of Serbia.

2. Materials and Methods

2.1. Sampling

As part of the research into the diversity of arthropods in Resava Cave, several expeditions have been carried out in recent years. The cave (Figure 4) has been studied biospeleologically in detail on six occasions. It was visited in 2020 (May and October), 2021 (May and September) and 2022 (September and November). The research was carried out on the basis of the authorization for research on strictly protected and protected wild species for scientific research and educational purposes granted by the Ministry of Environmental Protection of the Republic of Serbia for each of the three research years.

The expeditions were conducted by the team of the Institute of Zoology of the University of Belgrade—Faculty of Biology. The taxonomic identification was carried out mostly by experts from the same institution and to a lesser extent by experts from other institutions in Serbia and abroad (Bulgaria, China, Czech Republic, Poland, Romania and Slovakia). During the field trips, arthropods were collected manually in vials containing 70% ethanol. Empty pitfall traps (plastic cups with vaseline-coated walls) baited with rotten meat were also set up (Figure 4) during the visits to observe the arthropod fauna of the cave (they were checked after a few days). In addition to the traps set in search of terrestrial arthropods, all of the cave's halls and canals were thoroughly visually inspected during each research visit. For identification, we used available keys and original descriptions for different arthropod groups. For some species, the male genitalia were removed in order to identify them accurately. Most arthropod species were photographed *in situ* using a Nikon D5300 digital camera (Nikon Corp., Tokyo, Japan) equipped with a Tamron SP Di AF 90 mm F/2.8 macro lens (Tamron Co., Ltd., Saitama, Japan) and an EM-140 DG flash ring (Sigma Corp., Kawasaki, Japan) with a home-made light diffuser. Further identification and imaging using a Zeiss SteREO Discovery.V12 stereomicroscope (Carl Zeiss, Jena, Germany) equipped with a Flexacam C3 camera (Leica Microsystems, Wetzlar, Germany) was performed in the laboratory of the Institute of Zoology, University of Belgrade—Faculty of Biology. To obtain fully focused images, the frames were stacked using Zerene Stacker ver. 1.04. All figures were processed with Adobe Photoshop CS6 ver. 13.0.1.

All collected arthropod specimens were deposited in the collection of the Institute of Zoology, University of Belgrade—Faculty of Biology.

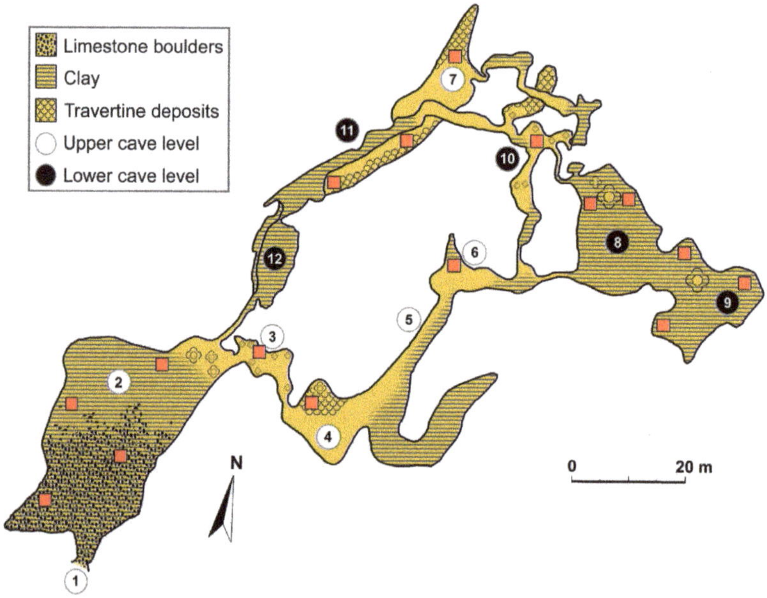

Figure 4. A plan of Resava Cave showing the sites where the arthropod fauna was sampled with pitfall traps (red squares) (modified after [15]). 1—entrance; 2—Entrance Hall; 3—Hall of Conjoined Columns; 4—Beehive Hall; 5—Canyon (=Dry) Canal; 6—Vestibule of History; 7—Crystal Hall; 8—Concert Hall (=Hall of Statues); 9—Boban's Hall; 10—Menza Gallery; 11—Coral Canals; 12—Kepa's (=Mud) Hall. Halls 6 and 7 are connected by an artificial tunnel.

2.2. Study Area

Resava Cave (=Resavska Pećina) is one of the best explored cave systems in Serbia. The original name was Divljakovac Cave (=Divljakovačka Pećina), named after the Divljakovac karst polje (=field), on the northern edge of which it is located. The cave is situated in the eastern part of the country, about 15 km southeast of the town of Despotovac (Figure 1) and about 4 km southeast of the village of Jelovac. The cave was formed in the limestones of the Babina Glava hill by the erosive activity of a sinking stream that flowed from the northwestern slopes of the Kučajske Planine Mts. [15].

The cave was discovered accidentally in 1962 by local shepherds and mountaineers, when they saw its entrance (44°04′22.1″ N, 21°37′47.7″ E), sitting on 485 m a.s.l. [15]. It was examined in detail by Serbian speleologists. Prof. Dr. Jovan Petrović and his colleagues were the first to carry out detailed speleological investigations in Resava Cave in the 1960s [26]. After these explorations, further speleological research was carried out by Prof. Dr. Radenko Lazarević [27] and Milorad Kličković [28]. Ten years after its discovery, on 22 April 1972, the cave was officially opened to visitors. It is visited by tens of thousands of tourists every year. The estimated age of the cave is 80 million years, while the oldest speleothems date back to 45 million years [26]. The total length of the cave halls, canals, galleries and a network of side passages is 4.5 km, of which 0.8 km have been made accessible to visitors. The depth of the cave is 80 m. Resava Cave has been protected since 1995, when it was declared a natural monument [29]. It has been a member of the International Show Caves Association (ISCA) since 2010. In 2017, LED lighting was installed to minimize the negative effects of light on the cave's microclimate and its wildlife.

The cave consists of two main levels, of which the upper level is on average drier and warmer, while the lower one is significantly wetter and colder. Limestone boulders are present in the Entrance Hall, while certain areas of the cave are covered with layers of clay and rich travertine deposits (Figures 3 and 4) [15].

2.3. Terminology

The classical Schiner-Racovitza terminology for subterranean terrestrial animals, which was widely used for a long time, comprised three ecological categories: troglobites, troglophiles and trogloxenes [30,31].

According to Sket [32], however, terrestrial cave dwellers should be divided into four ecological groups, depending on their degree of adaptation to the subterranean environment. Troglobites (=troglobionts) include species that inevitably complete their life cycle in caves. Their appearance and behavior are exclusively adapted to the subterranean way of life (e.g., loss of pigmentation and eyes, elongation of appendages). Eutroglophiles include species that can establish permanent subterranean populations, but are essentially epigean. Subtroglophiles include species that can only live in caves during a certain phase of their life cycle, but are dependent on the surface for some biological functions (e.g., feeding) and never establish permanent subterranean populations. Finally, trogloxenes include species that occur sporadically or accidentally in caves (they may be attracted to the cave by humidity, temperature or food) [32].

Although trogloxenes are not true cave dwellers (they are usually restricted to cave entrances) and their importance for the study of subterranean animals is low [32], their presence in caves should not be completely ignored, especially because they bring nutrients from epigean habitats into the caves. There is a possibility that some of them are actually subtroglophilic or even eutroglophilic when more is known about their ecology [33].

Similar to the terrestrial fauna, the aquatic subterranean fauna is also divided into corresponding ecological categories (stygobites, eustygophiles, substygophiles and stygoxenes).

In this paper, however, we have divided the subterranean terrestrial arthropods into three categories (troglobites, troglophiles and trogloxenes) according to the Schiner-Racovitza terminology, as it is difficult to distinguish between eutroglophiles and subtroglophiles and between subtroglophiles and trogloxenes in Resava Cave due to the paucity of data on the ecology of their populations in the explored cave. There is a possibility that some of the trogloxenic arthropods in Resava Cave are actually troglophilic, but there is no clear evidence for this.

3. Results

After three years of research, we found that Resava Cave provides permanent or temporary/occasional habitat for a total of 107 species of terrestrial arthropods. A total of four troglobitic species, 16 troglophilic species and 87 trogloxenic species were recorded in the cave.

3.1. Troglobitic Arthropods of Resava Cave

During the investigations, a total of four troglobitic species were registered (Table 1, Figure 5).

Table 1. A list of troglobitic arthropod species found in Resava Cave.

No.	Species	Family	Order	Class	Subphylum
1	*Centromerus serbicus* Deltshev, 2002	Linyphiidae	Araneae	Arachnida	Chelicerata
2	*Serbosoma kucajense* (B. Ćurčić & Makarov, 1998)	Anthroleucosomatidae	Chordeumatida	Diplopoda	Myriapoda
3	*Plusiocampa christiani* Condé & Bareth, 1996	Campodeidae	Diplura	Diplura	Hexapoda
4	*Duvalius petrovici* S. Ćurčić, Vrbica, Antić & B. Ćurčić, 2014	Carabidae	Coleoptera	Insecta	Hexapoda

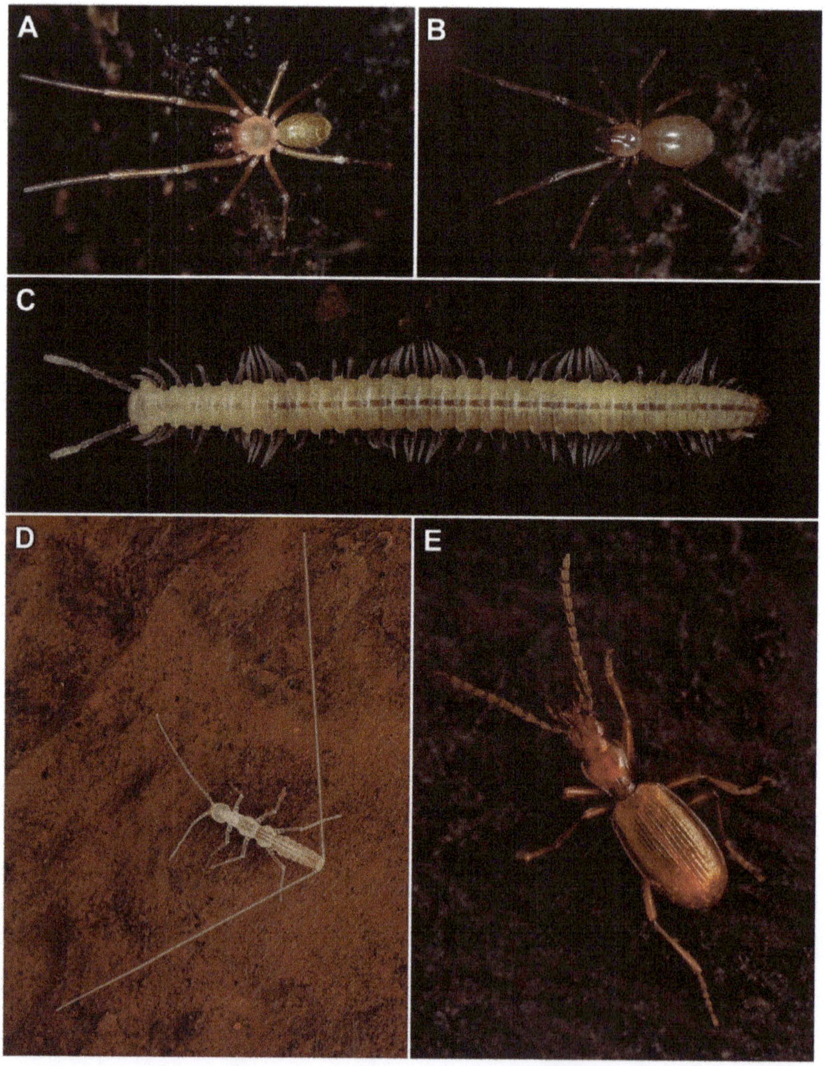

Figure 5. Troglobitic arthropods from Resava Cave. (**A**) A male of the spider *Centromerus serbicus*; (**B**) a female of *C. serbicus*; (**C**) the millipede *Serbosoma kucajense*; (**D**) the dipluran *Plusiocampa christiani*; (**E**) the ground beetle *Duvalius petrovici*. Photos: N. Vesović.

Of the troglobitic arachnids, only one species was recorded in Resava Cave—the spider *C. serbicus* (Figure 5A,B), which is endemic to caves in eastern Serbia [20,21]. Specimens of *C. serbicus* were observed in both levels of the cave (more frequently in the lower level), in complete darkness. The species was found in most of the halls of the cave with increased humidity both during visual inspection and in pitfall traps. We found its specimens most frequently under stones and decaying wood and in crevices in the walls.

Only one troglobitic myriapod has been recorded in Resava Cave, the millipede *S. kucajense* (Figure 5C), which is endemic to the former cave and Vrtačelje (=Ledena Pećina) Pit [19,25]. Live specimens of this species have been found in the cave in complete darkness during visual inspection and in pitfall traps (on the damp walls, on decaying wood and under stones).

As far as the troglobitic hexapods are concerned, two species live in Resava Cave. We have recorded the dipluran *P. christiani* (Figure 5D) and the ground beetle *D. petrovici* (Figure 5E), the former of which is endemic to caves in eastern Serbia [23], while the latter is a stenoendemic of Resava Cave [22]. Live specimens of the first species have been recorded during visual inspection and in pitfall traps in several halls of the cave (most frequently in the Concert Hall and the Boban's Hall). We found live specimens of the second species on the floor of the Menza Gallery, among decaying wood debris and in the damp clay. Specimens of this species were also found in pitfall traps set up on the damp walls and on the floor in the same gallery, but also in smaller numbers in the Coral Canals and the Concert Hall. Both troglobitic hexapod species were found in the lower level of the cave, in complete darkness and in similar microhabitats (on the damp floor and on the walls).

3.2. Troglophilic Arthropods of Resava Cave

During the investigations, a total of 16 troglophilic species were registered (Table 2, Figure 6).

Table 2. A list of troglophilic arthropod species found in Resava Cave.

No.	Species	Family	Order	Class	Subphylum
1	*Paranemastoma sillii* (Herman, 1871)	Nemastomatidae	Opiliones	Arachnida	Chelicerata
2	*Diplocephalus cristatus* (Blackwall, 1833)	Linyphiidae	Araneae	Arachnida	Chelicerata
3	*Porrhomma convexum* (Westring, 1851)	Linyphiidae	Araneae	Arachnida	Chelicerata
4	*Nesticus cellulanus* (Clerck, 1757)	Nesticidae	Araneae	Arachnida	Chelicerata
5	*Meta menardi* (Latreille, 1804)	Tetragnathidae	Araneae	Arachnida	Chelicerata
6	*Metellina merianae* (Scopoli, 1763)	Tetragnathidae	Araneae	Arachnida	Chelicerata
7	*Ixodes vespertilionis* Koch, 1844	Ixodidae	Ixodida	Arachnida	Chelicerata
8	*Hyloniscus* cf. *riparius* (C. Koch, 1838)	Trichoniscidae	Isopoda	Malacostraca	Crustacea
9	*Trachysphaera* sp.	Glomeridae	Glomerida	Diplopoda	Myriapoda
10	*Pygmarrhopalites pygmaeus* (Wankel, 1860)	Arrhopalitidae	Symphypleona	Collembola	Hexapoda
11	*Tomocerus vulgaris* (Tullberg, 1871)	Tomoceridae	Entomobryomorpha	Collembola	Hexapoda
12	*Troglophilus neglectus* Krauss, 1879	Rhaphidophoridae	Orthoptera	Insecta	Hexapoda
13	*Laemostenus punctatus* (Dejean, 1828)	Carabidae	Coleoptera	Insecta	Hexapoda
14	*Quedius mesomelinus* (Marsham, 1802)	Staphylinidae	Coleoptera	Insecta	Hexapoda
15	*Limonia nubeculosa* Meigen, 1804	Limoniidae	Diptera	Insecta	Hexapoda
16	*Speolepta leptogaster* (Winnertz, 1863)	Mycetophilidae	Diptera	Insecta	Hexapoda

Figure 6. Selected troglophilic arthropods of Resava Cave. (**A**) The harvestmen *Paranemastoma sillii*; (**B**) the spider *Meta menardi*; (**C**) the spider *Metellina merianae*; (**D**) the tick *Ixodes vespertilionis*; (**E**) the isopod *Hyloniscus* cf. *riparius*; (**F**) the millipede *Trachysphaera* sp.; (**G**) the springtail *Tomocerus vulgaris*; (**H**) the springtail *Pygmarrhopalites pygmaeus*; (**I**) the cricket *Troglophilus neglectus*; (**J**) the ground beetle *Laemostenus punctatus*; (**K**) the rove beetle *Quedius mesomelinus*; (**L**) the dipteran *Speolepta leptogaster*. Photos: N. Vesović.

A total of seven troglophilic arachnid species (one harvestmen, five spiders and one tick) are known from Resava Cave.

We found live specimens of the large troglophilic harvestmen *Paranemastoma sillii* (Hermann, 1871) (Figure 6A) in both levels of the cave, in complete darkness, mostly on the damp limestone surfaces (rocks, walls). We recorded a total of four troglophilic spider species in the cave: *Meta menardi* (Latreille, 1804) (Figure 6B), *Metellina merianae* (Scopoli, 1763) (Figure 6C), *Diplocephalus cristatus* (Blackwall, 1833) and *Nesticus cellulanus* (Clerck, 1757). The fifth troglophilic spider species, *P. convexum*, is only known from the literature [15] as an inhabitant of the cave, as we were unable to find it during our investigations. All the troglophilic spider species mentioned were observed mainly in the illuminated Entrance Hall (where there is the most food), and somewhat less in the inner parts of the cave, which belong to the upper level. The only tick species recorded in the cave, also a troglophile, was *Ixodes vespertilionis* Koch, 1844 (Figure 6D). We found live specimens of this species in both levels of the cave.

During our cave explorations, we found a single troglophilic crustacean—the isopod *Hyloniscus* cf. *riparius* (C. Koch, 1838) (Figure 6E). This is one of the most common species in the cave, occurring on very moist surfaces in both the upper and lower levels of the cave.

The only troglophilic millipede in the cave, *Trachysphaera* sp. (Figure 6F) was found in complete darkness in the lower level of the cave, but almost exclusively as single or few specimens on relatively dry, decaying wood.

A total of seven species of troglophilic hexapods (two taxa of the class Collembola and five taxa of the class Insecta) were recorded. These included two springtails, one orthopteran, two beetles and two dipterans. The springtail *Tomocerus vulgaris* (Tullberg, 1871) (Figure 6G) was found in both levels in almost all cave halls that are in complete darkness, but also on the floor of the illuminated Entrance Hall. It was observed in both humid and somewhat drier microhabitats. Another finding was that of a springtail belonging to the genus *Pygmarrhopalites* Vargovitsh, 2009—*Pygmarrhopalites pygmaeus* (Wankel, 1860) (Figure 6H). We found it in small numbers only in pitfall traps set up in the upper level of the cave (the Beehive Hall), in complete darkness.

The troglophilic cave cricket *Troglophilus neglectus* Krauss, 1879 (Figure 6I) was found in the Entrance Hall of the cave. Both recorded troglophilic beetle species [the ground beetle *Laemostenus punctatus* (Dejean, 1828) (Figure 6J) and the rove beetle *Quedius mesomelinus* (Marsham, 1802) (Figure 6K)] were found in both the upper and lower levels of the cave. We observed them near the entrance, in the partially illuminated part of the cave, but also in complete darkness in the Canyon Canal, the Coral Canals and the Kepa's Hall in smaller numbers, on relatively dry surfaces. The short-palped cranefly (*Limonia nubeculosa* Meigen, 1804) was observed in complete darkness in the upper level of the cave, as well as in the Entrance Hall. The fungus gnat *Speolepta leptogaster* (Winnertz, 1863) (Figure 6L) is a small troglophilic dipteran, which we observed in complete darkness on the damp walls in the upper level of the cave, as well as in pitfall traps set up in the lower level of the cave.

3.3. Trogloxenic Arthropods of Resava Cave

During the investigations, a total of 87 trogloxenic species were registered (Table 3, Figure 7).

Table 3. A list of trogloxenic arthropod species found in Resava Cave.

No.	Species	Family	Order	Class	Subphylum
1	*Neobisium carpaticum* Beier, 1935	Neobisiidae	Pseudoscorpiones	Arachnida	Chelicerata
2	*Lacinius dentiger* (C. L. Koch, 1848)	Phalangiidae	Opiliones	Arachnida	Chelicerata
3	*Opilio ruzickai* Šilhavý, 1938	Phalangiidae	Opiliones	Arachnida	Chelicerata
4	*Phalangium opilio* Linnaeus, 1761	Phalangiidae	Opiliones	Arachnida	Chelicerata
5	*Harpactea* sp.	Dysderidae	Araneae	Arachnida	Chelicerata
6	*Stemonyphantes lineatus* (Linnaeus, 1758)	Linyphiidae	Araneae	Arachnida	Chelicerata
7	*Agroeca cuprea* Menge, 1873	Liocranidae	Araneae	Arachnida	Chelicerata
8	*Liocranum rupicola* (Walckenaer, 1830)	Liocranidae	Araneae	Arachnida	Chelicerata
9	*Alopecosa trabalis* (Clerck, 1757)	Lycosidae	Araneae	Arachnida	Chelicerata
10	*Pardosa alacris* (C. L. Koch, 1833)	Lycosidae	Araneae	Arachnida	Chelicerata
11	*Piratula latitans* (Blackwall, 1841)	Lycosidae	Araneae	Arachnida	Chelicerata
12	*Trochosa terricola* Thorell, 1856	Lycosidae	Araneae	Arachnida	Chelicerata

Table 3. *Cont.*

No.	Species	Family	Order	Class	Subphylum
13	*Philodromus margaritatus* (Clerck, 1757)	Philodromidae	Araneae	Arachnida	Chelicerata
14	*Philodromus* sp.	Philodromidae	Araneae	Arachnida	Chelicerata
15	*Pisaura mirabilis* (Clerck, 1757)	Pisauridae	Araneae	Arachnida	Chelicerata
16	*Metellina segmentata* (Clerck, 1757)	Tetragnathidae	Araneae	Arachnida	Chelicerata
17	*Tmarus* sp.	Thomisidae	Araneae	Arachnida	Chelicerata
18	*Xysticus lanio* C. L. Koch, 1835	Thomisidae	Araneae	Arachnida	Chelicerata
19	*Xysticus* sp.	Thomisidae	Araneae	Arachnida	Chelicerata
20	*Ligidium* cf. *hypnorum* (Cuvier, 1792)	Ligiidae	Isopoda	Malacostraca	Crustacea
21	*Trachelipus rathkii* (Brandt, 1833)	Trachelipodidae	Isopoda	Malacostraca	Crustacea
22	*Lithobius forficatus* (Linnaeus, 1758)	Lithobiidae	Lithobiomorpha	Chilopoda	Myriapoda
23	*Lithobius muticus* C. L. Koch, 1847	Lithobiidae	Lithobiomorpha	Chilopoda	Myriapoda
24	*Lithobius nigripalpis* L. Koch, 1867	Lithobiidae	Lithobiomorpha	Chilopoda	Myriapoda
25	*Cryptops anomalans* Newport, 1844	Cryptopidae	Scolopendromorpha	Chilopoda	Myriapoda
26	*Callipodella fasciata* (Latzel, 1882)	Callipodidae	Callipodida	Diplopoda	Myriapoda
27	*Craspedosoma raulinsii* Leach, 1814	Craspedosomatidae	Chordeumatida	Diplopoda	Myriapoda
28	*Glomeris hexasticha* Brandt, 1833	Glomeridae	Glomerida	Diplopoda	Myriapoda
29	*Megaphyllum bosniense* (Verhoeff, 1897)	Julidae	Julida	Diplopoda	Myriapoda
30	*Polydesmus subscabratus* Latzel, 1884	Polydesmidae	Polydesmida	Diplopoda	Myriapoda
31	*Lepismachilis* sp.	Machilidae	Microcoryphia	Insecta	Hexapoda
32	*Poecilimon schmidtii* (Fieber, 1853)	Phaneropteridae	Orthoptera	Insecta	Hexapoda
33	*Pachytrachis gracilis* (Brunner von Wattenwyl, 1861)	Tettigoniidae	Orthoptera	Insecta	Hexapoda
34	*Forficula auricularia* Linnaeus, 1758	Forficulidae	Dermaptera	Insecta	Hexapoda
35	*Ectobius erythronotus* Burr, 1898	Blattellidae	Blattodea	Insecta	Hexapoda
36	*Bertkauia lucifuga* (Rambur, 1842)	Epipsocidae	Psocodea	Insecta	Hexapoda
37	*Leptoglossus occidentalis* Heidemann, 1910	Coreidae	Hemiptera	Insecta	Hexapoda
38	*Lygaeus equestris* (Linnaeus, 1758)	Lygaeidae	Hemiptera	Insecta	Hexapoda
39	*Nemoura cinerea* (Retzius, 1783)	Nemouridae	Plecoptera	Insecta	Hexapoda
40	*Pedilophorus auratus* (Duftschmid, 1825)	Byrrhidae	Coleoptera	Insecta	Hexapoda
41	*Abax carinatus* (Duftschmid, 1812)	Carabidae	Coleoptera	Insecta	Hexapoda
42	*Anchomenus dorsalis* (Pontoppidan, 1763)	Carabidae	Coleoptera	Insecta	Hexapoda
43	*Calathus fuscipes* (Goeze, 1777)	Carabidae	Coleoptera	Insecta	Hexapoda
44	*Carabus convexus* Fabricius, 1775	Carabidae	Coleoptera	Insecta	Hexapoda
45	*Carabus coriaceus* Linnaeus, 1758	Carabidae	Coleoptera	Insecta	Hexapoda
46	*Carabus montivagus* Palliardi, 1825	Carabidae	Coleoptera	Insecta	Hexapoda
47	*Carabus ullrichii* Germar, 1824	Carabidae	Coleoptera	Insecta	Hexapoda
48	*Carabus violaceus* Linnaeus, 1758	Carabidae	Coleoptera	Insecta	Hexapoda
49	*Harpalus rufipes* (De Geer, 1774)	Carabidae	Coleoptera	Insecta	Hexapoda
50	*Leistus rufomarginatus* (Duftschmid, 1812)	Carabidae	Coleoptera	Insecta	Hexapoda
51	*Molops piceus* (Panzer, 1793)	Carabidae	Coleoptera	Insecta	Hexapoda
52	*Platynus scrobiculatus* (Fabricius, 1801)	Carabidae	Coleoptera	Insecta	Hexapoda

Table 3. *Cont.*

No.	Species	Family	Order	Class	Subphylum
53	*Trechus quadristriatus* (Schrank, 1781)	Carabidae	Coleoptera	Insecta	Hexapoda
54	*Mesosa curculionoides* (Linnaeus, 1761)	Cerambycidae	Coleoptera	Insecta	Hexapoda
55	*Otiorhynchus perdix* (Olivier, 1807)	Curculionidae	Coleoptera	Insecta	Hexapoda
56	*Otiorhynchus raucus* (Fabricius, 1777)	Curculionidae	Coleoptera	Insecta	Hexapoda
57	*Ruteria hypocrita* (Boheman, 1837)	Curculionidae	Coleoptera	Insecta	Hexapoda
58	*Lampyris noctiluca* (Linnaeus, 1767)	Lampyridae	Coleoptera	Insecta	Hexapoda
59	*Oedemera femoralis* Olivier, 1803	Oedemeridae	Coleoptera	Insecta	Hexapoda
60	*Valgus hemipterus* (Linnaeus, 1758)	Scarabaeidae	Coleoptera	Insecta	Hexapoda
61	*Ocypus nitens* (Schrank, 1781)	Staphylinidae	Coleoptera	Insecta	Hexapoda
62	*Ocypus olens* (Müller, 1764)	Staphylinidae	Coleoptera	Insecta	Hexapoda
63	*Eilema sororcula* (Hufnagel, 1766)	Erebidae	Lepidoptera	Insecta	Hexapoda
64	*Scoliopteryx libatrix* (Linnaeus, 1758)	Erebidae	Lepidoptera	Insecta	Hexapoda
65	*Asthena albulata* (Hufnagel, 1767)	Geometridae	Lepidoptera	Insecta	Hexapoda
66	*Camptogramma bilineata* (Linnaeus, 1758)	Geometridae	Lepidoptera	Insecta	Hexapoda
67	*Colotois pennaria* (Linnaeus, 1761)	Geometridae	Lepidoptera	Insecta	Hexapoda
68	*Epirrhoe* sp.	Geometridae	Lepidoptera	Insecta	Hexapoda
69	*Epirrita* sp.	Geometridae	Lepidoptera	Insecta	Hexapoda
70	*Hypomecis punctinalis* (Scopoli, 1763)	Geometridae	Lepidoptera	Insecta	Hexapoda
71	*Operophtera fagata* (Scharfenberg, 1805)	Geometridae	Lepidoptera	Insecta	Hexapoda
72	*Rheumaptera cervinalis* (Scopoli, 1763)	Geometridae	Lepidoptera	Insecta	Hexapoda
73	*Micropteryx* sp.	Micropterigidae	Lepidoptera	Insecta	Hexapoda
74	*Diloba caeruleocephala* (Linnaeus, 1758)	Noctuidae	Lepidoptera	Insecta	Hexapoda
75	*Hedya* sp.	Tortricidae	Lepidoptera	Insecta	Hexapoda
76	*Micropterna nycterobia* McLachlan, 1875	Limnephilidae	Trichoptera	Insecta	Hexapoda
77	*Micropterna sequax* McLachlan, 1875	Limnephilidae	Trichoptera	Insecta	Hexapoda
78	*Stenophylax permistus* McLachlan, 1895	Limnephilidae	Trichoptera	Insecta	Hexapoda
79	*Camponotus ionius* Emery, 1920	Formicidae	Hymenoptera	Insecta	Hexapoda
80	*Diphyus quadripunctorius* (Müller, 1776)	Ichneumonidae	Hymenoptera	Insecta	Hexapoda
81	*Lymantrichneumon disparis* (Poda, 1761)	Ichneumonidae	Hymenoptera	Insecta	Hexapoda
82	*Calliphora vomitoria* (Linnaeus, 1758)	Calliphoridae	Diptera	Insecta	Hexapoda
83	*Culex pipiens* Linnaeus, 1758	Culicidae	Diptera	Insecta	Hexapoda
84	*Heleomyza serrata* (Linnaeus, 1758)	Heleomyzidae	Diptera	Insecta	Hexapoda
85	*Heteromyza atricornis* Meigen, 1830	Heleomyzidae	Diptera	Insecta	Hexapoda
86	*Tarnania fenestralis* (Meigen, 1838)	Mycetophilidae	Diptera	Insecta	Hexapoda
87	*Chorisops nagatomii* Rozkošný, 1979	Stratiomyidae	Diptera	Insecta	Hexapoda

Figure 7. Selected trogloxenic arthropods from Resava Cave. (**A**) The pseudoscorpion *Neobisium carpaticum;* (**B**) the harvestmen *Lacinius dentiger;* (**C**) the harvestmen *Opilio ruzickai;* (**D**) the spider *Liocranum rupicola;* (**E**) the spider *Philodromus margaritatus;* (**F**) the centipede *Lithobius forficatus;* (**G**) the millipede *Megaphyllum bosniense;* (**H**) the millipede *Polydesmus subscabratus;* (**I**) the psocodean *Bertkauia lucifuga;* (**J**) the ground beetle *Molops piceus;* (**K**) the herald moth—*Scoliopteryx libatrix;* (**L**) the yellow shell moth—*Camptogramma bilineata;* (**M**) the caddisfly *Stenophylax permistus;* (**N**) the ichneumonid wasp *Lymantrichneumon disparis;* (**O**) the dipteran *Heleomyza serrata.* Photos: N. Vesović.

The largest number of arachnids found in Resava Cave represent trogloxenic species. During the investigations, a total of 19 such species were registered (Table 3, Figure 7A–E). The trogloxenic arachnids from the cave belong to the orders Pseudoscorpiones (one species) (Figure 7A), Opiliones (three species) (Figure 7B,C) and Araneae (15 species) (Figure 7D,E). The arachnid families richest in trogloxenes are Lycosidae (wolf spiders) (with four species), Thomisidae and Phalangiidae (with three species each). Most of the trogloxenic arachnids were found in the Entrance Hall of the cave.

Among the Crustacea, we found only two trogloxenic species (Table 3), both belonging to the order Isopoda, but to two different families (Ligiidae and Trachelipodidae). We found both trogloxenic isopods in the Entrance Hall of the cave (in the leaf litter, on the floor and on rocks).

Most of the myriapods found in Resava Cave are also trogloxenic taxa. We found a total of nine such species in the cave (Table 3, Figure 7F–H). They mostly belong to the class Diplopoda (five species), i.e., to the orders Callipodida, Chordeumatida, Glomerida, Julida and Polydesmida (with one species each). Slightly fewer species (four) belong to the class Chilopoda and the orders Lithobiomorpha (with three species) and Scolopendromorpha (with one species). The family and genus of myriapods richest in trogloxenes in Resava Cave are Lithobiidae and *Lithobius* Leach, 1814, with three species each, while all other families and genera of myriapods were represented by a single trogloxenic species each. In the Entrance Hall of Resava Cave we found all trogloxenic species of myriapods (in the leaf litter, on the ground, on rocks and under stones). *Callipodella fasciata* (Latzel, 1882) and *Lithobius forficatus* (Linnaeus, 1758) (Figure 7F) are the most common of the trogloxenic myriapod species recorded. The species *Glomeris hexasticha* Brandt, 1833, *Megaphyllum bosniense* (Verhoeff, 1897) (Figure 7G) and *Lithobius muticus* C. L. Koch, 1847 are somewhat less numerous, while the other trogloxenic myriapods are represented by far fewer individuals.

The largest number of hexapods in Resava Cave represent trogloxenic species. During our investigations, we found a total of 57 such species (Table 3, Figure 7I–O). The trogloxenic hexapods we detected all belong to the class Insecta, i.e., to the orders Coleoptera (23 species), Lepidoptera (13 species), Diptera (six species), Hymenoptera, Trichoptera (with three species each), Orthoptera, Hemiptera (with two species each), Dermaptera, Blattodea, Microcoryphia, Plecoptera and Psocodea (with one species each) (Table 3). The insect families richest in trogloxenes are Carabidae and Geometridae, with 13 and eight species, respectively.

We found most of the trogloxenic species of hexapods in the partially illuminated Entrance Hall of the cave (on the walls, on the floor, on rocks or under stones). The most common was the psocodean *Bertkauia lucifuga* (Rambur, 1842) (Figure 7I) (it was particularly common on large limestone boulders). Other trogloxenic insect species common in the Entrance Hall of the cave were the ground beetles *Carabus montivagus* Palliardi, 1825 and *Molops piceus* (Panzer, 1793) (Figure 7J), the European earwig—*Forficula auricularia* Linnaeus, 1758, the herald moth—*Scoliopteryx libatrix* (Linnaeus, 1758) (Figure 7K), the yellow shell moth—*Camptogramma bilineata* (Linnaeus, 1758) (Figure 7L), the rove beetle *Ocypus nitens* (Schrank, 1781), the caddisfly *Stenophylax permistus* McLachlan, 1895 (Figure 7M), the common house mosquito—*Culex pipiens* Linnaeus, 1758 and the dipteran *Heteromyza atricornis* Meigen, 1830. The remaining trogloxenic insect taxa were less common in the cave (Figure 7N).

4. Discussion

As a result of strong geotectonic movements (e.g., folding and faulting) of the limestone masses in Serbia, numerous fissures and canals were formed, which penetrate deep into the limestone layers and enable both free circulation of groundwater and intensification of the karst process in the superficial and deep parts of the limestone masses [15].

Study of the hypogean inhabitants of the Serbian karst has provided further evidence of their great age and diverse origins [3]. Apart from this, it is apparent that specific aspects of the geomorphological and climatic events in the Balkans, together with the peculiarities of historical development of the fauna there, have led to the area becoming the main center of dispersal and colonization of species and species groups, i.e., the main source for the revival and emergence of biodiversity in the Mediterranean and Southeast Europe [3].

Summarizing the results of our investigations carried out in recent years, a total of 107 arthropod species (66 hexapods, 27 arachnids, 11 myriapods and three crustaceans) were found in Resava Cave. Four species (3.74%) are troglobites (Table 1, Figure 5),

16 species (14.95%) are troglophiles (Table 2, Figure 6), while the remaining 87 species (81.31%) are trogloxenes (Table 3, Figure 7).

As there is neither an underground stream nor stagnant water in Resava Cave, no aquatic subterranean organisms (stygobites, stygophiles or stygoxenes) were found there.

The following caves and pits in the surroundings of Resava Cave were also biospeleologically examined in the past (Table 4): Lazar's (=Zlot) Cave (with 26 species of terrestrial arthropods, of which nine troglobites) [34], Vernjikica Cave (with 15 species of terrestrial arthropods, of which six troglobites) [34], Golema Porica Pit (with 10 species of terrestrial arthropods, of which three troglobites) [35,36], Sesalac Cave (with six species of terrestrial arthropods, of which two troglobites) [24,37,38], Devojačka (=Gaura Fećilor) Cave (with five species of terrestrial arthropods, of which four troglobites) [39,40], Vrtačelje (=Ledena Pećina) Pit (with four species of terrestrial arthropods, of which two troglobites) [25] and Tupižnička Ledenica Pit (with four species of terrestrial arthropods, of which one troglobite) [41]. In addition to Resava Cave, Lazar's Cave, Vernjikica Cave and Sesalac Cave are also adapted for tourist visits. The terrestrial arthropod fauna of certain caves in other parts of Serbia, which are adapted for tourist visits, was also studied. According to the available literature data, 15 species of terrestrial arthropods live in Hadži-Prodan's Cave in southwestern Serbia, three of which are troglobitic [42], Stopića Cave in western Serbia is inhabited by eight taxa of terrestrial arthropods, one of which is troglobitic [43], while Ceremošnja Cave in eastern Serbia is home to five species of terrestrial arthropods, three of which are troglobitic [44,45]. Only in two of the caves mentioned (Lazar's Cave and Vernjikica Cave) a greater number of troglobitic terrestrial arthropod species was found than in Resava Cave, but the latter cave is characterized by the highest overall diversity of terrestrial arthropods (with 107 species) compared to all others in Serbia. One of the reasons for this assertion lies in the particular morphology of Resava Cave (it has a large Entrance Hall), which allows many trogloxenes to reside in it. It should also be noted that the biospeleological investigations in most of the caves mentioned lasted a year or even less, and they mainly focused not on the entire cave fauna, but only on troglobites.

Table 4. List of selected biospeleologically surveyed caves/pits in Serbia (with four or more arthropod species detected) with the number of troglobitic and total terrestrial arthropod species recorded.

No.	Cave/Pit Name	Troglobitic Arthropods	Total Terrestrial Arthropods
1	Lazar's (=Zlot) Cave	9	26
2	Vernjikica Cave	6	15
3	Resava Cave	4	107
4	Devojačka (=Gaura Fećilor) Cave	4	5
5	Golema Porica Pit	3	10
6	Hadži-Prodan's Cave	3	15
7	Ceremošnja Cave	3	5
8	Sesalac Cave	2	6
9	Vrtačelje (=Ledena Pećina) Pit	2	4
10	Stopića Cave	1	8
11	Tupižnička Ledenica Pit	1	4

Considering the medium size of Resava Cave and the lack of permanent water flow in the two main levels on the one hand, and the large number of arthropod species recorded on the other, the cave is relatively rich in hypogean terrestrial arthropod fauna compared to other caves in Serbia that have been biospeleologically studied so far. The terrestrial arthropod fauna of Resava Cave appears to be well preserved, and populations of endemic and protected arthropod species are constantly present despite the fact that it is a site visited by tourists, which speaks for its responsible management.

4.1. *Arachnida of Resava Cave*

The arachnid fauna of Resava Cave comprises 27 species (one troglobite, seven troglophiles and 19 trogloxenes). According to the literature [15,20,21], only three species (*C. serbicus*, *N.* cf. *carpaticum* and *P. convexum*) have been reported so far in Resava Cave, of which we were able to confirm two (*C. serbicus* and *N. carpaticum*), while the remaining 24 species we found were not yet known for Resava Cave. Representatives of four arachnid orders (Araneae, Ixodida, Opiliones and Pseudoscorpiones) were recorded. Most of the species we found belong to spiders (21) and harvestmen (four), while pseudoscorpions and ticks were represented by one species each. The families with the most species in Resava Cave are Linyphiidae and Lycosidae (with four species each), followed by Tetragnathidae, Thomisidae, Phalangiidae (with three species each), Liocranidae and Philodromidae (with two species each). Two species each were recorded within the genera *Metellina* Chamberlin & Ivie, 1941, *Philodromus* Walckenaer, 1826 and *Xysticus* C. L. Koch, 1835, while the other genera include one species each. During our investigations, we observed a large number of individuals of certain species of spiders (*C. serbicus*, *M. menardi* and *M. merianae*) and harvestmen (*O. ruzickai* and *P. sillii*), which indicates the presence of their stable populations in the cave. The presence of the pseudoscorpion *N. carpaticum* was confirmed in the Entrance Hall of the cave [15], where there are deposits of leaf litter that favor the development of this species.

Interestingly, despite several visits by our research team to Resava Cave in recent years, the spider *P. convexum* has not been recorded there since its discovery in 1998 [15]. On the contrary, the spider *C. serbicus* was found in large numbers, which confirms that the population of this specialized troglobitic species is quite stable in Resava Cave. This finding is very valuable as it is the only troglobitic arachnid in the cave. It is endemic to several caves in eastern Serbia (Vernjikica Cave, Resava Cave, Lazar's Cave, Izviđačka Cave and Manda's Cave) [20,21]. Considering the troglophiles, we found a stable population of the harvestmen *P. sillii*, an inhabitant of beech forests and caves, which is widely distributed in the Balkans and Southern Europe [46]. The troglophilic spiders *M. menardi* and *M. merianae* are common in Resava Cave and form stable populations there. They prefer somewhat drier habitats and we have often seen them on the walls, where they weave their webs. The troglophilic tick *I. vespertilionis* is a rather large species, which parasitizes bats and is usually found in caves with bat colonies [47].

4.2. *Crustacea of Resava Cave*

The crustacean fauna of Resava Cave includes three isopod species (one troglophile and two trogloxenes) that we found during our investigations, but which were not previously known to live in the cave. During our research, we observed a large number of the troglophilic species *H.* cf. *riparius*, suggesting its stable population in the cave. It is native to Central and Eastern Europe and can endure long periods under water [48]. Other species are trogloxenic and were observed sporadically.

The expected number of isopod species in Resava Cave is probably somewhat higher than we have indicated.

4.3. *Myriapoda of Resava Cave*

A total of 11 species of myriapods were recorded, which are divided into two classes: Diplopoda (with seven species) and Chilopoda (with four species). We found a significantly higher number of myriapod species in Resava Cave than indicated in the literature [19,24], where only two species of millipedes were reported to inhabit the cave (*S. kucajense* and *Trachysphaera* sp.). During our investigations, we found the two species of diplopods mentioned above and nine others that can be considered new to the fauna of Resava Cave. So far, no representatives of the class Chilopoda have been recorded in Resava Cave, but we found four centipede species there [*Cryptops anomalans* Newport, 1844, *Lithobius forficatus* (Linnaeus, 1758), *L. muticus* and *L. nigripalpis* L. Koch, 1867]. During our expeditions, representatives of seven orders and eight families of Myriapoda (five orders and six families of Diplopoda and two orders and two families of Chilopoda) were recorded in Resava Cave. Within the class Diplopoda, species of the orders Callipodida, Chordeumatida, Glomerida, Julida and Polydesmida and of the families Anthroleucosomatidae, Callipodidae, Craspedosomatidae, Glomeridae, Julidae and Polydesmidae were recorded. Within the class Chilopoda, taxa of the orders Lithobiomorpha and Scolopendromorpha and of the families Cryptopidae and Lithobiidae were found. Most of the recorded myriapod species belong to the order Lithobiomorpha (with three species), followed by the orders Chordeumatida and Glomerida (with two species each), while all other orders are represented by only one species each. The families with the most species are Lithobiidae (with three species) and Glomeridae (with two species), while all other families have only one species each. Of the eight genera of myriapods, *Lithobius* (with three species) is the most species-rich, while the other genera contained only one species each. Most species of Diplopoda and Chilopoda have stable populations in Resava Cave, as evidenced by the large number of specimens found during our research and their presence during all our visits.

The presence of the most important myriapod species, the troglobitic diplopod *S. kucajense*, was confirmed in Resava Cave mainly in its lower level (only one specimen was observed in the upper level), which is wetter and colder, favoring the development of this species. This finding is very valuable as it is the only troglobitic myriapod from Resava Cave. The increased humidity on the walls and on the floor of the cave (the presence of trickling water) and the presence of decaying wood as a food source are essential for the survival of this species. The population of this species in Resava Cave, although vulnerable and not so numerous, is stable, as evidenced by the presence of live specimens that we have observed during most of our visits. This information is very important for the monitoring and conservation of this species in its natural microhabitats in Resava Cave.

In addition, we confirmed the presence of *Trachysphaera* sp. in Resava Cave. We were unable to identify the specimens to species level as we only collected females and juveniles, whereas *Trachysphaera* species can only be accurately identified from the males. The taxon from Resava Cave most closely resembles the species *Trachysphaera costata* (Waga, 1857), *T. corcyraea* (Verhoeff, 1900) and *T. similicostata* (Radu & Ceuca, 1951), which were found in some other caves of the Carpatho-Balkanides of Serbia [24]. It is the only troglophilic myriapod from Resava Cave.

The findings of the millipede species *C. fasciata* and *Polydesmus subscabratus* Latzel, 1884 (Figure 7H), which are endemic to the Balkans, are also of great importance. Our record of *P. subscabratus* is significant as the species is very rare in Serbia and its previous findings in the country date back to the 1980s. Our discovery of the centipede species *L. nigripalpis* in Resava Cave is the first precise finding of this species in Serbia, which confirms its occurrence in this country. This species is distributed in the Balkans and the Middle East [49]. Other centipede species from Resava Cave are more widespread, some of which are subcosmopolitan (*L. forficatus*). The four species of centipedes mentioned were observed in the Entrance Hall of the cave (under stones and in the leaf litter).

The definite number of myriapod species in Resava Cave is probably somewhat higher than we have indicated.

4.4. Hexapoda of Resava Cave

We identified a total of 66 species of hexapods (two troglobites, seven troglophiles and 57 trogloxenes), which were assigned to the classes Collembola, Diplura and Insecta. Since only one species was previously known from Resava Cave [22], 65 species are new to the fauna of the cave. Representatives of 15 orders (three orders of Entognatha and 12 orders of Insecta) were recorded. Within the Entognatha, species of the classes Collembola and Diplura were recorded, while within the class Insecta, taxa of the orders Blattodea, Coleoptera, Diptera, Dermaptera, Hemiptera, Hymenoptera, Lepidoptera, Microcoryphia, Orthoptera, Plecoptera, Psocoptera and Trichoptera were found. Most of the hexapod species recorded belong to beetles (26), moths (13), dipterans (eight), caddisflies, hymenopterans and orthopterans (three species each). The families with the most hexapod species in Resava Cave are Carabidae (with 15 species), Geometridae (with eight species), Curculionidae, Staphylinidae and Limnephilidae (with three species each). The most species-rich hexapod genera were *Carabus* Linnaeus, 1758 (with five species), *Micropterna* Stein, 1873, *Ocypus* Leach, 1819 and *Otiorhynchus* Germar, 1822 (with two species each), while the other recorded genera contained only one species each. Certain species of springtails, rove beetles and psocodeans are very numerous in Resava Cave. These are *T. vulgaris*, *Q. mesomelinus* and *B. lucifuga*, respectively.

Apart from Resava Cave, the troglobitic dipluran *P. christiani* has been recorded in several other caves in the area, i.e., it is an endemic of the Kučajske Planine Mts. and Mt. Beljanica in eastern Serbia [23]. The carabid species *D. petrovici* is a stenoendemic of Resava Cave [22]. It is obvious that the increased humidity on the walls (the presence of trickling water) and on the floor of the cave, as well as a sufficient food supply are essential for the survival of these species.

The population of the troglophilic springtail *T. vulgaris* in Resava Cave is very large, as the species was very common in the cave. As far as its distribution is concerned, it is a widespread species in the Holarctic, often found in Europe and North America [50]. Another troglophilic species, the tiny collembolan *P. pygmaeus*, is actually the smallest arthropod in Resava Cave, but also one of the smallest in Serbia. Compared to *T. vulgaris*, we observed a much lower occurrence of *P. pygmaeus* in the cave, but it is also questionable whether it can be easily overlooked due to its small size. The number of Collembola species potentially inhabiting this cave is probably higher (possibly not less than 10), but they are difficult to find *in situ*. In contrast to most other Orthoptera, which feed on plants and live outside caves, the troglophilic *T. neglectus* is carnivorous and very common in European caves, especially in the Balkans [51].

Both troglophilic beetle species from Resava Cave are predators and are considered common in European caves, where they are found in bat guano [52]. The carabid *L. punctatus* is common in Serbian caves [52] and is also found outside caves. The rove beetle *Q. mesomelinus* is often found on a substrate consisting of bat guano and decaying wood [52]. Its population in Resava Cave is much larger than that of *L. punctatus*.

The troglophilic cranefly *L. nubeculosa* is widespread in European caves and is also found in forests, where its larvae feed on rotting wood [53]. The other troglophilic dipteran in the cave, *S. leptogaster*, lives in subterranean habitats of the Palaearctic and its larvae spin slimy hanging threads [54].

Interestingly, among the trogloxenes, the findings of the psocodean *B. lucifuga* and the dipteran *Heleomyza serrata* (Linnaeus, 1758) (Figure 7O) in Resava Cave are the first records of these two species on the territory of Serbia.

The actual number of hexapods in Resava Cave is probably much higher, especially among the trogloxenes, as their presence in the cave may also be accidental.

4.5. Distribution of Ecological Groups of Terrestrial Arthropods in Resava Cave

As far as the distribution of troglobites is concerned, they inhabit completely dark parts of Resava Cave, especially in the lower level of the cave (less often in the upper level), where the presence of moisture in the substrate (especially with trickling water) and air has been noted. We observed them mainly in the lower level of the cave on the walls, on decaying pieces of wood, on the floor, on rocks and speleothems and under stones, while they were much less present in the upper level of the cave. Most troglophiles were observed inside the cave on less moist substrate than that inhabited by troglobites (under stones, on the limestone walls, in guano). *Troglophilus neglectus* was only found in the illuminated Entrance Hall, where part of the populations of *T. vulgaris* and *L. punctatus* were also detected. Trogloxenes are mainly found in the Entrance Hall (under stones, in the rubble and leaf litter, on the walls), the driest part of Resava Cave, but with the greatest food supply.

The large number of arthropod species recorded indicates that this cave is a very important refuge for fauna. The greatest biological richness of Resava Cave is the occurrence of endemic troglobitic species, two of which (*C. serbicus* and *S. kucajense*) are strictly protected in Serbia [55]. The populations of troglobites in the cave, although vulnerable, are present and stable throughout the year, which is confirmed by the fact that we found live individuals of all four troglobitic species during all six visits, which is particularly important for the monitoring and conservation of these species in their natural microhabitats in the cave. The occurrence of *C. serbicus* is significantly high in the cave, similar to that of the dipluran *P. christiani* and the millipede *S. kucajense*. The occurrence of *D. petrovici* is significantly lower than that of the other three troglobites in the cave, which is not unusual as the former species is a relatively large subterranean predator.

It is certain that the actual number of terrestrial arthropod species living in Resava Cave is higher than currently known, as some species are easier to overlook. These are rare species, those with smaller populations, species with shorter activity periods and those from isolated microhabitats. Therefore, it is necessary to continue the biospeleological studies in Resava Cave in the future to record such taxa. It is desirable to conduct surveys using other collection methods to determine the actual state of arthropod diversity in the cave (including the status and abundance of their populations).

As far as sampling is concerned, possible biases that may have affected the results (representativeness of the arthropod samples collected) could certainly be reduced by spatial and temporal improvements. Therefore, as many different microhabitats as possible should be investigated in the cave, which would probably lead to the collection of additional taxa, especially from the class Collembola. On the other hand, sampling in each month of the year, repeated several times in the following years, would certainly provide new information on the terrestrial fauna of the cave. In temperate caves, flooding and cold air infiltration in winter and early spring can disrupt the relatively constant physical conditions of the cave environment [56].

Like other habitats, caves also have their own trophic webs, but in caves they are simpler and functionally less complex than in epigean ecosystems due to the lower species richness [57]. The arthropod species found in Resava Cave are detritivores or predators. Detritivores such as springtails and millipedes play a very important role among the arthropods. They feed on bat droppings and decompose other organic matter that can enter the cave from surface habitats in various ways (from the soil, with plant roots, through sinking streams, with trogloxenic animals) [56]. The detritivores are eaten by predators such as spiders, harvestmen and pseudoscorpions. Centipedes and beetles are often the main predators in subterranean habitats.

In addition, it is necessary to focus future studies on other groups of invertebrates (roundworms, snails, annelids, etc.), which are also very important in the trophic webs of the cave and surrounding habitats.

4.6. Human Influence on Subterranean Biodiversity

As far as the impact of human activities on subterranean biodiversity is concerned, each cave is unique due to its morphological characteristics, its gate system and its management measures. The relative humidity, temperature, CO_2 concentration, artificial light, noise, etc. can be affected by tourist traffic. To assess the risks, a multi-year study of the terrestrial fauna of a cave should be carried out, focusing on the troglobites. The results from a cave in France show that tourism does not necessarily have a negative impact on the biodiversity of the cave [58]. The authors demonstrated that neither the number of specimens observed nor the species richness at the sampling sites had decreased as a result of tourist visits to the cave [58]. If human influence does not cause significant microclimatic changes and does not disturb the microhabitats of invertebrates (suitable spaces such as rock crevices), the survival of cave-dwelling arthropods is not at risk [58]. Despite the high vulnerability of cave-dwelling organisms, disturbance of cave habitats does not necessarily lead to a decline in species diversity if the disturbance is limited in terms of duration, intensity and spatial extent in relation to the size of the cave [58]. In a case study of four Romanian caves, some of them were found to maintain normal levels of relative humidity despite visits by tourists [59]. In the unventilated caves, however, the temperature and CO_2 concentration showed large fluctuations during the peak tourist season [59].

In the same study, the authors proposed several management measures for the sustainable use of show caves [59]. First, control of lighting is required through a combination of early removal of lampenflora and adjustment of lighting levels (both light intensity and duration). Another measure is continuous monitoring of CO_2 levels and occasional ventilation through the airlocks during the peak season in caves without a natural entrance. In caves where the hazard varies seasonally, seasonal changes in visitor management can be made, benefiting both the cave environment and the visitor health [59]. Finally, microbiological monitoring of surfaces should be routinely carried out to control pathogens introduced by visitors [59].

It is difficult to discuss the impact of adaptation to tourism on the biodiversity of Resava Cave, as there is no comparable data before the cave was commercially opened to visitors. However, we can state that the wildlife we encountered in the cave seems to survive relatively undisturbed despite the introduction of stairs and LED lights in the cave. It seems like a good idea to bring rotting pieces of wood into the cave and place them on the floor of the damp halls in the lower level of the cave to give the millipedes a better chance of survival (it should be ensured that the wood provided does not contain any animals). This potential conservation measure would be particularly important for the troglophilic *Trachysphaera* sp., which is always present in the cave, albeit in small numbers. The provision of additional wood as a food source would certainly strengthen its population in the cave. Another advantage of this measure would be the possibility to collect male specimens for the final identification of the species.

If there are no new anthropogenic changes, the current management does not seem to jeopardize the survival of the fauna in Resava Cave, which has already been demonstrated for other show caves *in situ* [58]. Of course, further studies on ecological impacts would provide much more information on the status and prospects of terrestrial arthropod populations in the cave. Certainly, conservation efforts should focus on the obligate cave dwellers—the troglobites.

Author Contributions: Conceptualization, N.V. and S.Ć.; methodology, N.V. and S.Ć.; software, N.V.; validation, N.V. and S.Ć.; formal analysis, N.V. and S.Ć.; investigation, N.V. and S.Ć.; resources, N.V. and S.Ć.; data curation, N.V., C.D., P.M., D.A., D.Z.S., D.V.S., K.S., M.B., A.I.-Ć. and S.Ć.; writing—original draft preparation, N.V.; writing—review and editing, N.V. and S.Ć.; visualization, N.V.; supervision, S.Ć.; project administration, S.Ć.; funding acquisition, S.Ć. All authors have read and agreed to the published version of the manuscript.

Funding: This research was funded by the Serbian Ministry of Science, Technological Development and Innovation (contract numbers 451-03-65/2024-03/200178, 451-03-66/2024-03/200178, 451-03-66/2024-03/200197 and 451-03-65/2024-03/200117) and the Public Company "Resava Cave" (Despotovac, Serbia) (contract numbers 03-2035/2020, 03-1063/2021, 03-1037/2022 and 03-2556/2023).

Institutional Review Board Statement: Not applicable.

Data Availability Statement: Data are contained within the article.

Acknowledgments: We would like to thank Ľubomír Kováč (Pavol Jozef Šafárik University in Košice, Slovakia), Sára Ferenți (University of Oradea, Romania), Ljiljana Protić (Belgrade, Serbia), Dilian Georgiev (University of Plovdiv, Bulgaria), Andrzej Józef Woźnica (Wrocław University of Environmental and Life Sciences, Poland), Daoyuan Yu (Nanjing Agricultural University, China) and Jaroslav Boháč (University of South Bohemia in České Budějovice, Czech Republic) for their help in identifying certain arthropod species and for confirming our initial identifications. We also thank Vukašin Gojšina (University of Belgrade, Serbia) for his help in our field research activities. We are indebted to the Public Company "Resava Cave" (Despotovac, Serbia), especially to the director Ana

Bogdanović and the deputy director Vladan Gicić, for supporting our biological research in Resava Cave.

Conflicts of Interest: The authors declare no conflicts of interest.

References

1. Camacho, A.I.; Bello, E.; Becerra, J.M.; Vaticon, I. A natural history of the subterranean environment and its associated fauna. In *The Natural History of Biospeleology, Monografias*; Camacho, A.I., Ed.; Museo Nacional de Ciencias Naturales, Consejo Superior de Investigaciones Cientificas, Graficas Mar-Car, S.A.: Madrid, Spain, 1992; pp. 171–195.
2. Furon, R. La paléogéographie. In *Essai sur L'évolution des Continents et des Oceans*, 2nd ed.; Payot: Paris, France, 1959; pp. 1–405.
3. Ćurčić, B. The cave fauna in Serbia: Origin, historical development, and diversification. In *Speleological Atlas of Serbia*; Đurović, P., Ed.; Geographical Institute "Jovan Cvijić"; Serbian Academy of Sciences and Arts; Institute for Nature Conservation of Serbia; Faculty of Geography, University of Belgrade & Faculty of Biology, University of Belgrade: Belgrade, Serbia, 1998; pp. 75–83.
4. Polak, S.; Pipan, T. The subterranean fauna of Križna Jama, Slovenia. *Diversity* **2021**, *13*, 210. [CrossRef]
5. Zagmajster, M.; Polak, S.; Fišer, C. Postojna-Planina Cave System in Slovenia, a hotspot of subterranean biodiversity and a cradle of speleobiology. *Diversity* **2021**, *13*, 271. [CrossRef]
6. Lukić, M.; Fišer, C.; Delić, T.; Bilandžija, H.; Pavlek, M.; Komerički, A.; Dražina, T.; Jalžić, B.; Ozimec, R.; Slapnik, R.; et al. Subterranean fauna of the Lukina Jama-Trojama cave system in Croatia: The deepest cave in the Dinaric Karst. *Diversity* **2023**, *15*, 726. [CrossRef]
7. Delić, T.; Pipan, T.; Ozimec, R.; Culver, D.C.; Zagmajster, M. The subterranean species of the Vjetrenica Cave System in Bosnia and Herzegovina. *Diversity* **2023**, *15*, 912. [CrossRef]
8. Francke, O.F.; Monjaraz-Ruedas, R.; Cruz-López, J.A. Biodiversity of the Huautla Cave System, Oaxaca, Mexico. *Diversity* **2021**, *13*, 429. [CrossRef]
9. Ferreira, R.L.; Berbert-Born, M.; Souza-Silva, M. The Água Clara Cave System in northeastern Brazil: The richest hotspot of subterranean biodiversity in South America. *Diversity* **2023**, *15*, 761. [CrossRef]
10. Xiong, K.N. Karst and world natural heritage in southern China. In Proceedings of the Southwest Region Conference of Mountain Environment and Ecological Civilization Construction of the Chinese Geographical Society Annual Academic Conference, Kunming, China, 27 April 2013.
11. Huang, S.; Wei, G.; Wang, H.; Liu, W.; Bedos, A.; Deharveng, L.; Tian, M. Ganxiao Dong: A hotspot of cave biodiversity in northern Guangxi, China. *Diversity* **2021**, *13*, 355. [CrossRef]
12. Huang, S.; Zhao, M.; Luo, X.; Bedos, A.; Wang, Y.; Chocat, M.; Tian, M.; Liu, W. Feihu Dong, a new hotspot cave of subterranean biodiversity from China. *Diversity* **2023**, *15*, 902. [CrossRef]
13. Tian, M.; Huang, S.; Jia, X. A contribution to cavernicolous beetle diversity of South China Karst: Eight new genera and fourteen new species (Coleoptera: Carabidae: Trechini). *Zootaxa* **2023**, *5243*, 1–66. [CrossRef]
14. Tian, M.; Huang, S.; Wang, D. Discovery of a most remarkable cave-specialized trechine beetle from southern China (Coleoptera, Carabidae, Trechinae). *ZooKeys* **2017**, *725*, 37–47. [CrossRef]
15. Đurović, P. (Ed.) *Speleological Atlas of Serbia*; Geographical Institute "Jovan Cvijić", Serbian Academy of Sciences and Arts; Institute for Nature Conservation of Serbia; Faculty of Geography, University of Belgrade & Faculty of Biology, University of Belgrade: Belgrade, Serbia, 1998; pp. 1–290.
16. Grubić, A.; Jankičević, J. Carbonate paraplatform in the Upper Jurassic and Lower Cretaceous of eastern Serbia. *Mem. Serb. Geol. Soc.* **1973**, *10*, 73–85. (In Serbian)
17. Vasić, N.; Obradović, J.; Grubin, N. Characteristics of Jurassic System in the Danubicum-Pesača-Greben and Miroč area. In Proceedings of the International Symposium "Geology in the Danube Gorges", Special Edition No. 25, Donji Milanovac-Orșova, Yugoslavia-Romania, 23–26 September 1997; Grubić, A., Berza, T., Eds.; Geoinstitut: Belgrade, Serbia, 1997; pp. 29–35.
18. Gavrilović, D. The karst of Serbia. *Mem. Serb. Geol. Soc.* **1976**, *13*, 1–28. (In Serbian)
19. Ćurčić, B.P.M.; Makarov, S.E. Three new endemic species of the millipede genus *Bulgarosoma* Verhoeff (Diplopoda: Chordeumatida: Anthroleucosomatidae) from East Serbia. *Arch. Biol. Sci.* **1998**, *50*, 51–61.
20. Deltshev, C.; Ćurčić, B.P.M. A contribution to the study of the genus *Centromerus* Dahl (Araneae: Linyphiidae) in caves of the Balkan Peninsula. *Rev. suisse Zool.* **2002**, *109*, 167–176. [CrossRef]
21. Deltshev, C.; Ćurčić, B.; Wang, C.X.; Yao, Z.Y.; Antić, D.; Ćurčić, S.; Rađa, T. New data on the spiders (Araneae) in the caves of Balkan Peninsula. *Arch. Biol. Sci.* **2014**, *66*, 465–471. [CrossRef]
22. Ćurčić, S.; Vrbica, M.; Antić, D.; Petković, M.; Marković, Đ.; Ćurčić, B.; Vesović, N. *Duvalius* (*Paraduvalius*) *petrovici* n. sp. and *D.* (*P.*) *sotirovi* n. sp. (Carabidae: Trechinae: Trechini), two new troglobitic ground beetles from eastern and southeastern Serbia. *Arch. Biol. Sci.* **2014**, *66*, 889–900. [CrossRef]
23. Sendra, A.; Borko, Š.; Jiménez-Valverde, A.; Selfa, J.; Lukić, M.; Miculinić, K.; Rađa, T.; Antić, D. Cave-adapted campodeids (Hexapoda, Diplura, Campodeidae) from the Dinarides and adjacent karst regions. *Rev. suisse Zool.* **2021**, *128*, 15–52. [CrossRef]
24. Antić, D.; Šević, M.; Macek, O.; Akkari, N. Review of *Trachysphaera* Heller, 1858 (Diplopoda: Glomerida: Glomeridae) in Serbia, with taxonomic notes on the genus. *Zootaxa* **2021**, *5047*, 273–299. [CrossRef] [PubMed]

25. Nešić, D.; Pavićević, D.; Ognjenović, S. The results of speleomorphological and biospeleological studies of the pit Vrtačelje (Ledena Pećina). *Zašt. Prir.* **2008**, *58*, 15–25. (In Serbian)
26. Petrović, J. *Resava Caves*, 4th ed.; National Library "Resavska Škola" & Institute of Geography, Faculty of Science, University of Novi Sad: Novi Sad, Serbia, 2002; pp. 1–80. (In Serbian)
27. Lazarević, R. Resava Cave before the reconstruction of the lights. *Zašt. Prir.* **1991**, *43–44*, 58–79. (In Serbian)
28. Kličković, M. Misconceptions about Resavska Cave. *Zašt. Prir.* **2016**, *66*, 49–58. (In Serbian)
29. Anonymous. Decree on the Protection of the Natural Monument "Resava Cave". *Služb. Glasn. RS* **1995**, *9*. (In Serbian)
30. Schiner, J.R. Fauna der Adelsberger-, Lueger-und Magdalenen-Grotte. In *Die Grotten und Höhlen von Adelsberg, Lueg, Planina und Laas*; Schmidl, A., Ed.; Braumüller: Vienna, Austria, 1854; pp. 231–272.
31. Racovitza, E.G. L'essai sur les problèmes biospéologiques. *Arch. Zool. Exp. Gén.* **1907**, *6*, 371–488.
32. Sket, B. Can we agree on an ecological classification of subterranean animals? *J. Nat. Hist.* **2008**, *42*, 1549–1563. [CrossRef]
33. Deltshev, C. The faunistic diversity of cave-dwelling spiders (Arachnida, Araneae) of Greece. *Arachnol. Mitt.* **2011**, *40*, 23–32. [CrossRef]
34. Vesović, N.; Ćurčić, S. *Report on the Study of Insects and Other Invertebrates in the Caves of the Lazar's Canyon (Lazar's Cave and Vernjikica Cave)*; University of Belgrade—Faculty of Biology: Belgrade, Serbia, 2022; pp. 1–11. (In Serbian)
35. Nešić, D.; Pavićević, D.; Lazarević, P.; Zatezalo, A. Golema Porica on the Rtanj Mt.—A speleological object of important geological and biological diversity. *Zašt. Prir.* **2009**, *60*, 427–437. (In Serbian)
36. Ćurčić, B.P.M.; Dimitrijević, R.N.; Makarov, S.E.; Ćurčić, S.B.; Tomić, V.T.; Antić, D.Ž.; Ilić, B.S.; Ćurčić, N.B. *Roncus radgost* n. sp., *R. jarevid* n. sp. and *R. crnobog* n. sp.: Three new cave dwellers from eastern Serbia (Neobisiidae, Pseudoscorpiones). *Arch. Biol. Sci.* **2013**, *65*, 751–760. [CrossRef]
37. Nešić, D.; Jović, D.; Kličković, M.; Zatezalo, A.; Medenica, I. Results of the research of the Sesalačka Cave as a basis for its protection (Sokobanja structural basin, eastern Serbia). *Zašt. Prir.* **2022**, *72*, 39–51. (In Serbian) [CrossRef]
38. Antić, D.Ž.; Tomić, V.T.; Ćurčić, B.P.M.; Ćurčić, S.B.; Stamenković, S.P.; Lučić, L.R.; Makarov, S.E. Genus *Belbogosoma* Ćurčić & Makarov, 2008, with descriptions of new troglobitic species from East Serbia, Balkan Peninsula (Diplopoda: Chordeumatida: Anthroleucosomatidae). *Arch. Biol. Sci.* **2014**, *66*, 907–918. [CrossRef]
39. Nešić, D.; Pavićević, D. The results of a recent study of the Devojačka Pećina Cave (Gaura Fećilor). *Zašt. Prir.* **2009**, *60*, 419–426. (In Serbian)
40. Antić, D.Ž.; Ćurčić, B.P.M.; Tomić, V.T.; Ćurčić, S.B.; Stojanović, D.Z.; Dudić, B.; Makarov, S.E. One hundred millipede species in Serbia (Arthropoda: Myriapoda: Diplopoda). *Arch. Biol. Sci.* **2013**, *65*, 1559–1578. [CrossRef]
41. Nešić, D.; Pavićević, D.; Petrović, B.; Zatezalo, A. Results of recent studies of Tupižnička Ledenica. *Zašt. Prir.* **2008**, *59*, 67–79. (In Serbian)
42. Anonymous. *Natural Monument "Hadži-Prodan's Cave". Protection Study*; Institute for Nature Conservation of Serbia: Belgrade, Serbia, 2012. (In Serbian)
43. Anonymous. *Natural Monument "Stopića Cave". Proposal for Placing Under Protection as a Natural Asset of Outstanding Importance*; Institute for Nature Conservation of Serbia: Belgrade, Serbia, 2004. (In Serbian)
44. Pavićević, D.; Vesović, N.; Popović, M.; Ćurčić, S. A new troglobitic ground beetle (Carabidae: Trechinae) from eastern Serbia. *Zašt. Prir.* **2018**, *68*, 61–66. [CrossRef]
45. Antić, D.Ž.; Stojanović, D.Z.; Makarov, S.E. *Cornogonopus*—A new monotypic cave-dwelling genus of the family Anthroleucosomatidae (Diplopoda, Chordeumatida) from Serbia, Balkan Peninsula. *Biol. Serb.* **2020**, *42*, 32–47. [CrossRef]
46. Băncilă, R.I.; Pradel, R.; Choquet, R.; Plăiaşu, R.; Gimenez, O. Using temporary emigration to inform movement behaviour of cave-dwelling invertebrates: A case study of a cave harvestman species. *Ecol. Entomol.* **2018**, *43*, 551–559. [CrossRef]
47. Arthur, D.R. The *Ixodes* ticks of Chiroptera (Ixodoidea, Ixodidae). *J. Parasitol.* **1956**, *42*, 180–196. [CrossRef] [PubMed]
48. Wijnhoven, H. Landpissebedden van de Ooijpolder: Deel 1. verspreiding (Crustacea: Isopoda: Oniscidea). *Ned. Faun. Meded.* **2000**, *11*, 55–131. (In Dutch)
49. Bachvarova, D.; Doichinov, A.; Stoev, P.; Kalchev, K. Habitat preferences and effect of environmental factors on the seasonal activity of *Lithobius nigripalpis* L. Koch, 1867 (Chilopoda: Lithobiomorpha: Lithobiidae). *Int. Res. J. Nat. Sci.* **2015**, *3*, 1–13.
50. Babenko, A.; Stebaeva, S.; Turnbull, M.S. An updated checklist of Canadian and Alaskan Collembola. *Zootaxa* **2019**, *4592*, 1–125. [CrossRef]
51. Holuša, J.; Kočárek, P.; Vlk, R.; Marhoul, P. Annotated checklist of the grasshoppers and crickets (Orthoptera) of the Czech Republic. *Zootaxa* **2013**, *3616*, 437–460. [CrossRef]
52. Nešić, D.; Pavićević, D.; Zatezalo, A.; Mijatović, M.; Grubač, B. The results of a complex study of the Ogorelička Cave. *Zašt. Prir.* **2008**, *59*, 51–66. (In Serbian)
53. Košel, V.; Horváth, M. Spatial and temporal distribution of family Limoniidae (Insecta, Diptera) in a cave of the western Carpathians (Slovakia). *Dipterol. Bohemosl.* **1995**, *7*, 97–109.
54. Ševčík, J.; Kjærandsen, J.; Marshall, S.A. Revision of *Speolepta* (Diptera: Mycetophilidae), with descriptions of new Nearctic and Oriental species. *Can. Entomol.* **2012**, *144*, 93–107. [CrossRef]
55. Anonymous. The Rulebook on Proclamation and Protection of Strictly Protected and Protected Wild Species of Plants, Animals and Fungi. *Služb. Glasn. RS.* **2010**, *5/10*. (In Serbian)
56. Barr, T.C. Observations on the ecology of caves. *Am. Nat.* **1967**, *101*, 475–491. [CrossRef]

57. Parimuchová, A.; Dušátková, L.P.; Kováč, Ľ.; Macháčková, T.; Slabý, O.; Pekár, S. The food web in a subterranean ecosystem is driven by intraguild predation. *Sci. Rep.* **2021**, *11*, 4994. [CrossRef]
58. Faille, A.; Bourdeau, C.; Deharveng, L. Weak impact of tourism activities on biodiversity in a subterranean hotspot of endemism and its implications for the conservation of cave fauna. *Insect Conserv. Divers.* **2014**, *8*, 205–215. [CrossRef]
59. Constantin, S.; Mirea, I.C.; Petculescu, A.; Arghir, R.A.; Măntoiu, D.Ș.; Kenesz, M.; Robu, M.; Moldovan, O.T. Monitoring human impact in show caves. A study of four Romanian caves. *Sustainability* **2021**, *13*, 1619. [CrossRef]

diversity

MDPI

Article

Exploring Ice Cave Biodiversity in Northeastern Italy

Leonardo Latella [1] and Stefano Brighenti [2,*]

[1] Zoology Department, Museo di Storia Naturale di Verona, Lungadige Porta Vittoria 9, 37129 Verona, Italy; leonardo.latella@comune.verona.it
[2] Competence Centre for Mountain Innovation Ecosystems, Free University of Bozen/Bolzano, Piazza dell'Università 5, 39100 Bolzano, Italy
* Correspondence: stefano.brighenti@unibz.it

Abstract: The ice stored in caves is a widespread yet neglected cryospheric component. The cold-adapted biodiversity of ice caves has received very little attention from research, despite the potential abundance of endemic troglobiotic and cryophilic species and their consequent sensitivity to the changing climate. In this study, we investigated the invertebrate diversity of two ice caves in Northeastern Italy (Bus delle Taccole and Caverna del Sieson, Veneto Region). During 2022 and 2023, we sampled, using pitfall traps, the invertebrates dwelling at different locations in each cave: the shaft base, an intermediate hall, and the cave bottom. At each cave location, we also collected ice samples, on which we measured the stable isotopes of oxygen and hydrogen ($\delta^{18}O$, $\delta^{2}H$), and monitored the air temperature with data-loggers. The two caves had different invertebrate communities, both dominated by a combination of troglobiotic and cryophilic taxa. Despite a low taxonomic richness, which was higher at Taccole (15 taxa) than at Sieson (11 taxa), both caves hosted rare/endemic species, four of which are not described yet. At each cave, the ice water isotopic signatures differed among cave locations, suggesting the ice had formed under different climatic conditions, and/or resulted from different frequencies of thawing/freezing events. The occurrence of summer melt at both caves suggests that these unique ecosystems will quickly disappear, along with their specialized and unique biodiversity.

Keywords: ice cave; subterranean biology; speleology; glaciology; invertebrate communities; Veneto

1. Introduction

The recent special report from the Intergovernmental Panel on Climate Change (IPCC) on the ocean and cryosphere in a changing climate highlighted a global contraction of permafrost, glaciers, seasonal snow cover, and Arctic sea ice [1]. Rising temperatures are responsible for the current widespread cryospheric loss, and climate models suggest that warming will continue unabated, leading to the loss of most glaciers in the coming decades, paralleled by a widespread permafrost warming and degradation. Small glaciers are most affected by this recent melt, with climate scenarios suggesting losses of up to 80 percent by the end of this century [1]. However, the same report does not include information on the perennial ice hosted in caves, which is the least-visible and -studied component of the global cryosphere.

The ice hosted in caves can originate from different processes, strongly dependent on the conformation of the cave and the microclimatic context. These include snow metamorphism, the freezing of infiltrating spring/rain/snowmelt water, and/or water vapor condensation [2]. These processes can be inferred with the use of the stable isotopes of oxygen and hydrogen in the water molecule. Isotopic ratios ($\delta^{18}O$, $\delta^{2}H$) are a useful tool to investigate the formation and internal dynamics of cave ice [3–6]. The future of this ice will strongly depend on the percolation of warm spring/rainfall water within the karst system and/or on enhanced ablation caused by increased air temperature within the cave [7,8]. Therefore, a rapid reduction in cave ice is predicted to occur during the next few decades, since the rates of ablation are up to two–three times faster than those of formation and lead to persistently negative annual mass balances [9]. Cave ice deposits are facing the risk of disappearing completely within a decade, with the irreparable loss of historical data on climate, environmental conditions, and the life they host [8].

Most caves that host perennial ice are located in central and southeastern Europe, a region that has experienced some of the most rapid glacier ice losses in recent decades [8]. In Italy, more than 1600 caves are classified as cryo-caves, due to the presence of multiyear snow, firn, or ice. At least 10% of such caves can be included in the ice cave classification (i.e., have a perennial ice deposit), testifying how widespread these cryospheric features are [10]. In Italy, the heterogeneous geology has enabled not only caves formed in limestone, dolomite, and marble terrains, in the Alps as well as in the Apennines, but also in lava tubes on Mount Etna [10]. Most of the Italian ice caves are in the Alpine arc and no caves with perennial ice deposits are known in the Apennines, except for the Abisso Revel in Tuscany [11].

Despite their considerable number, there are only a few studies on ice caves in Italy, and none of these concern the fauna inhabiting such peculiar environments. Ice caves represent unique ecosystems characterized by extreme cold and low light conditions [12]. Troglobionts and stygobionts, already adapted to life in subterranean environments, implement strategies to survive in constantly frozen environments [13]. In addition to these taxa, species typical of glacial or nival environments and that can bear darkness, even if not fully adapted to subterranean life, are commonly found in ice caves, even in their deepest areas. Still, very little is known about the physiological and morphological adaptations of ice cave species, and given the rate at which these ecosystems are disappearing, research about these unique environments needs to be intensified.

For this reason, the Natural History Museum and speleological groups in Verona began to study the ecology and fauna of ice caves in the Alps and Pre-Alps [14–17], establishing the first comprehensive biological research on this kind of environment.

In this study, we explored the ice cave fauna in two caves in Northeastern Italy to test the hypothesis that these environments host a combination of troglobiotic (due to the cave habitat) and cryophilic (due to the ice presence) taxa, including endemic species due to the geographical isolation of such caves.

2. Materials and Methods

2.1. Study Area

The two investigated caves are located in the western part of the Veneto Prealps (Southeastern Italy; Figure 1) [18] in two different mountain ranges: the Baldo-Altissimo group (Verona Province) and the Asiago Plateau (Vicenza Province).

The Baldo-Altissimo group, on which Bus delle Taccole (Taccole Cave, called Taccole hereafter; Figure 2) is located, has a surface area of 398 km^2 and a maximum elevation of 2200 m above sea level (a.s.l.). This mountain range is predominantly composed of carbonatic rocks; Dolomia Principale, Calcari Grigi, Calcari oolitici di S. Vigilio, Rosso Ammonitico, Scaglia Rossa, and Maiolica [19,20]. Caverna del Sieson (Sieson Cavern, called Sieson hereafter; Figure 3) is located in the Asiago Plateau, which has a surface of about 1000 km^2. In the Asiago Plateau, four main carbonatic formations can be distinguished: the Dolomia Principale Formation (Norian–Rhaetian), the Mount Zugna Formation (Hettangian–Sinemurian), the Loppio Oolitic Limestone Formation (mid-Sinemurian), and the Rotzo Formation (Sinemurian–Pliensbachian). The last three formations belong to the Calcari Grigi Group [21].

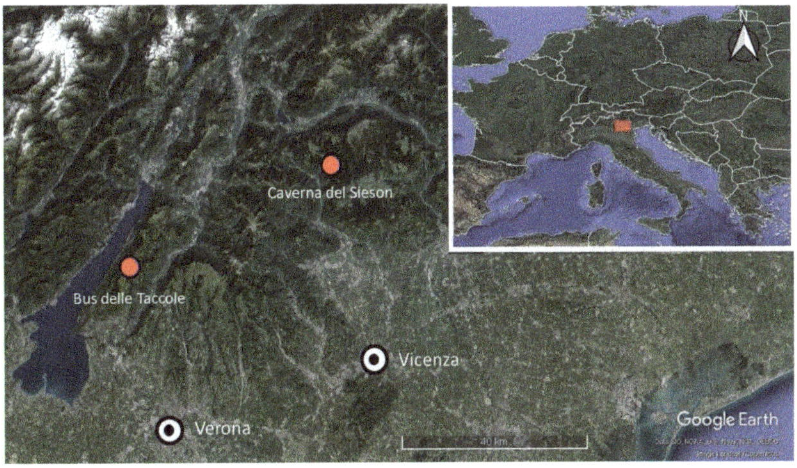

Figure 1. Location of the two studied caves (red dots) in the Lessinia Mountains (the red rectangle in the inlet box indicates the area in Italy).

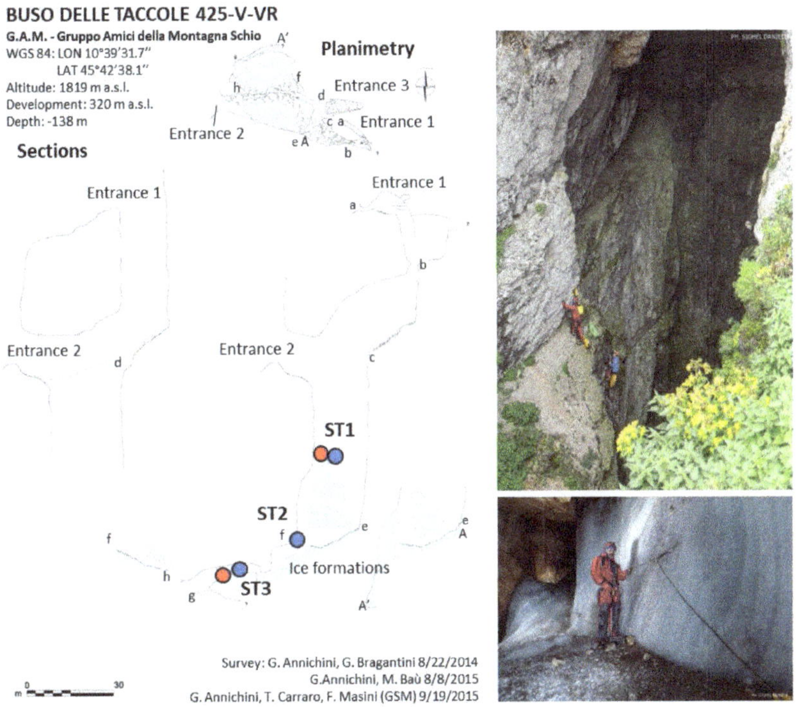

Figure 2. Cave map of Bus delle Taccole, with the locations of the sampling locations (blue dots) and the positions of dataloggers (red dots) used in the study. The extension of the permanent ice coverage inside the cave is illustrated in light blue. On the right: main entrance of the cave, and ice mass at the cave bottom (photos: speleoclick/Sighel D.). NOTE: letters in the map indicate the landmarks.

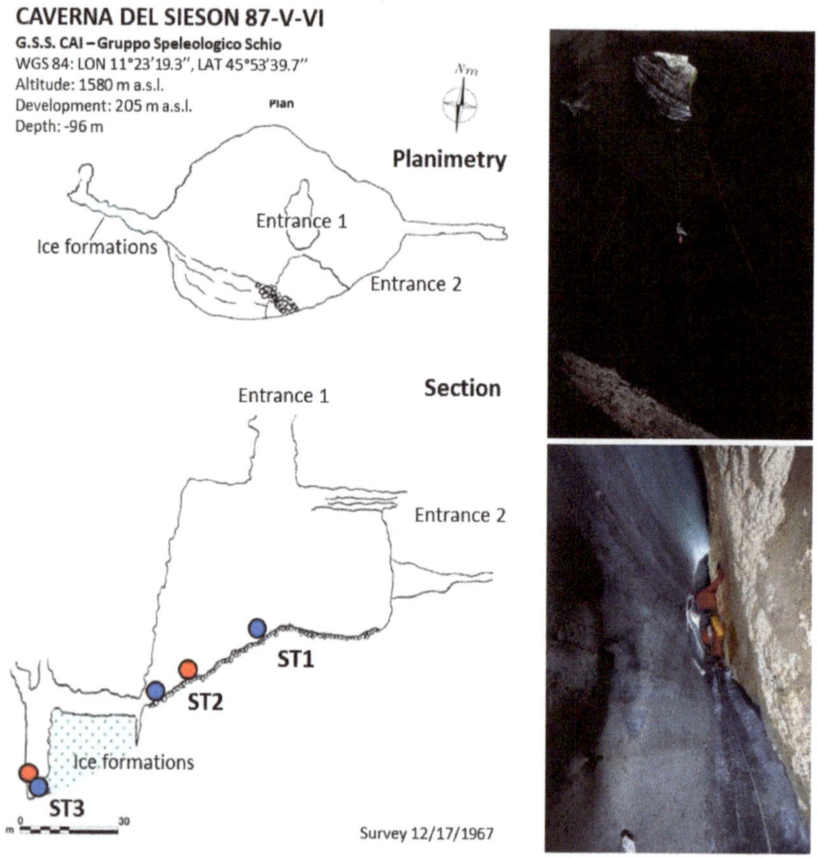

Figure 3. Cave map of Caverna del Sieson with the locations of the selected stations (blue dots) and the positions of dataloggers (red dots) used in the study. The extension of the permanent ice coverage inside the cave is illustrated in light blue. On the right: abseiling at the cave shaft base and descent along the cave ice. (Photos: speleoclick/Sighel D.).

2.1.1. Bus Delle Taccole

Cadastral number: 425 VVR
Locality: Baldo Mountain, Cima Telegrafo, Brenzone District, Verona Province
Coordinates WGS84: long. 10°89′31.7″–lat. 45°42′38.1″
Altitude: 1819 m a.s.l.
Development: 320 m
Depth: −138 m

The cave consists of three large shafts offset (Figure 2). A −40 m shaft departs from the upper entrance and connects to the top of the lower one, or enters a meander that, after about 15 m, past a landslide, enters a −40 m shaft, also in communication with the one below. Going up a few meters instead, it is possible to continue in the meander. The lower entrance is a large crack about 20 m high and 10 m wide that enters through a debris chute into a −65 m shaft, with a regular diameter of 20 × 15 m. The base widens considerably and is encumbered by a large accumulation of ice that bisects the shaft. Between the wall and the glacial mass, in a northerly direction, it is possible to descend for about −20 m between ice and rock. Continuing, it is possible to descend a few meters below the ice vault.

At the base of the entrance shaft, there are accumulations of soil and bird guano (a small colony of *Pyrrhocorax graculus* (Linnaeus, 1766) nests on the upper part of entrance shaft walls), which are an evident trophic supply for the cave.

2.1.2. Caverna del Sieson

Cadastral number: (87 V–VI)
Locality: Casare Campolongo, Rotzo District, province of Vicenza
Coordinates WGS84: long. 11°23′19.3″–lat. 45° 53′39.7″
Altitude: 1580 m a.s.l.
Development: 205 m
Depth: −96 m

The cave consists of a large shaft with an upper entrance of about ten meters that then enters an lower one, with a very wide opening that leads into a shaft about 50 m in diameter and 40 m deep (Figure 3). At the base of the shaft, one descends along a debris cone for another 30 m until reaching the beginning of the glacier, which one descends for 20 m in a narrow passage between rock and ice. The mass of ice in the cave has evidently been decreasing in recent years. There are no bat or bird colonies at the entrance or in the inner areas of the cave, so the trophic input from outside is mainly from organic material dropped on the bottom of the pit.

2.2. *Specimens Sampling*

For each cave, we selected three sampling locations: Station 1 (ST1) at the entrance shaft base, Station 2 (ST2) in a deeper area but not in direct contact with the ice, Station 3 (ST3) at the bottom of the cave in contact with the ice mass (Figures 2 and 3).

The trophic supply, in terms of animal or plant organic material, was very different among the sampling locations of both caves. It was abundant at the base of the entrance shaft, due to the presence of organic material from outside and from the top of the shaft, and was extremely limited in the deeper areas occupied by ice.

Each station was sampled using standard pitfall traps, consisting of a glass cup with an open diameter of 10 cm, filled with propylene glycol, in which was placed a tube containing an attracting bait of blue mold cheese. Traps were left on site for 13 months in total at Sieson (April 2022–May 2023) and for 16 months in total at Taccole (August–October 2020, June 2022–October 2023). The traps were changed every three months between spring and autumn and every six months in winter. The collected specimens were fixed in 75% ethanol for morphological study. All specimens sampled in this study are preserved in the collections of the Museo di Storia Naturale of Verona, Italy.

2.3. *Temperature Measurement*

Two Tinytag Plus temperature dataloggers (−30 °C to +50 °C, accuracy 0.01 °C) were positioned inside each cave, one at the base of the shaft and one at the cave bottom (Figures 2 and 3), where they recorded the air temperature (T_{air}, °C) at 12 h intervals.

2.4. *Ice Sampling and Isotopic Analyses*

During each sampling campaign, we used an ice screw to sample ice aliquots belonging to the external part of the ice mass (0–15 cm). Sampling spots included the cave trap locations, at the floor (0.5 m) and at breast height (1.5 m), where screws were inserted horizontally. The ice was collected into 50 mL polyethylene containers with double caps. Samples were transported in thermal bags to the laboratory of the Faculty of Science and Technology, University of Bolzano, where they were stored at 4 °C. At Sieson, we also collected samples of snow and of water dripping from the cave walls and from a pond at the cave bottom. The ratios of oxygen and hydrogen stable isotopes (δ^2H, $\delta^{18}O$) of the melted ice water were determined with a laser spectrometer (Cavity Ring-Down Spectroscopy Picarro L2130i, Santa Clara, CA, USA) after filtration (0.45 µm). The precision of the analyzer was 0.5 ‰ for δ^2H, and 0.25 ‰ for $\delta^{18}O$. The memory effect [22] was minimized, following the procedure reported in Penna et al. [23]. All isotopic results were referred to the Vienna Standard Mean Oceanic Water and expressed in ‰ notation.

2.5. Data Analysis

To visually represent the dissimilarities among the communities belonging to different cave locations, we used the "vegan" package [24] in R (version 4.2.2) to perform a non-metric multidimensional scaling (NMDS). We estimated community dissimilarity distances based on the Bray–Curtis index on log-transformed taxa abundances. Before the analyses, we discarded the larvae (identified at a broad taxonomic detail), and taxa found with only one specimen (*Hymenaphorura* sp., *Lepidocyrtus* sp., and *Pogonognathellus flavescens*). We identified the cryophilic and troglobiotic taxa (see Table 1) to plot them on a bidimensional ordination. Finally, we tested if communities from the two caves were significantly different based on the ANOSIM (analysis of similarity) test in "vegan".

Table 1. List of the known taxa found in the studied caves. Tb—troglobiont; Cy—cryophilic; 1—present; 0—absent.

Class	Order	Family	Genus/Species/Subspecies	Status	Tac. St 1	Tac. St 2	Tac. St 3	Sie. St 1	Sie. St 2	Sie. 3
Arachnida	Trombidiformes	Rhagidiidae	Gen. sp.	Tb	1	1	1	0	0	0
Arachnida	Trombidiformes	Fam.	Gen. sp.		1	1	0	0	0	1
Diplopoda	Chordeumatida	Neoatractosomatidae	*Osellasoma caoduroi* Mauries, 1984	Tb	1	1	1	0	0	0
Collembola	Poduromorpha	Onychiuridae	*Deutheraphorura* n. sp.	Tb	1	1	0	0	0	0
Collembola	Poduromorpha	Onychiuridae	*Onychiuroides* n.sp.	Tb	0	0	0	1	1	1
Collembola	Poduromorpha	Onychiuridae	*Hymenaphorura* sp.		0	1	0	0	0	0
Collembola	Poduromorpha	Hypogastruridae	*Ceratophysella* cf *macrocantha*		0	1	1	0	0	0
Collembola	Poduromorpha	Hypogastruridae	*Ceratophysella bengtssoni* (Agren, 1904)		0	0	1	0	0	0
Collembola	Poduromorpha	Neanuridae	*Pseudachorudina alpina* Stach, 1949	Cy	1	0	0	0	0	0
Collembola	Entomobryomorpha	Isotomiidae	*Folsomia nigrimaculata* Najt, 1981	Cy	1	0	0	0	0	0
Collembola	Entomobryomorpha	Isotomiidae	*Desoria* n. sp.	Cy	0	0	0	1	1	1
Collembola	Entomobryomorpha	Tomoceridae	*Pogonognathellus flavescens* (Tullberg, 1871)		0	0	0	0	0	1
Collembola	Entomobryomorpha	Entomobryidae	Gen. sp.		0	0	1	0	0	0
Collembola	Entomobryomorpha	Entomobryidae	*Lepidocyrtus* sp.		0	1	0	0	0	0
Collembola	Entomobryomorpha	Entomobryidae	*Pseudosinella concii* Gising, 1950	Tb	0	0	0	0	1	1
Collembola	Neelipleona	Neelidae	*Megalothorax carpaticus* Papáč and Kováč, 2013		0	0	0	0	1	0
Insecta	Coleoptera	Staphylinidae	Gen. sp.		0	1	0	1	0	0
Insecta	Coleoptera	Staphylinidae	Pselaphinae Gen. sp.		0	0	0	1	0	0
Insecta	Coleoptera	Leiodidae	*Halberria* n. sp.	Tb	1	1	1	0	0	0
Insecta	Coleoptera	Cucujidae	Gen. sp.					1		
Insecta	Diptera	Trichoceridae	*Trichocera maculipennis* Meigen, 1818	Cy	1	1	1	1	1	1
Insecta	Diptera	Limoniidae	*Chionea araneoides* (Dalman, 1816)	Cy	0	0	1	0	1	1

Based on the isotopic ratios, we calculated for each water sample the parameter d-excess [25], defined as follows:

$$\text{d-excess (‰)} = \delta^2\text{H} - 8\,\delta^{18}\text{O}$$

We estimated the correlation between δ^{18}O and d-excess with the non-parametric Spearman's rank correlation coefficient (ρ).

Since no isotopic data on local precipitation were available, we used as a reference those from a close-by pre-Alpine catchment, Ressi (721 m a.s.l., Vicenza province; reference period: 2012–2022 [26]), located 15 km and 35 km away from Sieson and Taccole, respectively. When producing a dual isotope plot of $\delta^{18}O$ and $\delta^{2}H$ with the collected data, we plotted the local meteoric water line (LMWL) of Ressi, as described by Marchina et al. [27] and Zuecco et al. [28].

3. Results

3.1. Subterranean Fauna

The fauna sampled in the two caves consisted of 22 taxa of invertebrates. Of these, 15 were found at Taccole and 11 were found at Sieson. Six troglobiotic species were sampled, four of them at Taccole and two of them at Sieson. Five species can be considered cryophilic, two of which were found in both caves (Table 1).

Regarding Diplopoda, in Taccole, we sampled the Neoatractosomatidae *Osellasoma caoduroi* Mauries, 1984 (Figure 4). Although present throughout the cave, it was particularly abundant at the bottom.

Figure 4. A specimen of the Diplopoda *Osellasoma caoduroi* collected in the Buso del Vallon (Photo: L-Latella).

Collembola were the most represented order within the two caves. Nine species were found at Taccole, with one troglobiotic and two cryophilic species. Inside Sieson, five Collembola species were present, one troglobiotic and one cryophilic (Table 1). Among Neanuridae, *Pseudachorudina alpina* Stach, 1949, was found at Taccole, with a few specimens. Onychiuridae included a new species belonging to the genus *Deutheraphorura* n. sp., which was abundant in Taccole, especially at the shaft base and in the intermediate zone of the cave; the troglobiotic *Onychiuroides* n. sp. was abundant at all sampling locations in the Sieson cave (B. Valle pers. com.). For the Isotomidae, one species belonging to the genus *Folsomia nigrimaculata* Najt, 1981 was found at Taccole. At Sieson, Isotomidae were represented by *Desoria* n. sp., a probable new species related to *D. fjellbergi* (Najt, 1981), a species typical of nival environments (B. Valle pers. com.). At Sieson, it was abundant at the cave bottom near the ice mass. To the Entomobryidae belongs *Pseudosinella concii*

Gisin, 1950, quite common in the innermost areas of Sieson. The troglobiotic Neelidae *Megalothorax carpaticus* Papáč and Kováč, 2013 was found only in ST2 of Sieson.

In addition to Collembola, the most interesting troglobiotic species was the Leiodidae Cholevinae *Halberria* n. sp. This is a new species related to *Halberria carlini* Vailati, 2017, present at all stations sampled within the Taccole cave.

Diptera were represented by two cold-adapted species. The Trichoceridae *Trichocera maculipennis* Meigen, 1818, was sampled in both caves at the adult stage, as well as at the larval stage in Taccole. The Limoniidae *Chionea araneoides* Dalman, 1816 (snow fly) was present in both caves as well.

3.2. Distinct Cave Communities

The NMDS (stress < 0.001, k = 2) showed distinct communities in the two caves, even though the ANOSIM test did not show significant differences ($R = 1$, $p = 0.09$), which was very likely due to the low sample size. The cryophilics *T. maculipennis*, *P. alpina*, and *F. nigrimaculata*, as well as the troglobiotics *Halberria* n. sp., Rhagidiidae, *O. caudoroi*, and *Deuteraphorura* n. sp. were more aligned with the Taccole samples, whereas the cryophilic *Desoria* n. sp. and the troglobiotics *Pseudosinella* concii and *Onychiuroides* sp. were more aligned with the Sieson ones (Figure 5). The communities belonging to different locations of the cave were more homogeneous at Sieson. Contrastingly, the abundance of *C. araneoides* and *T. maculipennis* increased, and that of *Desoria* n. sp. decreased, when moving from ST1 to ST2 and ST3 in Taccole (Figure 5).

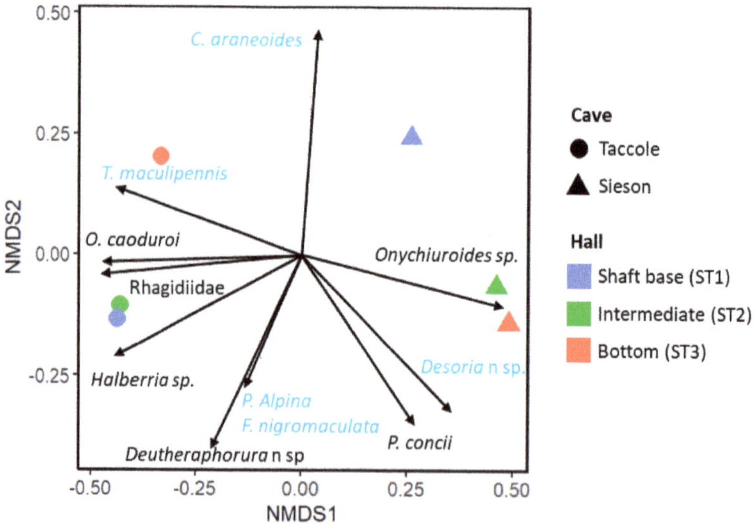

Figure 5. NMDS biplot, based on the Bray–Curtis index, of the communities belonging to different trap locations, focusing on cryophilic (light blue) and troglobite (black) taxa.

3.3. Environmental and Isotopic Conditions

The two caves had similarly cold air conditions (Table 2; Figure 6). Their bottoms had a larger thermal variability than the shaft base, and a lower T_{max}. Daily T_{air} generally exceeded 0 °C during the period of August–November at the cave bottom, and during July–November at the shaft base (Table 2). The collection of records during different periods did not allow the drawing of any comparisons of the thermal profiles at the two caves.

Table 2. Main thermal features of the two caves at their bottoms and shaft bases. For each cave, we refer here to the daily values of the period when dataloggers were recording at both locations. * Freezing index refers to the number of days during which T_{avg} was higher than 0 °C.

Thermal Parameter	Sieson Shaft Base	Sieson Bottom	Taccole Shaft Base	Taccole Bottom
Reference period	9/14/22–5/16/23	9/14/22–5/16/23	9/13/20–5/8/21 9/11/21–12/15/21	9/13/20–5/8/21 9/11/21–12/15/21
T_{min}/T_{max}	−5.5/4.2 °C	−2.9/4.3 °C	−3.1/1.2 °C	−2.5/1.0 °C
$T_{avg} \pm T_{sd}$	−0.9 ± 1.3 °C	−0.4 ± 0.8 °C	−0.7 ± 1.0 °C	−0.5 ± 0.7 °C
T_{med}	−0.8 °C	−0.5 °C	−0.3 °C	−0.2 °C
T_{aug}	na	na	0.14 ± 0.16 °C	0.01 ± 0.03 °C
Freezing index *	163	161	192	181

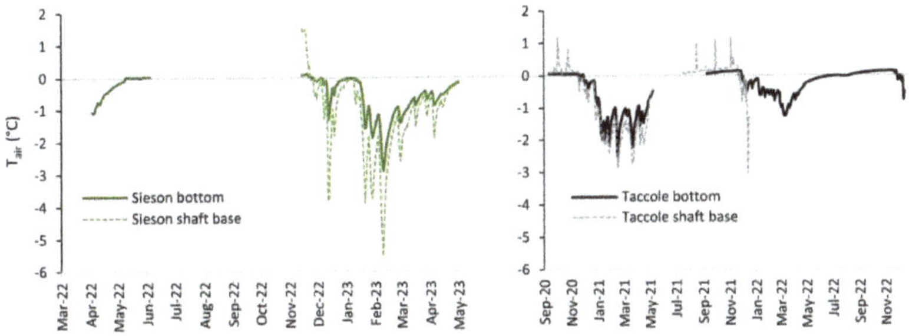

Figure 6. Time series of average daily air temperature (T_{air}) at the shaft base and bottom locations of the two caves. The 0 °C threshold is provided (dashed black line) to aid interpretation.

Overall, the two caves had similar isotopic conditions, with $\delta^{18}O$ in the range of −6/−9 ‰ and d-excess in the range of 9–14 ‰. At each cave, the isotopic composition of the ice generally differed among cave locations (Figure 7). At Sieson, there was a tendency towards a depletion in heavy isotopes when moving at depth and, at each location, the breast height ice was more depleted in heavy isotopes than the floor ice. By contrast, at Taccole, different locations had only slight isotopic differences, except at the shaft base, where the breast height ice was more enriched in heavy isotopes when compared with that of the other sampling locations/heights. D-excess only slightly differed between different locations at Sieson, where the larger differences were found between the floor and breast height ice (Figure 7). Indeed, at each location, breast height ice had higher d-excess when compared with the floor ice. In contrast, at Taccole, d-excess increased when moving from the shaft base to the bottom location, with slighter differences between floor and breast height when compared to Sieson. In the ice of both caves, we found a strong negative correlation between $\delta^{18}O$ and d-excess ($\rho = -0.8$, $p < 0.001$).

When plotted in the dual isotope plot (Figure 8), the ice samples shaped freezing lines with a lower slope and intercept than those of the global meteoric water line ($\delta^2H = 8\ \delta^{18}O + 10$), and of the local meteoric water line (LMWL) of the close-by Ressi catchment. The samples of water and snow that were collected at Sieson more evidently aligned along this LMWL than those of Taccole. The isotopic range of cave ice reflected the values of spring/autumn rainfall occurring in the Ressi catchment (highest affinity with the months of April and September), and it was isotopically more enriched with heavy isotopes than the snow collected above the Sieson (Figure 8).

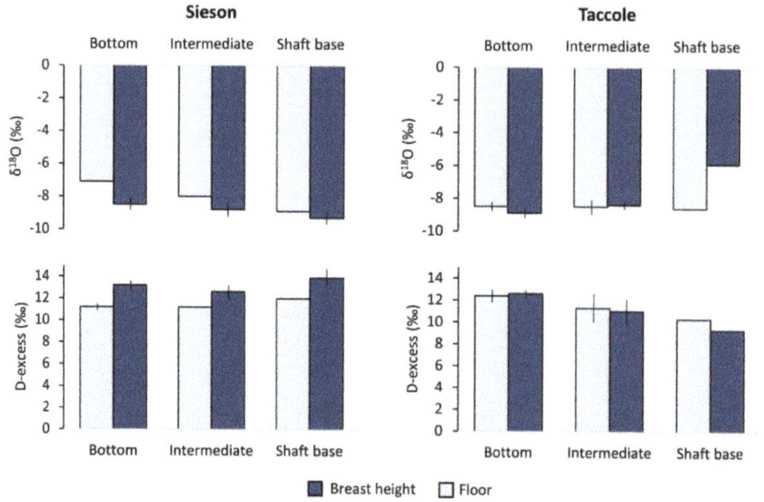

Figure 7. Bar plots of $\delta^{18}O$ and d-excess values of the ice at the two caves, at different locations and heights of the ice mass. Minimum and maximum values are displayed with vertical bars at each plot.

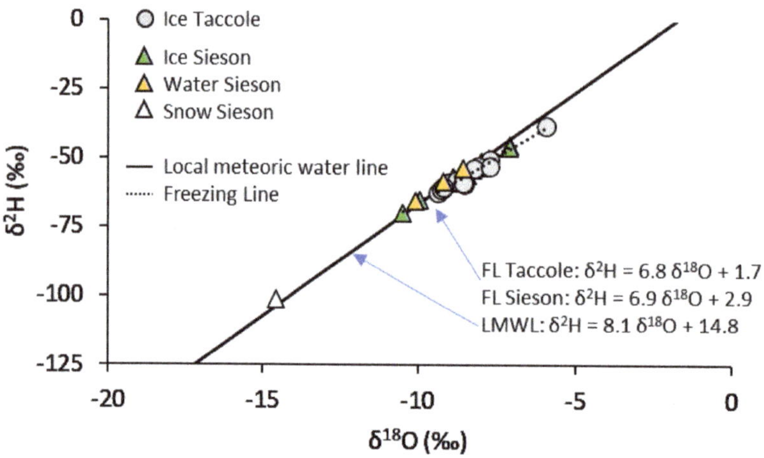

Figure 8. Dual-isotope plot showing the samples of ice at the two caves, the sample of snow that was collected immediately outside from the Sieson, and the water dripping from stalactites or belonging to a pond located at the cave bottom of Sieson. LMWL = local meteoric water line for the Ressi catchment [28]; FL = freezing line.

4. Discussion

4.1. *Ice Cave Fauna Composition*

In the two investigated caves, the troglobiotic biodiversity was low and limited to a few taxonomic groups that were mainly represented by Collembola and Coleoptera, as commonly observed in ice caves of temperate zones [13]. Overall, our hypothesis on the presence of a mixture of troglobiotic and cryophilic taxa, including endemic species, can be retained. In ice caves, typical subterranean-dwelling species co-dwell with a "cryophilic ipogean faunistic association" consisting of elements (mainly Diptera, such as the genera *Trichocera* and *Chionea*) that, although not exclusive of subterranean life, are well adapted to cold conditions [14,29]. At the Taccole cave, we found the Neanuridae *P. alpina*, a species typical of high mountain environments, and the Isotomidae *F. nigrimaculata*, a rather rare species typical of the cold environments of the Pyrenees [30]. The Trichoceridae *T. maculipennis* (winter crane fly) is a Holarctic cold-tolerant species, known from the Arctic to the southern regions of the Mediterranean zone and the Far East [31]. It has been reported in different cold caves in central Europe [31], and it was sampled in both the study caves at the adult stage, as well as at the larval stage at Taccole. The Limoniidae *C. araneoides* (snow fly) is another cryophilic species found at both caves. Adults of the genus Chionea are easily distinguishable due to the absence of the wings (aptery). In alpine and pre-Alpine environments, these flies are visible in winter walking on the snow cover and in caves with a low internal temperature [14].

While some cryophiles (e.g., *T. maculipennis* and *C. araneoides*) were found in both caves, all troglobiotic species were endemic. At Taccole, these included the Neoatractosomatidae *O. caodurol*. It is an endemic species for the Veneto and Lombardy pre-Alps, known at present only from another cave in Monte Baldo and on the pre-Alps of Bergamo (Lombardy). According to Mauries [32], it may be a Quaternary element related to cold climates that found refuge in the cave after the last glaciations. The Onychiuridae family was represented by a new species belonging to the genus *Deutheraphorura* n. sp., abundant at Taccole, where a second new species related to *Halberria carlini* Vailati, 2017, described from Monte Altissimo (Baldo-Altissimo Group), was found. At Sieson, we found a probable new species of *Desoria* n. sp. (Isotomidae), related to *D. fjellbergi* (Najt, 1981), which is typical of nival environments (B. Valle pers. com.).

More widespread cave taxa were found too. Several genera of rhagid mites include species adapted to life in subterranean habitats [33]. As in the case of the specimens sampled at Taccole, adaptations are usually manifested through the elongation of appendages and the progressive development of sensory organs, such as the increased length and number of rhagidial solenidia on the tarsi and tibiae of the first two pairs of legs [33]. The first pair of elongated legs acts as antennae, while the extremely elongated tactile setae, particularly on the palpi and first pair of legs, are useful for extending the tactile perception zone and create combs that help capture the prey [33]. To the Entomobryidae belongs *Pseudosinella concii* Gisin, 1950, a species distributed in different caves in Italy and Switzerland, and is quite common in the innermost areas of the Sieson. The troglobiotic Neelidae *M. carpaticus* was found only at Sieson.

Overall, the coexistence of troglobiotic and cryophilic taxa in ice caves can be explained both by the refugia hypothesis, according to which the caves were used as a refuge during the Plio-Pleistocene climatic alternations [34], and by the paleogeography of the two mountain ranges. Both the Baldo-Altissimo group and the Asiago Plateau were in fact isolated during the climatic oscillations of the Quaternary, and are therefore rich in species that were isolated and thus became endemic, even not in subterranean environments [35].

4.2. Relation to Environmental Variables

The inputs of organic matter can be a key food resource for cave fauna [13]. At Taccole, abundant trophic inputs from the outside (mainly soil, plant, and animal organic matter) slightly influenced cave communities. Indeed, taxa abundance and diversity were higher at the intermediate depth, where the organic matter was also more abundant. At Sieson, where the external organic inputs were much lower than at Taccole (only a few plant fragments were found at the base of the entrance shaft), we did not find different diversity and abundances among sampling locations.

The air temperature was very low within the two caves, with average values below 0 °C at all investigated locations. This explains why cryophilic species did not show preference for a particular cave location and, instead, were found at all depths.

4.3. Heterogeneous Isotopic Fingerprint of Cave Ice

The investigated ice had a large spatial variability of isotopic composition in the two caves. This variability was observed not only among different locations but also within the same locations, between the upper and the floor samples of the ice mass profile. This spatial pattern was particularly evident for $\delta^{18}O$ at Sieson.

Since lower $\delta^{18}O$ values generally indicate the origin of the ice water under colder weather conditions than higher values [5], the positive trend of $\delta^{18}O$ with depth found for Sieson may suggest that the ice at the bottom had formed in warmer periods (i.e., season and/or years), when compared with that of the shaft base. The wide opening of the Sieson entrance suggests that winter precipitation and snowmelt can more effectively enter the shaft base and quickly freeze, as personally observed. At the bottom of the cave, higher isotopic values of the ice suggest a "seasonality bias" [3] of formation during a warmer season when compared with the ice of the shaft base. Differently, at Taccole, $\delta^{18}O$ values decreased with depth, but only at breast height. In this cave, the formation of the basal layers may have resulted from the congelation of water with a consistent origin for all depths, whereas the breast height layers (i.e., the younger ones) may have resulted from an inverse "seasonality bias", when compared with Sieson. Thus, a different internal structure may be responsible for the different vertical isotopic pattern of the ice in the two caves. The upper part of the Taccole shaft base ice might have formed more recently, given the higher isotopic values when compared with those of the other locations in the cave, i.e., the shaft base floor, the intermediate locations, and the cave bottom. According to the thermal profiles of this work, at the Taccole shaft base, the air temperature dropped below 0 °C earlier during autumn, when compared with the cave bottom. Thus, rainfall occurring in this period may freeze more easily at the shaft base than at the bottom, and this may explain the enriched isotopic values observed at the shallower (and likely younger) layer (breast height) of the shaft base.

Despite these spatial differences, similar isotopic ranges suggest comparable seasonal formation of the ice at Sieson and Taccole. From springtime to autumn, rainwater can more easily penetrate at depth before freezing, given the warm conditions found at the shaft base of the caves. Indeed, when compared with decadal data on precipitation collected at the close-by Ressi catchment, the $\delta^{18}O$ values of the ice were compatible with those of precipitation fallen during springtime and autumn. By contrast, the d-excess values were generally higher in these precipitation samples when compared with those of the ice collected in this study. This suggests ice formation under colder climates than nowadays and/or the occurrence of fractionation processes during the ice formation. Overall, the positive relationship between $\delta^{18}O$ and d-excess suggests the occurrence of thawing/freezing events prior to the ice formation [6] at both caves.

We suggest that the ice at both caves did not originate from the condensation of water vapor, which would have returned freezing lines with higher intercepts [3,36], nor from snow metamorphism, which would have resulted in much lower isotopic values than those observed. We hypothesize that the ice formation in these caves mainly resulted from cold air traps ("static caves with congelation ice") [2], rather than from chimney effects. This was also testified by the lack of evidence of thermal anomalies during wintertime, when the shaft bases were colder than the cave bottoms because of the relevant influence from outside temperature.

In general, the limited number of ice samples collected only allowed for a qualitative estimation of a very likely more complex isotopic profile of the cave ice. Furthermore, very little local-scale information is available on the isotopic conditions of precipitation and snowmelt at both sites. Also, Taccole is located at a further distance from the Ressi catchment, compared to Sieson, leading to weaker isotopic comparisons of ice and precipitation.

5. Conclusions

Ice caves are overlooked cold environments at imminent risk of extinction. Life in temperate subterranean environments is driven by certain environmental factors (darkness, constant temperature, lack of primary production, a scarcity of food resources, etc.). In ice caves, these factors are combined with extremely low temperatures and dry air. These environmental constraints allow for the existence of species-poor communities mostly composed of a few, highly adapted species, which are therefore extremely sensitive to environmental changes.

Our study on two ice caves in northeastern Italy confirms the uniqueness of these environments, hosting a combination of euryoecious, troglobiotic, and cryophilic species. The low local diversity of invertebrates is outpaced by the high conservation value of cave communities, including endemic taxa, three of which are new to science. The ice of these caves likely results from thawing/freezing events after the infiltration of springtime waters, and during different historical periods, depending on the cave location. The rapid pace at which this ice is shrinking was personally observed during the different field visits, and it was testified with temperature records by the unexpectedly large number of days at which melt likely occurs. This raises concerns about the potential loss of a unique and unexplored biodiversity that will be paralleled by the vanishment of cave ice during the upcoming years, in the investigated ice caves as well as those of temperate regions globally.

Our ongoing research in several other ice caves will provide a better understanding of the biodiversity of these cryospheric ecosystems, and of the relationship between ecological features and the isotopic signature of the cave ice. We will also perform radiocarbon dating of the ice to estimate its age and correlate it with paleogeographic events and faunal composition.

Author Contributions: Conceptualization, L.L. and S.B.; methodology, L.L. and S.B.; software, L.L. and S.B.; validation, L.L. and S.B.; formal analysis, L.L. and S.B.; investigation, L.L. and S.B.; resources, L.L.; data curation, L.L. and S.B.; writing—original draft preparation, L.L. and S.B.; writing—review and editing L.L. and S.B.; visualization, L.L. and S.B.; supervision, L.L. and S.B. All authors have read and agreed to the published version of the manuscript.

Funding: This research has been carried out within the PNRR research activities of the consortium iNEST (Interconnected North-East Innovation Ecosystem) funded by the European Union Next-GenerationEU (Piano Nazionale di Ripresa e Resilienza (PNRR) Missione 4 Componente 2, Investimento 1.5 D.D. 1058 23/06/2022, ECS_00000043 –Spoke1, RT1B, CUP I43C22000250006). This manuscript reflects only the Authors views and opinions, neither the European Union nor the European Commission can be considered responsible for them.

Institutional Review Board Statement: Not applicable.

Data Availability Statement: The data supporting this work will be made available upon request.

Acknowledgments: We would like to thank Dragan Antić (Belgrade, Serbia) and Barbara Valle (Milan, Italy) for their taxonomic identification and suggestions, Daniele Sighel (Trento, Italy) for allowing the use of photographs of the two caves, and Gaia Padovani Quarella (Verona, Italy) for the graphic elaboration of some figures. We are also grateful to all the speleologists who shared the cave surveys with us and the speleological clubs that supported the research, in particular GAM Verona, GSM Malo, GGT Treviso, Giorgio Annichini (GAM Verona, Italy), the enthusiastic organizer of the surveys to the Taccole, Marcello Manea (GSM Malo, Italy), who managed part of the sampling at Sieson, and Giacomo Troisi (GSM Malo, Italy) for his support in the isotopic sampling. A special thanks to Alessandro Tenca, manager of the Refuge Telegrafo (Baldo Mountain, Verona), and all his staff, for the hospitality. We also thank Chiara Marchina and Giulia Zuecco (University of Padova) for the long-term isotopic monitoring of the Ressi catchment that helped to better contextualize our isotopic data.

Conflicts of Interest: The authors declare no conflicts of interest.

References

1. Intergovernmental Panel on Climate Change [IPCC]. *Special Report on the Ocean and Cryosphere in a Changing Climate*; IPCC: Paris, France, 2019.
2. Luetscher, M.; Jeannin, P.-Y. A process-based classification of alpine ice caves. *Theor. Appl. Karstol.* **2007**, *17*, 5–10.
3. Kern, Z.; Fórizs, I.; Pavuza, R.; Molnár, M.; Nagy, B. Isotope Hydrological Studies of the Perennial Ice deposit of Saarhalle, Mammuthöhle, Dachstein Mts, Austria. *Cryosphere* **2011**, *5*, 291–298. [CrossRef]
4. Perşoiu, A.; Onac, B.P.; Wynn, J.G.; Bojar, A.-V.; Holmgren, K. Stable isotope behavior during cave ice formation by water freezing in Scărişoara Ice Cave, Romania. *J. Geophys. Res.* **2011**, *116*, D02111. [CrossRef]

5. Bădăluţă, C.-A.; Perşoiu, A.; Ionita, M.; Piotrowska, N. Stable Isotopes in Cave Ice Suggest Summer Temperatures in East-central Europe Are Linked to Atlantic Multidecadal Oscillation Variability. *Clim. Past* **2020**, *16*, 2445–2458. [CrossRef]
6. Smith, D.F.; Lyons, W.B.; Welch, S.A.; Zorn, M.; Tičar, J.; Lipar, M.; Carey, A.E. The Chemistry of Cave Ice: Two Examples from Slovenia. *Environ. Eng. Geosci.* **2023**, *29*, 291–307.
7. Holmlund, P.; Onac, B.P.; Hansson, M.; Holmgren, K.; Mörth, M.; Nyman, M.; Perşoiu, A. Assessing the palaeoclimate potential of cave glaciers: The example of the Scărişoara ice cave (Romania). *Geogr. Ann. A* **2005**, *87*, 193–201. [CrossRef]
8. Perşoiu, A.; Buzjak, N.; Onaca, A.; Pennos, C.; Sotiriadis, Y.; Ionita, M.; Zachariadis, S.; Styllas, M.; Kosutnik, J.; Hegyi, A.; et al. Record summer rains in 2019 led to massive loss of surface and cave ice in SE Europe. *Cryosphere* **2021**, *15*, 2383–2399. [CrossRef]
9. Colucci, R.R.; Fontana, D.; Forte, E.; Potleca, M.; Guglielmin, M. Response of ice caves to weather extremes in the southeastern Alps, Europe. *Geomorphology* **2016**, *261*, 1–11. [CrossRef]
10. Maggi, V.; Colucci, R.R.; Scoto, F.; Giudice, G.; Randazzo, L. Ice caves in Italy. In *Ice Caves*; Elsevier: Amsterdam, The Netherland, 2018; pp. 399–423.
11. Sivelli, M.; Vianelli, M. *Abissi delle Alpi Apuane: Guida Speleologica*; Società Speleologica Italiana: Bologna, Italy, 1982.
12. Perșoiu, A. Ice caves climate. In *Ice Caves*; Elsevier: Amsterdam, The Netherlands, 2018.
13. Iepure, S. Chapter 7—Ice Cave Fauna. In *Ice Caves*; Elsevier: Amsterdam, The Netherlands, 2018.
14. Avesani, D.; Latella, L. Spatio-temporal distribution of the genus Chionea (Diptera, Limoniidae) in the Buso del Valon ice cave and other caves in the Lessini Mountains (Northern Italy). *Boll. Mus. Civ. St. Nat. Verona* **2016**, *40*, 11–16.
15. Latella, L.; Zorzin, R.; Annichini, G.; Pellegrini, B. Il progetto "Buso del Vallon" negli Alti Lessini Veronesi. *Speleo. Ven.* **2019**, *23*, 175–179.
16. Petri, I.; Ballarin, F.; Latella, L. Seasonal abundance and spatio-temporal distribution of the troglophylic harvestman Ischyropsalis ravasinii (Arachnida, Opiliones, Ischyropsalididae) in the Buso del Valon ice cave, Eastern Italian Prealps. *Subt. Biol.* **2022**, *42*, 151–164. [CrossRef]
17. Latella, L. Analisi preliminare della fauna sotterranea delle Dolomiti di Brenta (TN). In Proceedings of the 17th Regional Symposium of Speleology of Trentino Alto Adige 2021, Trento, Italy, 20–21 November 2022; GS SAT: Trento, Italy, 2022.
18. Marazzi, S. *Atlante Orografico Delle Alpi—SOIUSA*; Priuli and Verlucca: Torino, Italy, 2005.
19. Pasa, A. *Carsismo e Idrografia Carsica nel Gruppo del M. Baldo e nei Lessini Veronesi. C.N.R.; Centro Studi Geografia Fisica, Ricerche Sulla Morfologia e Idrografiacarsica, 5*; Tipografia Mareggiani: Bologna, Italy, 1954; p. 150.
20. Zorzin, R.; Agostini, L.; Montecchi, M.C.; Torri, P.; Accorsi, C.A. Tanella cave (Monte Baldo—Verona, Italy): A record of environmental data on the last glacial period. *Acta Carsol.* **2011**, *40*, 189–205. [CrossRef]
21. Di Cuia, R.; Riva, A.; Scifoni, A.; Moretti, A.; Spötl, C.; Caline, B. Dolomite characteristics and diagenetic model of the Calcari Grigi Group (Asiago Plateau, Southern Alps–Italy): An example of multiphase dolomitization. *Sedimentology* **2011**, *58*, 1347–1369. [CrossRef]
22. Cui, J.; Tian, L.; Gerlein-Safdi, C.; Qu, D. The influence of memory, sample size effects, and filter paper material on online laser-based plant and soil water isotope measurements. *Rapid Commun. Mass Spectrom.* **2017**, *31*, 509–522. [CrossRef]
23. Penna, D.; Stenni, B.; Šanda, M.; Wrede, S.; Bogaard, T.A.; Michelini, M.; Fischer, B.M.C.; Gobbi, A.; Mantese, N.; Zuecco, G.; et al. Technical Note: Evaluation of between-sample memory effects in the analysis of δ2H and δ18O of water samples measured by laser spectroscopes. *Hydrol. Earth Syst. Sci.* **2012**, *16*, 3925–3933. [CrossRef]
24. Oksanen, J.; Blanchet, F.G.; Kindt, R.; Legendre, P.; O'hara, R.B.; Simpson, G.L.; Solymos, P.; Stevens, M.H.H.; Wagner, H. Vegan: Community Ecology Package. 2022 R Package Version 2.6-4. Available online: https://CRAN.R-project.org/package=vegan (accessed on 20 March 2024).
25. Dansgaard, W. Stable isotopes in precipitation. *Tellus* **1964**, *16*, 436–468. [CrossRef]
26. Zuecco, G.; Marchina, C. Hydrological Data from Ressi Experimental Catchment (Italy). 2021. Available online: https://osf.io/n24dg/ (accessed on 12 March 2024).
27. Marchina, C.; Zuecco, G.; Chiogna, G.; Bianchini, G.; Carturan, L.; Comiti, F.; Engel, M.; Natali, C.; Borga, M.; Penna, D. Alternative methods to determine the δ^2H-$\delta^{18}O$ relationship: An application to different water types. *J. Hydrol.* **2020**, *587*, 124951. [CrossRef]
28. Zuecco, G.; Marchina, C.; Gelmini, Y.; Amin, A.; van Meerveld, H.J.; Penna, D.; Borga, M. Ressi experimental catchment: Ecohydrological research in the Italian pre-Alps. *Hydrol. Process.* **2021**, *35*, e14095. [CrossRef]
29. Plachter, H. Cave-dwelling flies in Central Europe: Adaptation to environment, especially to low temperatures (Diptera, Nematocera: Trichoceridae et Sciaridae). *Oecologia* **1983**, *58*, 367–372. [CrossRef]
30. Lauga-Reyrel, F.; Lauga, J. Collembola of Cold Pyrenean habitats. *Eur. J. Soil Biol.* **1995**, *31*, 217–229.
31. Dahal, C.; Krzemińska, E.; Baez, M. Trichoceridae. In *Hjorth Andersen MCT (Coordinator) Catalogo de los Diptera de España, Portugal y Andorra (Insecta)*; Monografias SEA 8: Zaragoza, Spain, 2002; p. 82.
32. Mauries, J.P. Osellasoma, un remarquable nouveau genre de Diplopode cavernicole du Monte Baldo (Trentin) (Myriapoda: Diplopoda: Craspedosomida: Neoatractosomidae). Studi trentini di scienze naturali. *Acta Biol.* **1984**, *61*, 189–196.
33. Zacharda, M.; Isaia, M.; Piva, E. New troglobitic species of the genus Troglocheles (Acari: Prostigmata: Rhagidiidae) from caves in northern Italy and Austria, with a key to adult species of the genus. *J. Nat. Hist.* **2011**, *45*, 641–666. [CrossRef]
34. Humphreys, W.F. Relict fauna and their derivation. In *Ecosystems of the World*; Wilkens, H., Culver, D.C., Humphreys, W.F., Eds.; Subterranean Ecosystems; Elsevier: Amsterdam, The Netherlands, 2000; Volume 30, pp. 417–432.

35. Latella, L.; Verdari, N.; Gobbi, M. Distribution of terrestrial cave-dwelling arthropods in two adjacent Prealpine Italian areas with different glacial histories. *Zool. Stud.* **2012**, *51*, 1113–1121.
36. Lacelle, D.; Lauriol, B.; Clark, I.D. Formation of seasonal ice bodies and associated cryogenic carbonates in Caverne de l'Ours, Quebec, Canada: Kinetic isotope effects and pseudo-biogenic crystal structures. *J. Cave. Karst. Stud.* **2009**, *71*, 48–62.

diversity

MDPI

Article

Redescription of *Euscorpius studentium* Based on Adult Specimens; Updated Classification of Cavernicolous Euscorpiidae; and Review of Cavernicolous Scorpions in the Balkans

Javier Blasco-Aróstegui [1,2,*] and Lorenzo Prendini [2]

[1] Centre for Ecology, Evolution and Environmental Changes, Departamento de Biologia Animal, Faculdade de Ciências, Universidade de Lisboa, Campo Grande 016, 1749-016 Lisbon, Portugal

[2] Arachnology Lab and Scorpion Systematics Research Group, Division of Invertebrate Zoology, American Museum of Natural History, Central Park West at 79th Street, New York, NY 10024-5192, USA; lorenzo@amnh.org

* Correspondence: javierblasco77@gmail.com

Abstract: Cavernicolous scorpions are difficult to collect and study due to their often inaccessible habitats. Some have evolved unique morphological adaptations, known as troglomorphies, including reduced sclerotization and pigmentation, reduction and/or loss of eyes, attenuation and elongation of the appendages, which assist them to thrive in dark, humid and low-energy input environments. Cavernicolous scorpions are classified into accidentals, trogloxenes, troglophiles, and troglobites. The Balkans, and particularly the Dinaric Karst region, host a diverse cave-adapted fauna, including scorpions. Despite an 1895 report of a blind scorpion from Bosnia and Herzegovina, the first truly troglobitic European scorpion, *Euscorpius studentium* Karaman, 2020, was only described a few years ago, based on two immature specimens. In the present contribution, this unique species is redescribed based on the first adult specimens; the ecological classification of all currently known cavernicolous Euscorpiidae Laurie, 1896, is updated; a key to the identification of cavernicolous scorpions occurring in the Dinaric Karst is provided; and the historical and geographical factors affecting the distribution and conservation of cavernicolous scorpions in the Balkans is reviewed.

Keywords: Racovitzan impediment; subterranean environment; Arachnida; scorpions; ecological niches; taxonomy; Montenegro; climate change; habitat degradation

1. Introduction

Over millions of years, cavernicolous taxa adapted to the unique conditions of the subterranean environment, especially the absence of light, stable microclimate, and low energy input [1–3], by evolving specialized behavior, physiological responses, and ecomorphological adaptations known as troglomorphies. Among scorpions, such troglomorphies include anophthalmia (i.e., loss or reduction of median and lateral ocelli), reduction of sclerotization and pigmentation, and attenuation of the appendages [4–8].

Several classifications, building on the original Schiner-Racovitza System [9,10], have been proposed for cavernicolous and troglomorphic biota, depending on the phenotype, ecological niche and degree of specialization (e.g., [11–16]). Prendini et al. [8] reviewed these classifications as applied to scorpions and presented a key to classify cavernicolous and troglomorphic species according to their ecology and morphology. The key, which was based on Trajano and Carvalho's [15] redefinition of the three classical Schiner-Racovitza categories, trogloxenes, troglophiles and troglobites, and a fourth category, accidentals, differentiated between epigean, endogean, and hypogean scorpions, corresponding to those living on the surface, within the soil or leaf litter, and in subterranean environments, respectively. The differences between the redefined Schiner-Racovitza categories for scorpions, as outlined by Prendini et al. [8], can be summarized as follows: accidentals typically inhabit epigean habitats but occasionally enter caves either by mishap or in search of a milder climate; trogloxenes can thrive and establish stable populations in both epigean and hypogean habitats; troglophiles occur exclusively within hypogean habitats but exhibit few or no troglomorphic traits; troglobites or troglobionts are exclusively found within hypogean habitats and generally exhibit marked troglomorphies (i.e., loss or reduction of ocelli and pigmentation).

The Balkans, a group of nations in southeastern Europe, including Albania, Bosnia and Herzegovina, Bulgaria, Croatia, Greece, Kosovo, Montenegro, North Macedonia, Serbia, Slovenia and Romania, contain extensive cave systems, representing global hotspots of troglobitic fauna [17–23]. The complex and dynamic geoclimatic history of the Dinaric Karst region (Figures 1–3) in the western Balkans has resulted in unprecedented speciation through the colonization and isolation of different taxa, especially invertebrates [18,23–26]. This area acted as a refugium during the Miocene Climatic Transition and Plio-Pleistocene glacial cycles, sheltering relictual taxa, e.g., the descendants of hygrophilous and/or humicolous epigean fauna from the Tertiary [19,27,28] and promoted the diversification of younger, karst-adapted lineages, generating pronounced geographical structure in several groups. Subterranean radiations were fostered by the complex geotectonics, orogeny, fluvial networks, vast cave systems, and localized pits and cavities of the region [19,23,29].

Despite an early report of a blind scorpion from Bosnia and Herzegovina ("Les cavernes du sud de la Bosnie ... on y trouve ... et une petite espèce de Scorpion aveugle" (p. 23: [30]), no cavernicolous scorpions were described from the Dinarides until the last decade [31–36]. Like some other subterranean fauna, cavernicolous scorpions may occupy the *milieu souterrain superficiel* [37,38], i.e., not only the large chambers of caves but the network of narrow crevices and fissures interconnecting them. These restricted cavities are usually impossible to access, greatly limiting biospeleologists from obtaining comprehensive series of cavernicolous taxa for study [5,6,33]. The challenges involved in collecting cavernicolous taxa may create a false impression of rarity and provide a mere glimpse of their true diversity, a phenomenon referred to as the "Racovitzan impediment" [39].

The Dinaric Karst is now the only part of Europe, aside from Greece and Sardinia, known to harbor trogloxene, troglophile, and troglobite scorpions, and it contains the highest diversity among the three regions, with at least five species in two genera of the family Euscorpiidae Laurie, 1896, *Alpiscorpius* Gantenbein et al., 1999, and *Euscorpius* Thorell, 1876, occurring exclusively within caves (Table 1).

Euscorpius studentium Karaman, 2020, the first truly troglobitic European scorpion (Figure 4), displaying several marked troglomorphies (e.g., absence of median ocelli, reduction of lateral ocelli, and reduction of pigmentation), was described based on immature specimens from a cave in Montenegro (Figures 1–3). No further data were published about the morphology and biology of this species in the ensuing years. In the present contribution, this unique species is redescribed based on the first adult specimens; the ecological classification of all currently known cavernicolous Euscorpiidae is updated; a key to the identification of cavernicolous scorpions occurring in the Dinaric Karst is provided; and the historical and geographical factors affecting the distribution and conservation of cavernicolous scorpions in the Balkans is reviewed.

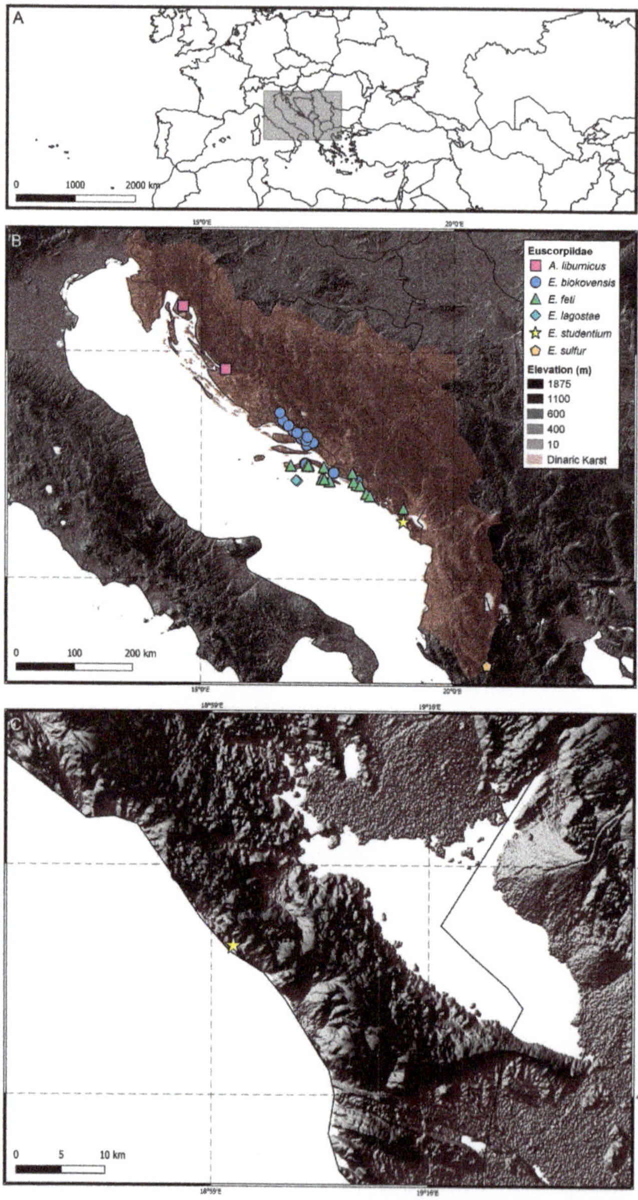

Figure 1. (**A**). Map of the Western Palearctic with international borders. Rectangle indicates area in (**B**). (**B**). Map illustrating extent of the Dinaric Karst from southern Slovenia across southern Croatia, Bosnia and Herzegovina and Montenegro, to Albania, and cavernicolous species of Euscorpiidae Laurie, 1896, occurring in the area. (**C**). Map indicating location of Spila Skožnica Cave, type locality of *Euscorpius studentium* Karaman, 2020, amid mountains surrounding Skadar Lake. Stars (**B**,**C**) indicate cave locality of *E. studentium*; squares, cave localities of *Alpiscorpius liburnicus* Tvrtković & Rebrina, 2022; circles, cave localities of *Euscorpius biokovensis* Tropea & Ozimec, 2020; triangles, cave localities of *Euscorpius feti* Tropea, 2013; diamonds, cave localities of *Euscorpius lagostae* Caporiacco, 1950; pentagon, cave locality of *Euscorpius sulfur* Kovařík et al., 2023.

Table 1. Genera and species of cavernicolous, troglobitic, and troglomorphic Euscorpiidae Laurie, 1896, with countries of occurrence, cave records, summary of troglomorphies, and ecological classification based on Prendini et al. [8].

***Alpiscorpius liburnicus* Tvrtković & Rebrina, 2022:** CROATIA. Habitat: inside cave. Troglomorphies: pigmentation and sclerotization reduced; pedipalps, legs, and metasoma attenuate; telson vesicle enlarged. Classification: hypogean: troglophile. Citations: [8,33–35,40–43].

***Alpiscorpius pavicevici* Tropea, 2021:** SERBIA. Habitat: inside cave; surface habitats, in leaf litter or under stones. Classification: endogean: accidental. Troglomorphies: none. Citations: [44].

***Euscorpius aquilejensis* (C.L. Koch, 1837):** CROATIA, ITALY, SLOVENIA. Habitat: inside cave; surface habitats, in rock crevices or under tree bark. Troglomorphies: pedipalps, legs, and metasoma attenuate. Classification: epigean, hypogean: accidental. Citations: [8,32,34–36,45–49].

***Euscorpius biokovensis* Tropea & Ozimec, 2020:** BOSNIA AND HERZEGOVINA, CROATIA. Habitat: inside caves. Troglomorphies: pigmentation and sclerotization reduced; pedipalps, legs, and metasoma attenuate; telson vesicle enlarged. Classification: hypogean: troglophile. Citations: [8,32–36,49].

***Euscorpius birulai* Fet et al., 2014:** GREECE. Habitat: inside cave. Troglomorphies: median ocelli reduced; pigmentation and sclerotization reduced; pedipalps, legs, and metasoma attenuate; telson vesicle enlarged. Classification: hypogean: troglophile. Citations: [8,32,33,36,48,50,51].

***Euscorpius canestrinii* (Fanzago, 1872):** ITALY. Habitat: inside cave; surface habitats, in rock crevices. Troglomorphies: none. Classification: epigean, hypogean: accidental. Citations: [52].

***Euscorpius concinnus* (C.L. Koch, 1837):** ITALY. Habitat: inside cave; surface habitats, in rock crevices, leaf litter, or under stones. Troglomorphies: none. Classification: epigean: accidental. Citations: [8,45,53–55].

***Euscorpius deltshevi* Fet et al., 2014:** BULGARIA, SERBIA. Habitat: inside cave; surface habitats, in rock crevices, leaf litter or under stones. Troglomorphies: none. Classification: epigean: accidental. Citations: [8,56,57].

***Euscorpius feti* Tropea, 2013:** BOSNIA AND HERZEGOVINA, CROATIA, MONTENEGRO. Habitat: both inside and outside cave. Troglomorphies: pigmentation and sclerotization slightly reduced; pedipalps, legs, and metasoma attenuate. Classification: epigean, hypogean: trogloxene. Citations: [8,31–36,47–49].

***Euscorpius giachinoi* Tropea & Fet, 2015:** GREECE. Habitat: both inside caves and outside. Troglomorphies: median ocelli reduced; pigmentation and sclerotization reduced; pedipalps, legs, and metasoma attenuate; telson vesicle enlarged. Classification: epigean, hypogean: trogloxene. Citations: [8,33,36,47,51].

***Euscorpius lagostae* Caporiacco, 1950:** CROATIA. Habitat: both inside and outside caves. Troglomorphies: pedipalps, legs, and metasoma slightly attenuate. Classification: epigean, hypogean: trogloxene. Citations: [34,58].

***Euscorpius sulfur* Kovařík et al., 2023:** ALBANIA, GREECE. Habitat: inside cave. Troglomorphies: pigmentation and sclerotization slightly reduced. Classification: hypogean: troglophile. Citations: [36].

***Euscorpius studentium* Karaman, 2020:** MONTENEGRO. Habitat: inside cave. Troglomorphies: median and lateral ocelli reduced; pigmentation and sclerotization reduced; pectinal teeth count reduced; pedipalps, legs, and metasoma attenuate; telson vesicle enlarged. Classification: hypogean: troglobite. Citations: [8,33–36].

Figure 2. *Euscorpius studentium* Karaman, 2020, habitat at type locality. (**A**). View of mountain range concealing Spila Skožnica Cave, as seen from Skadar Lake. (**B**). Entrance to Spila Skožnica Cave. (**C**). Twilight zone in Spila Skožnica Cave, with guano-coated soil and stones.

2. Materials and Methods

Specimens were collected inside the cave, by shining on the walls with ultraviolet (UV) lamps [59] and turning stones in loose and guano-coated soil. Adult specimens were preserved in 70% ethanol for morphological examination and a single juvenile in 95% ethanol for DNA isolation. Material is deposited in the collections of the American Museum of Natural History (AMNH), including the Ambrose Monell Cryocollection for Molecular and Microbial Research (AMCC), and the zoological collection of the Department of Biology and Ecology, University of Novi Sad, Serbia (ZZDBE). The newly collected material of *E. studentium* was compared morphologically with closely related species, using material deposited at the AMNH, as follows:

Euscorpius biokovensis Tropea & Ozimec, 2020: **CROATIA: Dalmatia Region:** *Splitsko-Dalmatinska County*: Bartulovići, Drinova II Cave, vicinity of Biokovo Nature Park, 43°24′28″ N 16°56′34.5″ E, 557 m, 11.viii.2022, J. Blasco-Aróstegui and P. Vicent, on walls and under stones inside cave, 1 ♂, 1 ♀(AMNH), 1 ♀(AMCC [LP 18574]); Župa, ca. 5.5 km NW on road 62 to Zagvozd, 43°20′51.6″ N 17°05′57.5″ E, 11.v.2023, A. Ullrich, 1 ♀(AMCC [LP 19477]); Biokovsko Selo, unnamed cave, 43°20′51.8″ N 17°05′58″ E, 780 m, 21.v.2024, J. Blasco-Aróstegui and H. Tahirović, inside cave, in crevices and under stones, 2 ♂, 2 ♀(AMNH), 2 ♀(AMCC [LP 20616, 20617]).

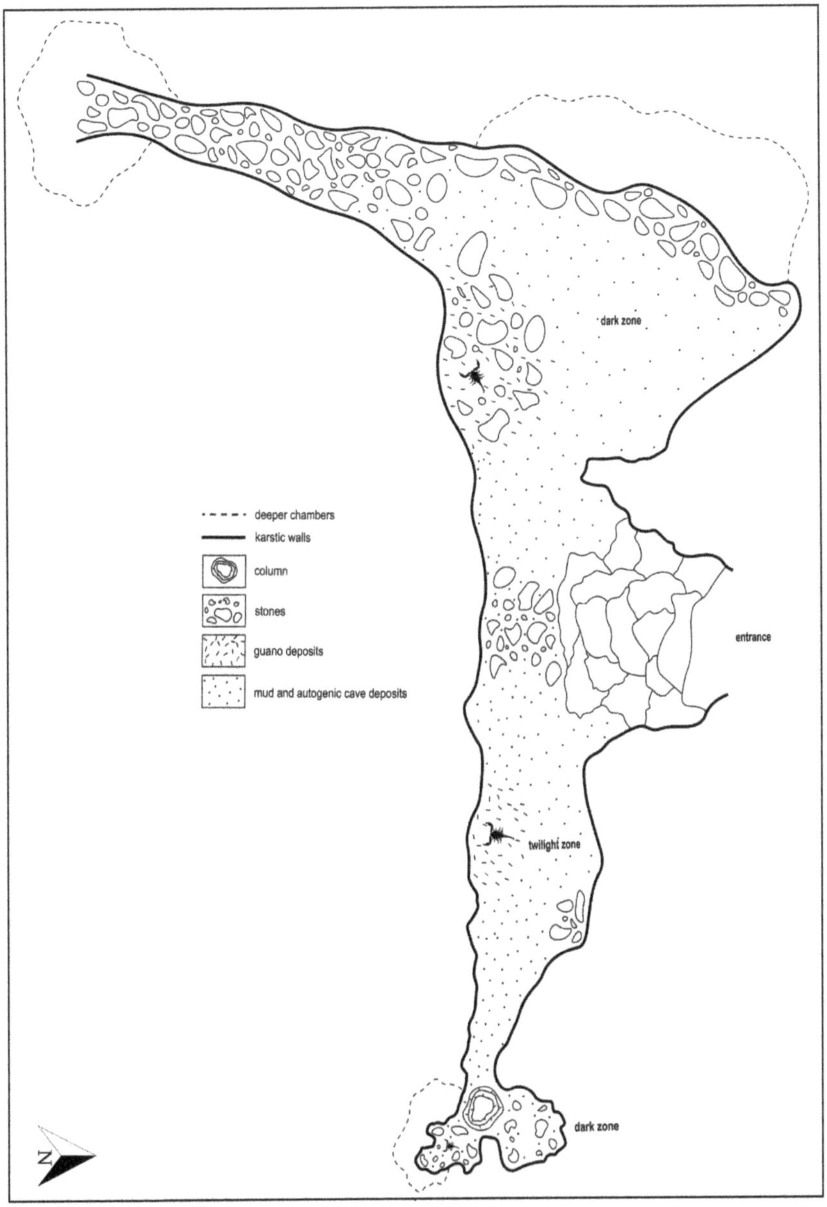

Figure 3. Schematic drawing of Spila Skožnica Cave layout, indicating zones in which scorpions were collected.

Euscorpius feti Tropea, 2013: **CROATIA: Dalmatia Region:** *Dubrovačko-Neretvanska County*: Pelješac Peninsula: Janjina, Gorska Jama Pit, 42°55′37.3″ N 17°25′29.7″ E, 243 m, 12.viii.2022, J. Blasco-Aróstegui and P. Vicent, in crevice of depression in steep rock, 1 ♀(AMCC [LP 18284]); Korčula, in Pišurka Cave, 42°57′34.3″ N 17°07′45.1″ E, 105 m, 12.viii.2022, J. Blasco-Aróstegui and P. Vicent, crevices at entrance, on floor and under stones in first chambers of cave, 5 ♂(AMNH), 3 ♀(AMCC [LP 18290, 18523, 18591]).

Figure 4. *Euscorpius studentium* Karaman, 2020, live habitus, dorsal aspect. (**A**). ♂(AMNH). (**B**). ♀(AMNH).

Live habitus photographs were taken inside the cave using a Google Pixel Pro 7 camera. Digital images of preserved adult specimens were taken in visible and long-wave UV light with a Microptics^{TM} ML-1000 digital photomicrography system at the AMNH. Morphological examination of specimens was conducted using a Nikon SMZ1500 stereoscope. Measurements (Tables 2 and 3) were taken following Stahnke [60] and Sissom et al. [61], using the ocular micrometer of a Nikon SMZ1500 stereoscope.

The cave location was georeferenced using a portable Garmin 64s GPS Navigation System. However, the exact coordinates have been withheld due to the sensitive nature of the site, which hosts a bat colony and a potentially endangered scorpion species. A map showing the approximate location and vicinity of the cave was created from a digital elevation layer available at DIVA-GIS using QGIS v.3.4. An approximate reconstruction of the cave's internal layout was prepared using Adobe Illustrator 2020.

The style of the species description follows previous works on *Euscorpius* systematics by the authors [49]. Morphological terminology follows Prendini et al. [62] for carapace topography and surface ornamentation; Loria and Prendini [63] for lateral ocelli; Vachon [64] and Prendini [65] for carinae and surfaces of pedipalps and legs, replacing "external" and "internal" with "retrolateral" and "prolateral", respectively; Prendini [65] for the patellar process, a spiniform apophysis on the prolateral surface of the pedipalp patella, teleologically referred to as the "patellar spur" in some literature; Vachon [66] for cheliceral dentition, replacing "external" and "internal" with "retrolateral" and "prolateral", respectively; Prendini and Loria [67] and Blasco-Aróstegui and Prendini [49] for lobes and notches on the pedipalp chela fingers; Vachon [64,68] for trichobothrial notation, with trichobothrial homology following Stockwell [69], in part; a modified version of Prendini [70] for tergal, sternal, and metasomal carinae; Volschenk and Prendini [5], Vignoli and Prendini [6], and Prendini et al. [7,8] for troglomorphies; Vachon [64] and Stahnke [60] for other characters.

Table 2. Measurements (mm) of adult male and female specimens of *Euscorpius studentium* Karaman, 2020, deposited in the American Museum of Natural History (AMNH), New York. Abbreviations: H, height; L, length; W, width. Notes: [1] sum of prosoma, mesosoma, and metasoma; [2] sum of tergites I–VII measured along midline; [3] sum of metasomal segments I–V and telson; [4] sum of pedipalp femur, patella, and chela; [5] measured from movable finger condyle to fingertip; [6] measured at median notch; [7] measured at median lobe.

		AMNH				AMNH	
		♂	♀			♂	♀
Total length [1]		41.9	40.28	Telson vesicle	L	4.5	3.62
Carapace	L	5.53	5.47		W	2.56	1.65
	Anterior W	3.87	3.63		H	2.22	1.93
	Posterior W	5.07	4.9	Aculeus	L	0.88	1.2
Mesosoma	L [2]	13.03	12.67	Pedipalp	L [4]	23.24	21.6
Metasoma	L [3]	23.34	22.14	Femur	L	5.55	5.26
Segment I	L	2.62	2.49		W	1.79	1.75
	W	1.96	1.64		H	1.12	1.1
	H	1.38	1.22	Patella	L	5.94	5.62
Segment II	L	3.08	2.96		W	1.9	1.85
	W	1.78	1.49		H	1.64	1.61
	H	1.37	1.29	Chela	L [5]	11.75	10.72
Segment III	L	3.22	3.12	Manus	L	6.7	6.18
	W	1.64	1.43		W	3.38	3.21
	H	1.38	1.25		H	2.67	2.49
Segment IV	L	3.82	3.46	Fixed finger	L	5.05	4.54
	W	1.57	1.34		W [6]	0.61	0.65
	H	1.38	1.25		H [6]	0.94	0.89
Segment V	L	6.1	5.29	Movable finger	L	5.93	5.48
	W	1.51	1.26		W [7]	0.58	0.58
	H	1.45	1.39		H [7]	1.1	0.81
Telson	L	5.38	4.82	Pectines	L	2.45	1.84

Table 3. Meristic data for known specimens of *Euscorpius studentium* Karaman, 2020, deposited in the University of Novi Sad (ZZDBE), Serbia, and the American Museum of Natural History (AMNH), New York. Counts (sinistral/dextral) of median denticle subrows, prolateral accessory denticles (PAD) and retrolateral accessory denticles (RAD) on fixed and movable fingers of pedipalp chela; retrolateral and ventral trichobothria on pedipalp chela and patella; ventromedian spinules (VMS) on telotarsi of legs I–IV; and pectinal teeth; - indicates missing data.

			♂ AMNH	♀ AMNH	Holotype subad. ♂ ZZDBE SC1/01	juv. ♂ AMNH	Paratype juv. ♀ ZZDBE SC2/02
Pedipalp chela	Fixed finger subrows		7/7	7/7	7/7	7/7	-/-
	Fixed finger PAD		11/11	11/11	-/11	11/11	-/-
	Fixed finger RAD		7/7	7/7	7/7	7/7	-/-
	Movable finger subrows		7/7	7/7	7/7	7/7	-/-
	Movable finger PAD		11/11	11/11	-/11	11/11	-/-
	Movable finger RAD		7/7	7/7	-//7	7/7	-/-
	External trichobothria	*Et*	5/5	5/5	5/5	5/5	5/5
		Est	1/1	1/1	1/1	1/1	1/1
		Esb	1/1	1/1	1/1	1/1	1/1
		Eb	3/3	3/3	3/3	3/3	3/3
		et	1/1	1/1	1/1	1/1	1/1
		est	1/1	1/1	1/1	1/1	1/1
		esb	1/1	1/1	1/1	1/1	1/1
		eb	1/1	1/1	1/1	1/1	1/1
	Ventral trichobothria	*V*	4/4	4/4	4/4	4/4	4/4
Pedipalp patella	External trichobothria	*et*	6/6	6/6	6/6	6/6	6/6
		est	4/4	4/4	4/4	4/4	4/4
		em	4/4	4/4	4/4	4/4	4/4
		esb	2/2	2/2	2/2	2/2	2/2
		esb_a	4/4	4/4	4/4	4/4	4/4
		eb	4/4	4/4	4/4	4/4	4/4
	Ventral trichobothria	*v*	7/7	7/7	7/7	7/7	7/7
Telotarsal VMS	Leg I		7/7	6/6	-/-	-/-	-/-
	Leg II		7/6	7/7	-/-	-/-	-/-
	Leg III		8/7	5/7	-/7	-/-	-/-
	Leg IV		9/9	-/8	-/-	-/-	-/-
Pectines	Teeth		7/7	6/6	7/7	7/7	7/7

3. Key to Identification of Cavernicolous Scorpions in the Dinaric Karst

1. Pedipalp patella retrolateral surface with three trichobothria in *em* series .... *Alpiscorpius liburnicus*

- Pedipalp patella retrolateral surface with four trichobothria in *em* series 2

2. Median ocelli absent, lateral ocelli reduced .... *Euscorpius studentium*

- Median ocelli present, lateral ocelli well-developed 3

3. Pedipalp patella retrolateral surface with four trichobothria in *eb* series and four trichobothria in eb_a series 4

- Pedipalp patella retrolateral surface with five trichobothria in *eb* series and 6–8 trichobothria in eb_a series .... *Euscorpius lagostae*

4. Pedipalp patella retrolateral surface with six trichobothria in *et* series; ventral surface with 7–9 trichobothria in *v* series 5

- Pedipalp patella retrolateral surface with more than six trichobothria in *et* series; ventral surface with 11–12 trichobothria in *v* series .... *Euscorpius feti*

5. Metasomal segment V, ventrolateral and ventromedian carinae obsolete, intercarinal surfaces smooth .... *Euscorpius biokovensis*

- Metasomal segment V, ventrolateral and ventromedian carinae distinct, intercarinal surfaces finely granular .... *Euscorpius sulfur*

4. Systematics

Family Euscorpiidae Laurie, 1896
Genus *Euscorpius* Thorell, 1876

Euscorpius studentium Karaman, 2020
Figures 1–12; Tables 1–3
Euscorpius studentium Karaman, 2020: 1–18, Figures 1–7, 11A and 15 [33].

Type Material. MONTENEGRO: Coastal Region: *Bar Municipality*: Čanj, Spila Skožnica Cave [42°10′ N 19°01′ E], 7.v.2017, I. Karaman, 1 subad. ♂holotype (ZZDBE SC1/01), 2.vi.2017, I. Karaman, 1 juv. ♀paratype (ZZDBE SC1/02). According to I. Karaman, the paratype was donated to G. Tropea (Rome, Italy).

Diagnosis. *Euscorpius studentium* can be separated from all other European species of Euscorpiidae, in which the ocelli are present and fully developed, by the absence of median ocelli and reduction of lateral ocelli (Figure 6). Additionally, the second-most proximal denticle in the first median denticle subrow of the movable finger is larger than the other denticles in *E. studentium* (Figures 8B and 9B), unlike other European species of Euscorpiidae, in which these denticles are similar in size.

Euscorpius studentium most closely resembles *E. biokovensis* and *E. feti*, two troglomorphic species inhabiting humid caves in the Dinaric Karst of Bosnia and Herzegovina, Croatia, and Montenegro, which also exhibit attenuation and elongation of the pedipalps and marked dorsoventral compression (Figures 4–12). However, in addition to the absence of median ocelli and reduction of lateral ocelli (Figure 6), *E. studentium* can be separated from these species by the following characters. Although the base coloration is similar (dark or reddish-brown to lighter brown, with carapace and pedipalps darker than mesosoma and metasoma) in all three species, *E. studentium* is notably paler (Figure 4). The pedipalp patella is also more elongate and slender in *E. studentium* (width 32.4% of length; Table 2) than in *E. biokovensis* (width 37.3% of length) and *E. feti* (width 37.1% of length) with the patellar process notably more developed, especially in the male, in *studentium* (height 135.1% of width; Figure 7) than in *E. biokovensis* (height 116.3% of width). and *E. feti* (height 113.8% of width) [31,32,48]. Whereas *E. studentium* possesses six trichobothria in the *et* series of the pedipalp patellar retrolateral surface and seven trichobothria in the *v* series of the patellar ventral surface (Figure 7), *E. biokovensis* possesses six and 7–9 trichobothria in these series, respectively, and *E. feti*, seven or eight and 11–12 trichobothria in these series, respectively [31,32,48]. Whereas the pedipalp chela is slender and compressed in all three species, the manus is more elongate in *E. studentium* (manus width 29.4% of chela length; Table 2; Figures 8 and 9) than in *E. biokovensis* (manus width 31.1% of chela length) and *E. feti* (manus width 33.6% of chela length). Unlike *E. studentium*, however, the chela fingers are more elongate and attenuate in the male of *E. biokovensis* and especially *E. feti*, which exhibits a more pronounced gap in the fixed finger and a wider median lobe on the movable finger. The pectinal tooth counts of *E. biokovensis* and *E. studentium* are generally lower, 7/7 (♂) or 6/6 (♀) (Table 3; Figure 6), than those of *E. feti*, 8–9/8–9 (♂) or 7–8/7–8 (♀) [31,32,48]. The legs, especially the terminal segments (i.e., basitarsi and telotarsi), exhibit similar attenuation in all three species, but this is more pronounced in *E. studentium* (Figure 10). The telotarsal ungues of *E. feti* and *E. studentium* (Figure 10) are more elongated and curved than those of *E. biokovensis*, which are slightly shorter and more robust. The ventrolateral and ventromedian carinae of metasomal segment V are distinct in *E. feti*, obsolete in *E. biokovensis*, and absent in *E. studentium* (Figure 11). The telson vesicle of the adult male is less globose in *E. biokovensis* and *E. studentium* (Figure 12A,B) than in *E. feti*.

Description. The following description is based on an adult male and an adult female (Figure 5), the first adults reported for the species (see Tables 2 and 3 for measurements and meristic data).

Total length: Medium-sized, 41.9 mm (♂; n = 1) or 40.3 mm (♀; n = 1) (Table 2).

Coloration: Uniformly depigmented (Figure 4). Carapace reddish-brown, darker in female, with posterior and posterolateral margins lighter. Pedipalps medium brown, slightly darker in female. Pedipalp carinae dark, blackish. Legs pale brown, darker in female. Tergites light brown, with posterior and lateral margins pale. Telson vesicle pale, with three slightly darker stripes longitudinally, one pair along ventrolateral sulci and one between anterodorsal sulci, more noticeable in female; aculeus black.

Chelicerae: Manus dorsal and retrolateral surfaces entirely smooth, with few scattered setae; prolateral and, to a lesser extent, ventral surfaces densely setose (Figure 6A,B). Fixed finger dorsal margin with four teeth: distal, subdistal, median and basal; distal and subdistal teeth well separated, median and basal fused into a bicuspid. Movable finger dorsal margin with five teeth: one prodistal, two subdistal, one median, and one basal; prodistal (DI) and retrodistal (DE) teeth asymmetric, DI twice length of DE; subdistal teeth small, separated; median teeth large; basal teeth small.

Carapace: Carapace anterior width of posterior width, 76.3% (♂; n = 1) or 74.1% (♀; n = 1); posterior width of length, 91.7% (♂; n = 1) or 89.6% (♀; n = 1) (Table 2). Carapace anterior margin sublinear, with obsolete anteromedian notch; anteromedian sulcus distinct, narrow, shallow; posteromedian sulcus shallow, more pronounced than anteromedian sulcus; posterolateral sulci shallow. Median ocelli absent; two pairs of small lateral ocelli, posterolateral major (PLMa) and median lateral major (MLMa), similar in size. Median ocular tubercle and superciliary carinae absent (Figure 6A,B). Anteromedian and posteromedian surfaces smooth; lateral surfaces very finely and sparsely granular; margins costate-granular. Carapace surfaces almost asetose; anterior margin with three or four microsetae; location of median ocular tubercle demarcated by two microsetae.

Sternum: Shape subpentagonal (Figure 6C,D) with marked posterior depression. Surface with four macrosetae and few additional microsetae.

Pedipalps: Femur width of length, 32.3% (♂; n = 1) or 33.3% (♀; n = 1) (Table 2). Prodorsal and proventral carinae complete, costate-granular, each with several large spiniform and subspiniform granules (Figure 7A,B). Retrodorsal carina complete, granular, discontinuous. Promedian and retroventral carinae complete, costate-granular, each with discontinuous spiniform granules and few macro- and microsetae. Ventromedian carinae partial, granular, becoming obsolete distally. Other carinae absent. Dorsal and ventral intercarinal surfaces finely granular in male, almost smooth in female. Prolateral and retrolateral intercarinal surfaces smooth or nearly so.

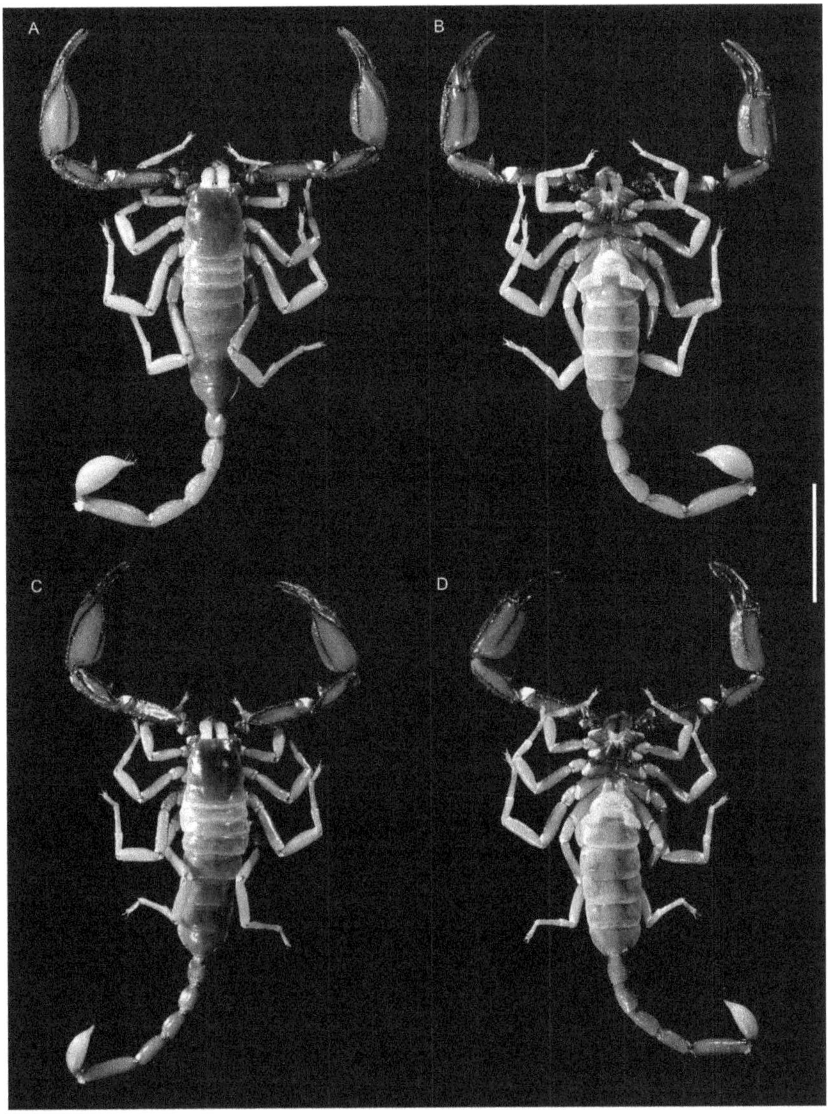

Figure 5. *Euscorpius studentium* Karaman, 2020, habitus, dorsal (**A**,**C**) and ventral (**B**,**D**) aspects. (**A**,**B**). ♂(AMNH). (**C**,**D**). ♀(AMNH). Scale bar = 10 mm.

Patella elongate, width of length, 32% (♂; $n = 1$) or 32.9% (♀; $n = 1$) (Table 2). Proventral carina complete, granular to costate-granular (Figure 7G,H). Prodorsal carina reduced to series of spiniform granules (Figure 7C,D). Retrodorsal and retroventral carinae complete, costate-granular (♂; Figure 7C,E,G) or granular (♀; Figure 7D,F,H), except at proximal and distal margins of segment. Retromedian carina partial, reduced to discontinuous row of granules (Figure 7E,F). Other carinae absent. Dorsal and ventral intercarinal surfaces finely granular to smooth (Figure 7C,D,G,H). Prolateral surface with distinct, elongate patellar process, more spiniform in male (Figure 7C,G) than in female (Figure 7D,H); surface of process smooth, with few microsetae and two macrosetae, one situated basally, the other apically (Figure 7C,D,G,H). Retrolateral intercarinal surface smooth (Figure 7E,F).

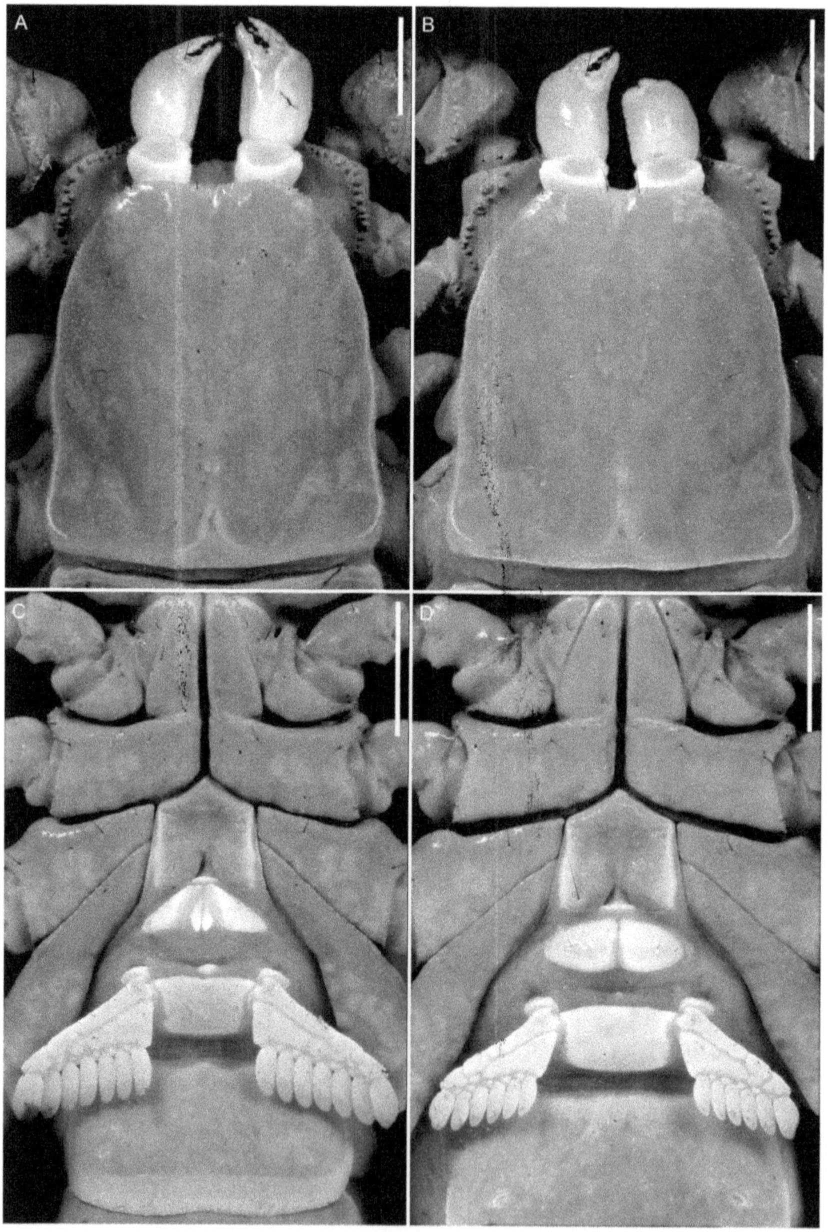

Figure 6. *Euscorpius studentium* Karaman, 2020, carapace, dorsal aspect (**A**,**B**) and sternum and pectines, ventral aspect (**C**,**D**). (**A**,**C**). ♂(AMNH). (**B**,**D**). ♀(AMNH). Scale bars = 2 mm.

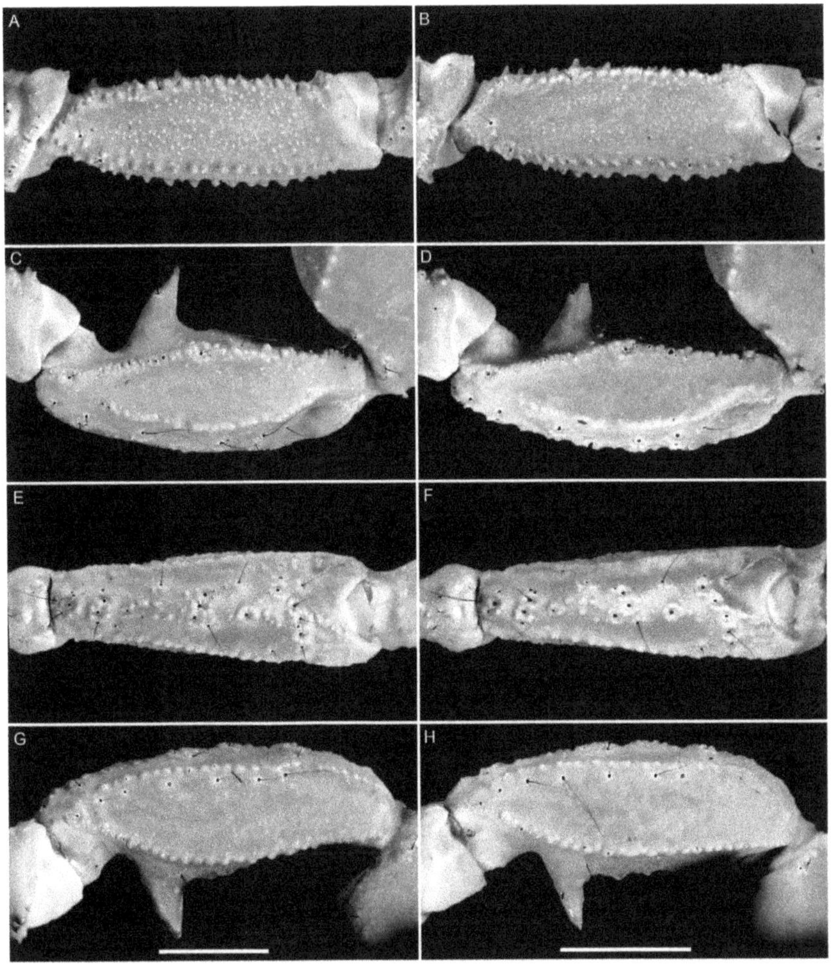

Figure 7. *Euscorpius studentium* Karaman, 2020, dextral pedipalp femur, dorsal aspect (**A**,**B**), and patella, dorsal (**C**,**D**), retrolateral (**E**,**F**), and ventral (**G**,**H**) aspects. (**A**,**C**,**E**,**G**). ♂(AMNH). (**B**,**D**,**F**,**H**). ♀(AMNH). Scale bars = 2 mm.

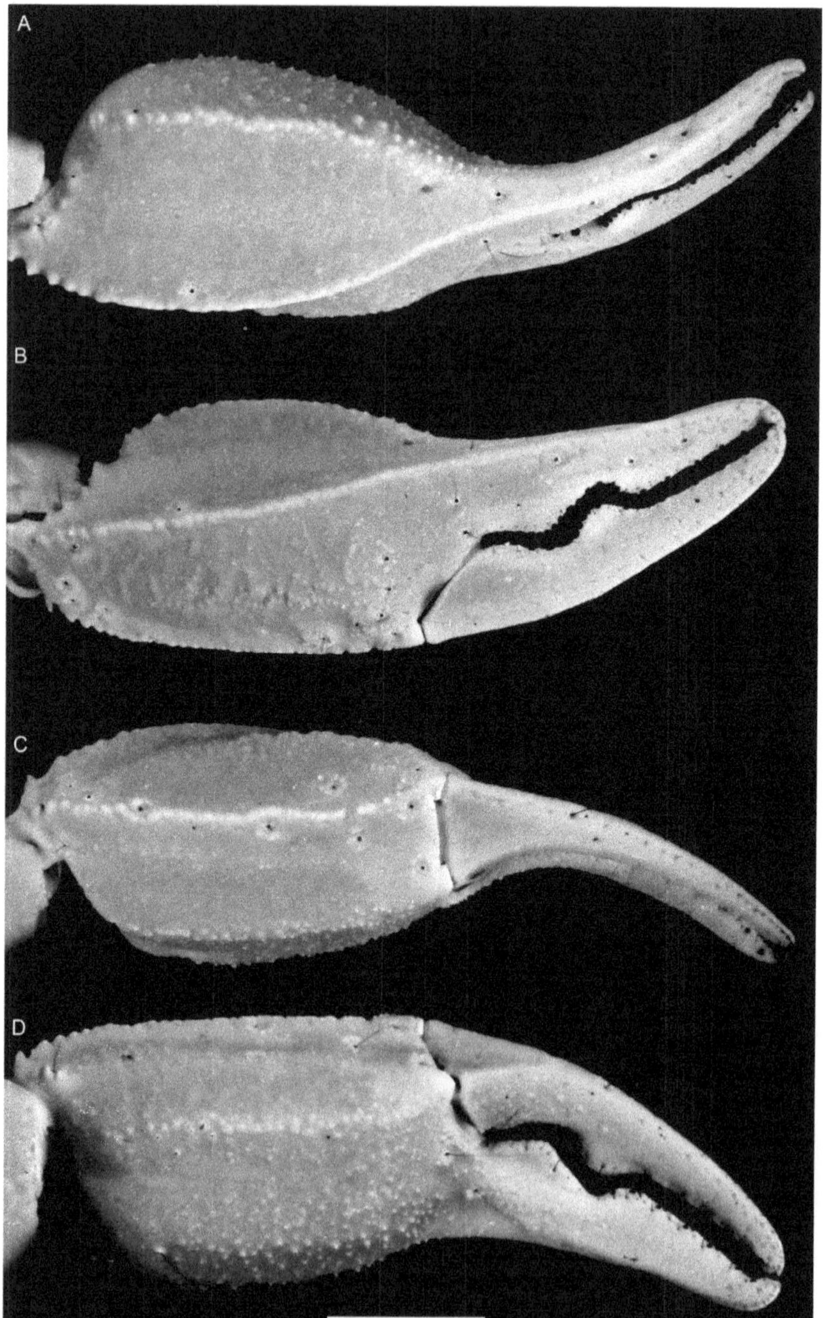

Figure 8. *Euscorpius studentium* Karaman, 2020, ♂(AMNH), dextral pedipalp chela, dorsal (**A**), retrolateral (**B**), ventral (**C**), and prolateral (**D**) aspects. Scale bar = 2 mm.

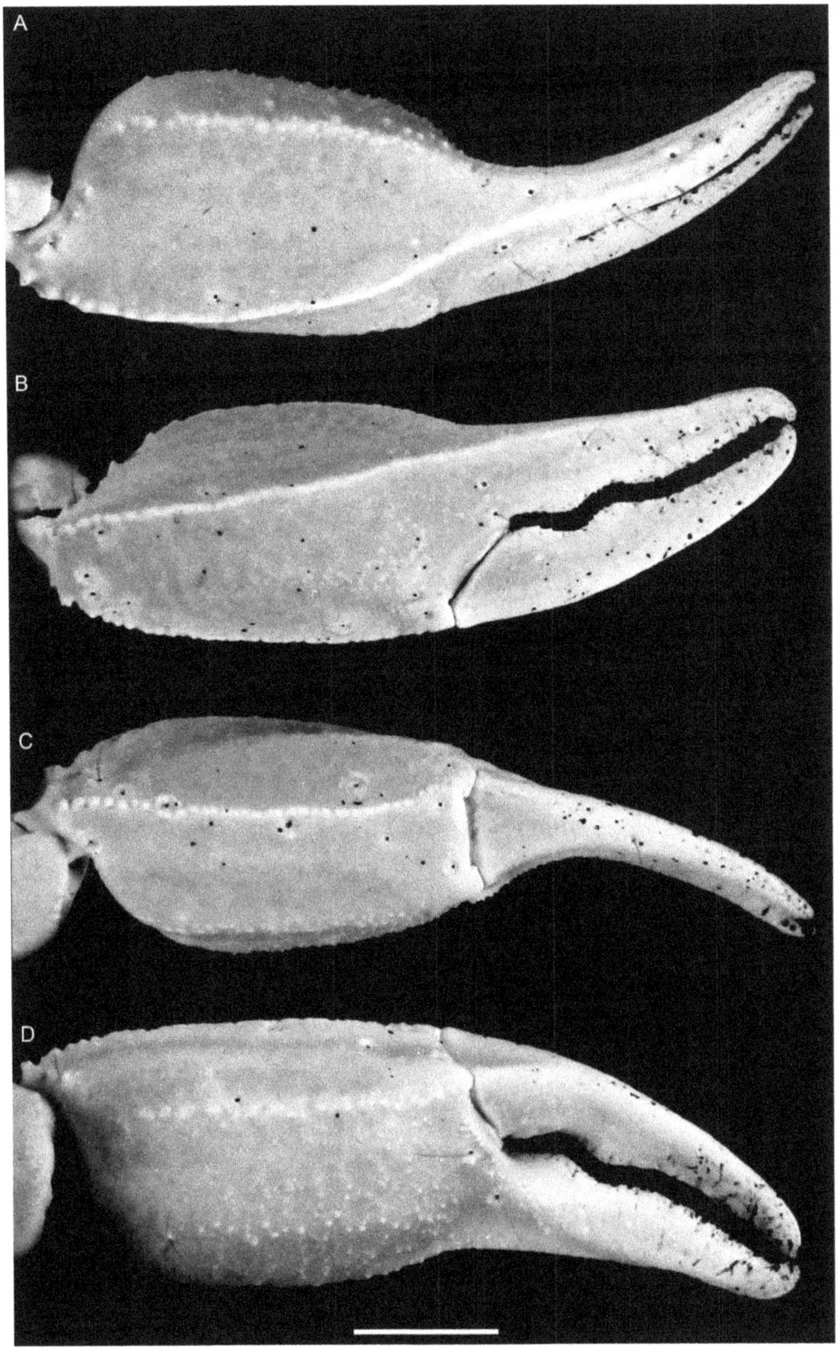

Figure 9. *Euscorpius studentium* Karaman, 2020, ♀(AMNH), dextral pedipalp chela, dorsal (**A**), retrolateral (**B**), ventral (**C**), and prolateral (**D**) aspects. Scale bar = 2 mm.

Chela slender, with elongate manus, fingers curved and narrower in male (Figure 8) than female (Figure 9); chela length of manus width, 28.8% (♂; n = 1) or 29.9% (♀; n = 1); manus width of length, 50.4% (♂; n = 1) or 51.9% (♀; n = 1); manus height of width, 79% (♂; n = 1) or 77.6% (♀; n = 1); movable finger length of manus length, 89% (♂; n = 1) or 88.7% (♀; n = 1); fixed finger width of length, 12.7% (♂; n = 1) and 14.3% (♀; n = 1); movable finger width of length, 9.8%, (♂; n = 1) or 10.6%, (♀; n = 1). Chela dorsal surface flat (Figures 8A,B and 9A,B), sloping slightly from proximal to distal margin, more so in male. Prolateral surface convex proximally, moderately (♀; Figure 9A,D) to markedly (♂; Figure 8A,D) concave distally (proximal to fixed finger). Retrolateral surface slightly (♀; Figure 9B) to markedly (♂; Figure 8B) convex medially. Ventral surface flat, sloping slightly from proximal to distal margin in female; proximal margin curved (Figures 8C and 9C). Proventral carina granular, incomplete, absent in proximal part of segment but notably more developed than ventromedian carina, which is obsolete (Figures 8C,D and 9C,D). Promedian carina obsolete, finely granular, extending to base of fixed finger (Figures 8D and 9D). Prodorsal carina obsolete, granular, comprising discontinuous row of granules in distal part of segment (Figures 8A,B and 9A,B). Dorsomedian carina costate-granular, partial, reduced to discontinuous row of subspiniform granules becoming obsolete in distal part of segment (Figures 8A,B and 9A,B). Dorsal secondary carinae vestigial. Digital carina costate-granular, reduced to discontinuous row of subspiniform granules distally (Figures 8A,B and 9A,B). Subdigital carina vestigial, reduced to few granules at proximal margin of segment (Figures 8A,B and 9A,B). Retromedian carina incomplete, comprising obsolete row of granules decreasing in size distally (Figures 8B,C and 9B,C). Retrolateral secondary and secondary accessory carinae vestigial, reduced to distal quarter of segment, proximal to movable finger condyle (Figures 8B,C and 9B,C). Retroventral carina complete, costate-granular, comprising subspiniform granules in proximal part of segment segment (Figures 8B–D and 9B–D). Ventromedian carina vestigial, reduced to fine granules proximally (Figures 8C,D and 9C,D). Dorsal and ventral intercarinal surfaces sparsely setose and smooth to glabrous. Prolateral and retrolateral intercarinal surfaces sparsely setose and finely granulo-reticulate (Figures 8 and 9). Fixed and movable fingers with pronounced proximal and medial lobes, respectively, and correspondingly deep proximal and medial notches in male (Figure 8B,D); proximal and medial lobes wider than high, but medial lobe higher and medial notch deeper; medial lobe fits evenly with median notch, leaving little to no gap between fingers, when closed. Proximal and medial lobes, and corresponding notches, of fixed and movable fingers shallower in female (Figure 9B,D). Median denticle rows of fixed and movable fingers each comprising seven subrows (Table 3) forming sublinear row, discontinuous at accessory denticles (Figures 8 and 9); second-most proximal denticle of first subrow on movable finger, distal to condyle, noticeably larger than other denticles (Figures 8B and 9B); fixed and movable fingers each with eleven prolateral and seven retrolateral accessory denticles (n = 4; Table 3) and single terminal denticle, interlocking unevenly, such that movable finger moderately displaced retrolaterally when closed (Figures 8 and 9). Intercarinal surfaces of fingers smooth, fairly setose (Figures 8 and 9).

Femur with three full-sized trichobothria (Figure 7A,B), two on dorsal surface (d_1, d_2), one on prolateral surface (i). Patella with 34 trichobothria, one petite (esb_2), 33 full sized (Table 3; Figure 7C–H): seven on ventral surface (v_1–v_7); 24 on retrolateral surface (et_1–et_6, est_1–est_4, em_1–em_4, esb_1, esb_2, esb_{a1}– esb_{a4}, eb_1–eb_4); two on dorsal surface (d_1, d_2); one on prolateral surface (i). Chela with 26 trichobothria, two petite (Et_4, Esb), 24 full sized (Table 3; Figures 8 and 9): eighteen on manus, four on ventral surface (V_1–V_4), ten on retrolateral surface (Et_1–Et_5, Est, Esb, Eb_1–Eb_3,), two on dorsal surface (Dt, Db), two on prolateral surface (it, ib); eight on fixed finger, four on dorsal surface (dt, dst, dsb, db), four on retrolateral surface (et, est, esb, eb).

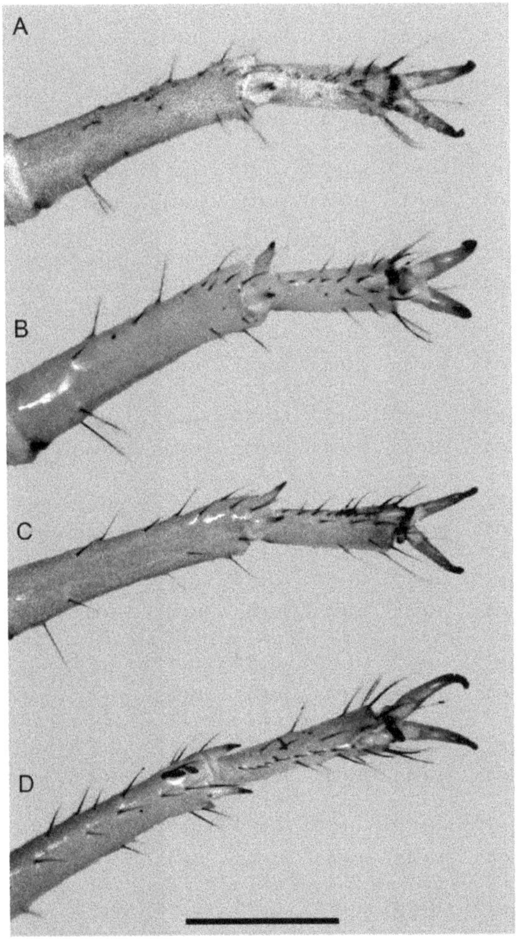

Figure 10. *Euscorpius studentium* Karaman, 2020, ♂(AMNH), dextral legs I–IV, basitarsi and telotarsi, proventral aspect (**A–D**). Scale bar = 1 mm.

Legs: Femora with costate-granular proventral carina complete on legs I–III, partial, restricted to proximal three-quarters on IV, and granular retrodorsal carina partial, restricted to proximal three-quarters, on I and IV, complete on legs II and III, other carinae absent; prolateral and retrolateral surfaces finely granular; few macrosetae on carinae and other surfaces. Patellae with costate-granular prodorsal carina on legs I and II, acarinate on III and IV, other carinae absent; few macrosetae on carina and other surfaces; pro- and retrolateral surfaces finely granular to smooth. Tibiae acarinate, with few setiform macrosetae on prolateral and retrolateral surfaces; tibial spurs absent. Basitarsi acarinate; prolateral and retrolateral surfaces sparsely setose; prolateral and retrolateral pedal spurs present distally. Telotarsi attenuate, each with single ventromedian row of six or seven elongate spinules on legs I and II, seven or eight on III, and eight or nine on IV (n = 2; Table 3); ungues elongate, narrow, curved, and equal in length; dactyl pronounced (Figure 10).

Genital operculum: Opercula suboval, sclerites completely separated longitudinally (Figure 6C,D). Genital papillae well developed, protruding medially from below opercula (♂; Figure 6C) or absent (♀; Figure 6D).

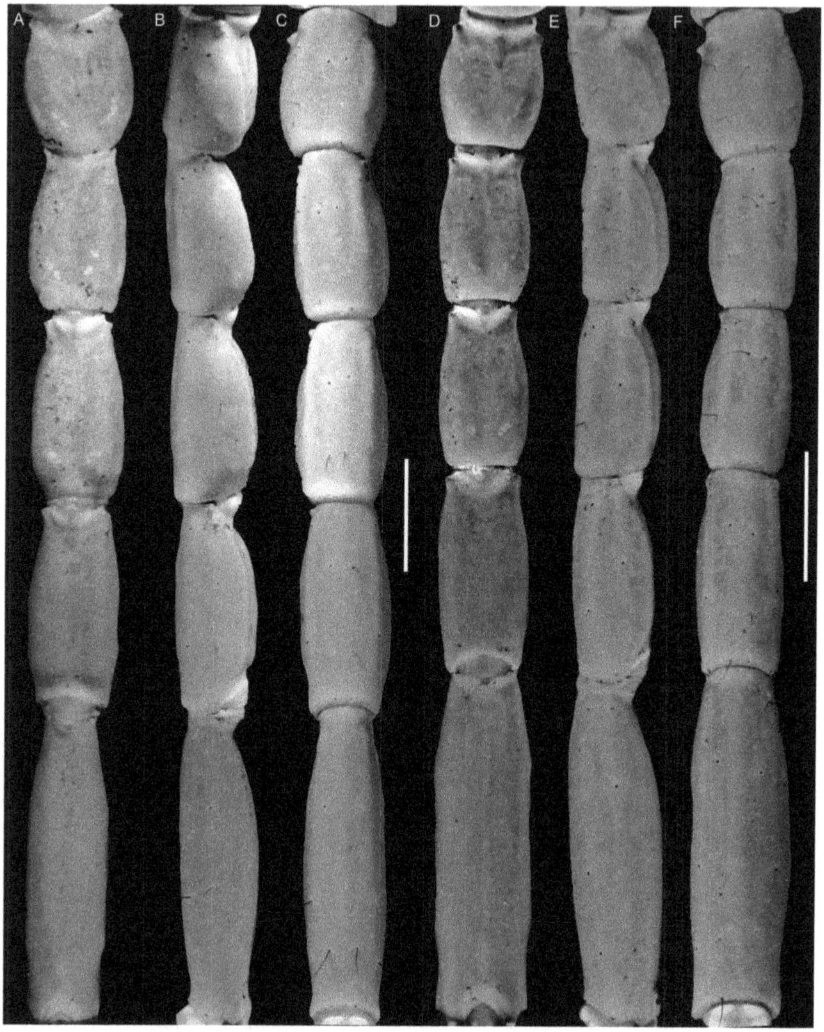

Figure 11. *Euscorpius studentium* Karaman, 2020 (AMNH), metasomal segments I–V, dorsal (**A**,**D**), lateral (**B**,**E**), and ventral (**C**,**F**) aspects. (**A**–**C**). ♂(AMNH). (**D**–**F**). ♀(AMNH). Scale bars = 2 mm.

Pectines: Basal plate rectangular, slightly oval; shallow anteromedian invagination in male, absent in female (Figure 6C,D). Three short marginal lamellae, proximal sclerite considerably longer than medial and distal sclerites; median lamellae count, 4/4. Pectinal tooth count, 7/7 (♂; n = 3) or 6–7/6–7 (♀; n = 2) (Table 3). Surfaces with macro- and microsetae, mostly on marginal lamellae but also on teeth.

Tergites: Pre-tergites and post-tergites progressively increasing in length. Pre-tergites glabrous; post-tergites III–VII surfaces acarinate and smooth medially, finely granular submedially and along posterior margins.

Sternites: Sternites III–VII surfaces acarinate, entirely smooth, and sparsely setose medially, each with several microsetae along lateral and posterior margins; III–VI each with pair of kidney-shaped spiracles sublaterally.

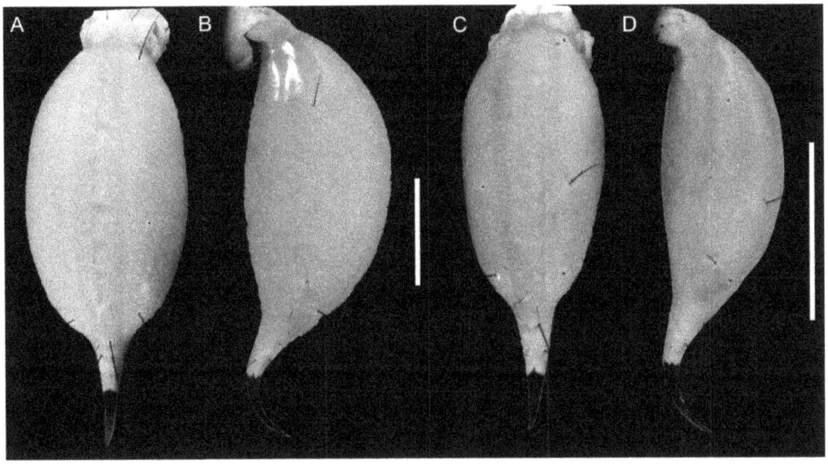

Figure 12. *Euscorpius studentium* Karaman, 2020, telson, ventral (**A**,**C**) and lateral (**B**,**D**) aspects. (**A**,**B**). ♂(AMNH). (**C**,**D**). ♀(AMNH). Scale bars = 2 mm.

Metasoma: Segments I–V narrow and slender, progressively increasing in length and decreasing in width (Table 2). All segments acarinate, except for few obsolete granules posteriorly on segments I–III (Figure 11). Anal arch of segment V with ten obsolete ventral lobes, two pairs of lateral lobes and rounded dorsal depression (Figure 11). Dorsal, lateral and ventral surfaces smooth on segments I–IV (Figure 11), almost completely smooth on V, except dorsolateral surfaces, finely granular (Figure 11B,E). Macrosetae arranged in groups of eight or nine per segment, situated ventrolaterally on segments I–IV, with few additional setae on V (Figure 11B,C,E,F).

Telson: Vesicle globose in male, elongate and slightly flattened ventrodistally in female (Figure 12); with anterodorsal and paired ventrolateral sulci; surfaces smooth with scattered macro- and microsetae mostly anterior to aculeus. Aculeus short, curved, forming obtuse angle with vesicle, angle broader in female (Figure 12); base of aculeus aligned diagonally to ventral surface of vesicle, more prominent in male (Figure 12).

DISTRIBUTION: Only known from Spila Skožnica Cave near Čanj, in the Coastal Region of Montenegro. An unidentified blind scorpion from a cave in Bosnia and Herzegovina (p. 23: [30]) may be closely related.

ECOLOGY: *Euscorpius studentium* is an obligate troglobite, restricted to a single cave in the Dinaric Karst (Figures 1–4). It displays pronounced troglomorphies, including absence of the median ocelli and reduction of the lateral ocelli, reduction in sclerotization and pigmentation, and attenuation of the pedipalps and legs (Figures 4–10 and 12). *Euscorpius biokovensis*, a closely related species with a wider distribution in the Dinaric Karst, has also only been found inside caves, but presents moderate to weak troglomorphies and is considered a troglophile (Table 1). *Euscorpius feti*, another species found both inside and outside caves in the Dinaric Karst, exhibits moderate to weak troglomorphies and is considered a trogloxene (Table 1).

The three specimens of *E. studentium* reported in the present contribution were collected with UV light detection, during daytime. The adult male of *E. studentium* was found walking on a humid limestone wall, in the twilight zone of the cave, near moist muddy soil covered with guano (Figure 3). The adult female and juvenile were taken from under stones situated directly beneath bat roosts, where the soil was moist and covered with guano, in the dark zones of the cave (Figure 3). Despite extensive searches, including the initial expeditions of I. Karaman and a subsequent visit to the cave in August 2022 by the first author, no additional specimens were collected (although exuviae were observed), suggesting that these scorpions are seasonally active and require high humidity. All specimens were collected in late spring, from May to early June, coinciding with the presence of bats and fresh guano deposits in the cave. The apparently greater activity of these scorpions in the upper chambers of the cave is probably associated with an increased energy input to the cave, resulting in a higher abundance of guanivores and other invertebrates, as well as higher levels of humidity during the rainy season.

An unidentified species of *Euscorpius* was collected in crevices near the cave entrance and surrounding forest in the foothills of the mountain.

CONSERVATION STATUS: *Euscorpius studentium* inhabits a small, humid cave in a low-elevation hill, known as Golo Brdo, on the coast of Montenegro (Figures 1–3). The cave is situated on a slope, surrounded by karren bedrock with patches of vegetation [33]. It houses a medium-sized colony of bats that become active in spring and depart the cave in summer, suggesting it serves as a winter roost. The cave interior, as well as the sub-Mediterranean forest on the foothills of the mountain in which it is located, are relatively undisturbed. However, the proximity of the cave to the tourist settlement of Čanj poses a potential threat to the cave and its occupants.

Euscorpius studentium is a highly stenotopic species, presently known from only three juveniles and two adults, collected in a single cave. Due to its very limited extent of occurrence, specific habitat requirements, and possibly low abundance, *E. studentium* may be vulnerable to anthropogenic environmental impacts, including climate change and habitat destruction or degradation. As this species is potentially at risk of extinction in an uncertain future, it would be advisable to list it as Endangered on the IUCN Red List of Threatened Species (https://www.iucnredlist.org, accessed on 27 November 2024). As a further recommendation to ensure the preservation of the scorpion and the bat colony, the cave should be designated a National Protected Area by the Environmental Protection Agency of Montenegro (https://epa.org.me, accessed on 27 November 2024).

MATERIAL EXAMINED: **MONTENEGRO: Coastal Region:** *Bar Municipality*: Čanj, Spila Skožnica Cave, 42°10′ N 19°01′ E, 149 m, 22.v.2024, J. Blasco-Aróstegui and H. Tahirović, inside karstic cave, in humid and guano covered area, walking along karstic wall and under stones, 1 ♂, 1 ♀(AMNH), 1 juv. ♂(AMCC [LP 20621]).

5. Discussion

The evolution of troglobitism has long interested biologists. Many researchers argue that it represents an evolutionary dead-end, as the extreme specialization and isolation of cave-dwelling species, which often leads to limited genetic diversity and reduced adaptability outside the cave environment, are presumed to be difficult to overcome [33,71–73]. Competing with epigean species that are already well adapted to surface habitats may further disadvantage hypogean species [73]. However, Humphreys [74] suggested that the removal of potential competitors might allow troglobitic species to recolonize endogean and epigean habitats—assuming suitable external conditions such as a highly humid environment—challenging the notion that troglobitism is a definitive evolutionary dead-end. More recent evidence from troglomorphic scorpions supports this view. Prendini et al. [7] found that endogean, humicolous scorpions of the North American family Typhlochactidae Mitchell, 1971, were evolutionarily younger than their hypogean counterparts. The hypogean condition was recovered as ancestral, indicating that endogean species evolved from hypogean ancestors on more than one occasion. Volschenk and Prendini [5] proposed that lithophily, the adaptation to rocky habitats, could be an evolutionary precursor to troglophily. This raises the question as to whether humicolous (endogean) habitats could represent a transitional stage between cave (hypogean) habitats and surface (epigean) habitats in some taxa. Leaf litter or soil microhabitats often share ecological characteristics with cave environments, e.g., higher levels of humidity and more stable temperatures than other surface microhabitats [75,76].

The troglobitic fauna of the Balkans is singularly ancient. It is assumed that the region's past tropical climate with an abundance of humid habitats promoted the diversification of an early humicolous and hygrophylous fauna during the Tertiary [19,27,28]. Part of this "proto-fauna" survived successive climatic transitions, including Miocene aridification and the Quaternary glaciations at the end of the Pliocene, by retreating to hypogean refugia, i.e., as climate relicts [19,26,77–79]. The presence of deep limestone sediments and subsequent processes of karstification (Figure 1B) facilitated the divergence of subterranean taxa in isolated hypogean niches, ultimately creating a remarkably diverse troglophilic and troglobitic fauna [18,23,26]. The extant cavernicolous species of the Balkans are, in effect, Darwin's (p. 136: [80]) "wrecks of ancient life", remnants of an ancient biota that evolved and adapted to subterranean environments over millions of years, diversifying through niche divergence and allopatric speciation.

Among Balkan cavernicolous taxa, scorpions are a rarity. Eleven of the thirteen euscorpiid species recorded from inside caves have been reported from the Balkans (Table 1), and little is known about their biology, ecology or genetic relationships. Five of these species are exclusively cavernicolous, each with varying degrees of specialization to subterranean environments. Noteworthy examples of differential adaptation to cave habitats include *E. biokovensis*, *E. feti*, and *E. studentium*. Using the ecological key of Prendini et al. [8], *E. feti* was classified as a trogloxene, which occasionally ventures into caves but is not entirely dependent on them. *Euscorpius feti* has often been recorded outside of caves [31–33,48], including in the present study, indicating a broader ecological tolerance (eurytopy). In contrast, *E. biokovensis* and *E. studentium* are considered a troglophile and troglobite, respectively. Both species are restricted to karstic systems, to which they exhibit different degrees of specialization [32–34]. Whereas *E. biokovensis* only displays attenuation of the appendages and slight reduction in sclerotization and pigmentation, *E. studentium* is an obligate troglobite, exhibiting pronounced troglomorphies (i.e., anophthalmia, reduced sclerotization and pigmentation, and attenuation of the appendages). Other remarkable Balkan cavernicolous *Euscorpius* species include the troglophilous *Euscorpius birulai* Fet et al., 2014, which appears to be restricted to a single cave on the Greek island of Euboea [50] and *Euscorpius sulfur* Kovařík et al., 2023, discovered within a sulfur cave on the border between Albania and Greece [36]. The other species in Table 1 represent sporadic records of typically epigean species in cave environments, displaying broad ecological tolerances and an ability to thrive in diverse environments. The opportunistic ability of euscorpiids to exploit a range of ecological niches probably contributed to their dispersal across Europe and beyond [81–83]. The only two cave-restricted species which have been studied phylogenetically appear to represent the most basal branch of their clade, in the case of *E. feti*, or to lack epigean relatives, in the case of *E. biokovensis* [34], consistent with the "Climate Relict" model [1,84,85]. Although the phylogenetic position of the troglobitic *E. studentium* has yet to be tested, it may represent another ancient lineage of Euscorpiidae, along with the montane species of the predominantly humicolous genus *Alpiscorpius* and the widely distributed genus *Tetratrichobothrius* Birula, 1917. Further evidence supporting the Climate Relict model in the Balkans is provided by several radiations of Dinaric Karst invertebrates in which unique, ancient lineages coexist with evolutionary younger, karst-adapted taxa [86–89]. For example, molecular analyses of pseudoscorpions inhabiting the Dinaric Karst identified several relictual species and revealed that at least 65% of the species in the region are endemic [23].

The restricted distributions, specific habitat requirements, and typically low abundance of cavernicolous taxa, either due to their occurrence in deeper cave chambers or because they follow a K-selected life history strategy, common to many troglobionts [33], render these species particularly susceptible to anthropogenic environmental impacts, including climate change and habitat destruction or degradation. Anthropogenic pressure on subterranean ecosystems and species is intensifying [90]. Climate change is the main factor altering the microclimate within caves, whereas habitat degradation, driven by pollution, mining, recreational tourism, and other forms of land use, threatens their structural integrity [90–92]. Troglobitic organisms frequently possess low thermal tolerance and limited vagility, rendering them particularly vulnerable [1,2,16]. Without prompt and effective conservation strategies, unique subterranean habitats and the specialized species they shelter face an inevitable decline [90]. The present contribution adds to a growing body of knowledge about the unique diversity of cavernicolous invertebrates, highlighting the need for continued exploration and conservation of these fragile ecosystems before it is too late.

Author Contributions: Conceptualization, J.B.-A. and L.P.; methodology, J.B.-A. and L.P.; software, J.B.-A.; validation, J.B.-A. and L.P.; formal analysis, J.B.-A.; investigation, J.B.-A. and L.P.; resources, L.P.; data curation, L.P.; writing—original draft preparation, J.B.-A.; writing—review and editing, J.B.-A. and L.P.; visualization, J.B.-A. and L.P.; supervision, L.P.; project administration, L.P.; funding acquisition, J.B.-A. and L.P. All authors have read and agreed to the published version of the manuscript.

Funding: J.B.-A. was supported by student scholarship 2021.04967.BD from the Fundação para Ciência e Tecnologia, Portugal. Fieldwork was supported by a grant from the Vincent Roth Fund for Systematics Research of the American Arachnological Society (AAS) and a Graduate Student Research Award from the Society of Systematic Biologists to J.B.-A. and by U.S. National Science Foundation grant DEB 2003382 to L.P.

Institutional Review Board Statement: Ethical review and approval were waived for this study due to experimentation on non-living invertebrate museum specimens.

Data Availability Statement: Data are contained within the article.

Acknowledgments: Material was collected on scientific research permits UP/I-352-04/22-08/77—517-10-1-1-22-4 (2022) and UP/I-352-04/24-08/85—517-10-1-2-24-2 (2024) from the Ministry of Economy and Sustainable Development and the Administration for the Conservation of Nature of Croatia and scientific research permit 03-UPI-558/6 (2024) from the Environmental Protection Agency of Montenegro. The authors thank the following: Brent E. Hendrixson (AAS Student Research Grants Chair) for guidance during the grant proposal application; Ivo Karaman (University of Novi Sad, Serbia), Pablo Vicent (University of Porto, Portugal) and Hamid Tahirović (University of Sarajevo, Bosnia and Herzegovina) for assisting the first author in the field; Boris Lauš (Association Hyla, Croatia), Snežana Dragićević (Montenegrin Academy of Sciences and Arts, Montenegro) and Vladimir Pavićević (Environmental Protection Agency of Montenegro) for assistance with collecting permits and in the field; Alex Ullrich (Munich, Germany) for donating some material examined in the study; Rebecca Johnson (AMNH) for assisting the first author with acquiring a U.S. visa; Pio A. Colmenares (AMNH) for logistical assistance; Steve Thurston (AMNH) for assistance with the figures for this contribution; Mario García París (Museo Nacional de Ciencias Naturales de Madrid, Spain) for providing some fieldwork supplies; Emilio Blasco and Juana María Aróstegui for assistance with shipping and organizing the fieldwork materials; Ivo Karaman, two anonymous reviewers, and the editor for constructive comments on the manuscript.

Conflicts of Interest: The authors declare no conflicts of interest.

References

1. Barr, T.C. Cave Ecology and the Evolution of Troglobites. In *Evolutionary Biology*; Dobzhansky, T., Hecht, M.K., Steere, W.C., Eds.; Springer: Boston, MA, USA, 1968; pp. 35–102.
2. Poulson, T.L.; White, W.B. The cave environment. *Science* **1969**, *165*, 971–981. [CrossRef] [PubMed]
3. Culver, D.C.; Sket, B. Hotspots of subterranean biodiversity in caves and wells. *J. Cave Karst Stud.* **2000**, *62*, 11–17.
4. Prendini, L. Substratum specialization and speciation in southern African scorpions: The Effect Hypothesis revisited. In *Scorpions 2001: In Memoriam Gary A. Polis*; Fet, V., Selden, P.A., Eds.; British Arachnological Society: Burnham Beeches, UK, 2001; pp. 113–138.
5. Volschenk, E.S.; Prendini, L. *Aops oncodactylus* gen. et sp. nov., the first troglobitic urodacid (Urodacidae: Scorpiones), with a re-assessment of cavernicolous, troglobitic and troglomorphic scorpions. *Invertebr. Syst.* **2008**, *22*, 235–257. [CrossRef]
6. Vignoli, V.; Prendini, L. Systematic revision of the troglomorphic North American scorpion family Typhlochactidae (Scorpiones: Chactoidea). *Bull. Am. Mus. Nat. Hist.* **2009**, *326*, 1–94. [CrossRef]
7. Prendini, L.; Francke, O.F.; Vignoli, V. Troglomorphism, trichobothriotaxy and typhlochactid phylogeny (Scorpiones, Chactoidea): More evidence that troglobitism is not an evolutionary dead-end. *Cladistics* **2010**, *26*, 117–142. [CrossRef]
8. Prendini, L.; Ehrenthal, V.L.; Loria, S.F. Systematics of the relictual Asian scorpion family Pseudochactidae Gromov, 1998, with a review of cavernicolous, troglobitic, and troglomorphic scorpions. *Bull. Am. Mus. Nat. Hist.* **2021**, *453*, 1–149. [CrossRef]
9. Schiner, J.R. Fauna der Adelsberger-, Lueger- und Magdalenen Grotte. In *Zur Höhlenkunde des Karstes. Die Grotten und Höhlen von Adelsberg, Lueg, Planina und Laas*; Schmidl, A., Ed.; Braumüller: Vienna, Austria, 1854; pp. 231–272.

10. Racovitza, E.G. Essai sur les problèmes biospéologiques. *Arch. Zool. Exp. Gén.* **1907**, *6*, 371–488.
11. Christiansen, K. Proposition pour la classification des animaux cavernicoles. *Spelunca* **1962**, *2*, 76–78.
12. Poulson, T.L. Cave adaptation in ambliopsid fishes. *Am. Midl. Nat.* **1963**, *70*, 257–290. [CrossRef]
13. Sket, B. Can we agree on an ecological classification of subterranean animals? *J. Nat. Hist.* **2008**, *42*, 1549–1563. [CrossRef]
14. Trajano, E. Ecological Classification of Subterranean Organisms. In *Encyclopedia of Caves*; White, W.B., Culver, D.C., Eds.; Academic Press: Waltham, MA, USA, 2012; pp. 275–277.
15. Trajano, E.; de Carvalho, M.R. Towards a biologically meaningful classification of subterranean organisms: A critical analysis of the Schiner-Racovitza system from a historical perspective, difficulties of its application and implications for conservation. *Subterr. Biol.* **2017**, *22*, 1–26. [CrossRef]
16. Mammola, S. Finding answers in the dark: Caves as models in ecology fifty years after Poulson and White. *Ecography* **2019**, *42*, 1331–1351. [CrossRef]
17. Hamann, O. *Europäische Höhlenfauna*; Hermann Costenoble: Jena, Germany, 1896; 319p.
18. Ćurčić, B.P.M. The Cave Fauna in Serbia: Origin, Historical Development, and Diversification. In *Speleological Atlas of Serbia*; Đurović, P., Ed.; Geographical Institute "Jovan Cvijić", Serbian Academy of Sciences and Arts: Belgrade, Serbia, 1998; pp. 75–83.
19. Ćurčić, B.P.M.; Jovanović, V.M. The cave-dwelling fauna of the Balkan Peninsula: Its origin and diversification. *Acta Entomol. Slov.* **2004**, *12*, 35–56.
20. Griffiths, H.I.; Kryštufek, B.; Reed, J.M. *Balkan Biodiversity: Pattern and Process in the European Hotspot*; Springer: Dordrecht, The Netherlands, 2004; 358p.
21. Sket, B. Diversity Patterns in the Dinaric Karst. In *Encyclopedia of Caves*; White, W.B., Culver, D.C., Eds.; Academic Press: Waltham, MA, USA, 2012; pp. 228–238.
22. Zagmajster, M.; Malard, F.; Eme, D.; Culver, D.C. Subterranean Biodiversity Patterns from Global to Regional Scales. In *Ecological Studies Series: Cave Ecology*; Moldovan, O.T., Kováč, L., Halse, S., Eds.; Springer: Cham, Switzerland, 2018; pp. 195–227.
23. Hlebec, D.; Podnar, M.; Kučinić, M.; Harms, D. Molecular analyses of pseudoscorpions in a subterranean biodiversity hotspot reveal cryptic diversity and microendemism. *Sci. Rep.* **2023**, *13*, 430. [CrossRef] [PubMed]
24. Deltshev, C. A faunistic and zoogeographical review of the spiders (Araneae) of the Balkan Peninsula. *J. Arachnol.* **1999**, *25*, 255–261.
25. Lukić, M.; Fišer, C.; Delić, T.; Bilandžija, H.; Pavlek, M.; Komerički, A.; Dražina, T.; Jalžić, B.; Ozimec, R.; Slapnik, R.; et al. Subterranean fauna of the Lukina Jama–Trojama Cave System in Croatia: The deepest cave in the Dinaric Karst. *Diversity* **2023**, *15*, 726. [CrossRef]
26. Vesović, N.; Deltshev, C.; Mitov, P.; Antić, D.; Stojanović, D.Z.; Stojanović, D.V.; Stojanović, K.; Božanić, M.; Ignjatović-Ćupina, A.; Ćurčić, S. The diversity of subterranean terrestrial arthropods in Resava Cave (eastern Serbia). *Diversity* **2024**, *16*, 234. [CrossRef]
27. Deeleman-Reinhold, C.L. Revision of the cave-dwelling and related spiders of the genus *Troglohyphantes* Joseph (Linyphiidae), with special reference to the Yugoslav species. In *Academia Scientiarum et Artium Slovenica, Classis IV: Historia Naturalis*; Opera; The Academy: Ljubljana, Yugoslavia, 1978; Volume 23, pp. 1–221.
28. Ćurčić, B.P.M. Cave-dwelling pseudoscorpions of the Dinaric Karst. In *Academia Scientiarum et Artium Slovenica, Classis IV: Historia Naturalis*; Opera; The Academy: Ljubljana, Yugoslavia, 1988; Volume 26, pp. 1–191.
29. Savić, I.R. Diversification of the Balkan fauna: Its origin, historical development and present status. *Adv. Arachnol. Dev. Biol.* **2008**, *12*, 57–78.
30. Apfelbeck, V. Sur la faune des cavernes de Bosnie et Herzégovine. *Spelunca* **1895**, *1*, 23–24.
31. Tropea, G. A new species of *Euscorpius* Thorell, 1876 from the western Balkans (Scorpiones: Euscorpiidae). *Euscorpius* **2013**, *174*, 1–10. [CrossRef]
32. Tropea, G.; Ozimec, R. Another new species of *Euscorpius* Thorell, 1876 from the caves of Croatia and Bosnia-Herzegovina (Scorpiones: Euscorpiidae) with notes on biogeography and cave ecology. *Euscorpius* **2020**, *308*, 1–13.
33. Karaman, I. A new *Euscorpius* species (Scorpiones: Euscorpiidae) from a Dinaric cave—The first record of troglobite scorpion in European fauna. *Biol. Serb.* **2020**, *42*, 14–31.
34. Podnar, M.; Grbac, I.; Tvrtković, N.; Hörweg, C.; Haring, E. Hidden diversity, ancient divergences, and tentative Pleistocene microrefugia of European scorpions (Euscorpiidae: Euscorpiinae) in the eastern Adriatic region. *Zool. Syst. Evol. Res.* **2021**, *59*, 1824–1849. [CrossRef]
35. Podnar, M.; Tvrtković, N.; Vuković, M.; Rebrina, F.; Grbac, I.; Hörweg, C. *Alpiscorpius liburnicus* sp. n. with a note on the "*Alpiscorpius croaticus* group" (Scorpiones: Euscorpiidae) in Croatia. *Nat. Croat.* **2022**, *31*, 265–282. [CrossRef]
36. Kovařík, F.; Audy, M.; Sarbu, S.M.; Fet, V. *Euscorpius sulfur* sp. n. (Scorpiones: Euscorpiidae), a new cave scorpion from Albania and northwestern Greece. *Euscorpius* **2023**, *376*, 1–14.
37. Juberthie, C.; Delay, B.; Bouillon, M. Sur l'existence d'un milieu souterrain superficiel en zone non calcaire. *C. R. De L'académie Des Sci.* **1980**, *290*, 49–52.
38. Mammola, S.; Giachino, P.M.; Piano, E.; Jones, A.; Barberis, M.; Badino, G.; Isaia, M. Ecology and sampling techniques of an understudied subterranean habitat: The Milieu Souterrain Superficiel (MSS). *Sci. Nat.* **2016**, *103*, 88. [CrossRef]
39. Ficetola, G.F.; Canedoli, C.; Stoch, F. The Racovitzan impediment and the hidden biodiversity of unexplored environments. *Conserv. Biol.* **2019**, *33*, 214–216. [CrossRef]

40. Potočnik, F. *Alpioniscus (Illyrionethes) christiani* spec. nov., eine neue Trichoniscinae-Art (Isopoda terrestrial) aus Jugoslawien. *Ann. Naturhist. Mus. Wien.* **1983**, *84*, 389–395.
41. Christian, E.; Potočnik, F. Ein Beitrag zur Kenntnis der Höhlenfauna der Insel Krk. *Biološki Vestn.* **1985**, *33*, 13–19.
42. Graham, M.R.; Webber, M.M.; Blagoev, G.; Ivanova, N.; Fet, V. Elevation of *Euscorpius germanus croaticus* Di Caporiacco, 1950. *Rev. Iber. Aracnol.* **2012**, *21*, 41–50.
43. Fet, V.; Graham, M.R.; Blagoev, G.; Karataş, A.; Karataş, A. DNA barcoding indicates hidden diversity of *Euscorpius* (Scorpiones: Euscorpiidae) in Turkey. *Euscorpius* **2016**, *216*, 1–12. [CrossRef]
44. Tropea, G. Concerning some Balkan Euscorpiidae populations (Scorpiones: Euscorpiidae). *Biol. Serb.* **2021**, *43*, 22–54.
45. Boldori, L. Cavernicola Italica. I. Dalle Alpi Occidentali alla Valle del Brenta, a nord del Po. Parte 1. Dai Protozoa ai Crustacea. *Ann. Mus. Civ. Stor. Nat. Brescia* **1977**, *14*, 127–172.
46. Tropea, G. Reconsideration of the taxonomy of *Euscorpius tergestinus* (Scorpiones: Euscorpiidae). *Euscorpius* **2013**, *162*, 1–23. [CrossRef]
47. Tropea, G.; Fet, V. Two new *Euscorpius* species from central-western Greece (Scorpiones: Euscorpiidae). *Euscorpius* **2015**, *199*, 1–16. [CrossRef]
48. Tropea, G.; Ozimec, R. Description of the adult male of *Euscorpius feti* Tropea, 2013 (Scorpiones: Euscorpiidae), with notes on cave ecology of this species. *Euscorpius* **2019**, *291*, 1–10. [CrossRef]
49. Blasco-Aróstegui, J.; Prendini, L. Glacial relicts? A new scorpion from Mount Olympus, Greece (Euscorpiidae: *Euscorpius*). *Am. Mus. Novit.* **2023**, *4003*, 1–36. [CrossRef]
50. Fet, V.; Soleglad, M.E.; Parmakelis, A.; Kotsakiozi, P.; Stathi, I. Two new species of *Euscorpius* from Euboea Island, Greece (Scorpiones: Euscorpiidae). *Arthropoda Sel.* **2014**, *23*, 111–126. [CrossRef]
51. Fet, V.; Parmakelis, A.; Stathi, I.; Tropea, G.; Kotsakiozi, P.; Kardaki, L.; Nikolakakis, M. Fauna and Zoogeography of Scorpions in Greece. In *Biogeography and Biodiversity of the Aegean*; Sfenthourakis, S., Pafilis, P., Parmakelis, A., Poulakakis, N., Triantis, K.A., Eds.; Broken Hill Publishers Ltd.: Nicosia, Cyprus, 2018; pp. 123–134.
52. Tropea, G.; Onnis, C. A remarkable discovery of a new scorpion genus and species from Sardinia (Scorpiones: Chactoidea: Belisariidae). *Arachn. Riv. Aracnol. Ital.* **2020**, *26*, 3–25.
53. Wolf, B. Ordnung Scorpiones. In *Animalium Cavernarum Catalogus*; Dr. W. Junk: The Hague, The Netherlands, 1937; Volume III, p. 534.
54. di Caporiacco, L. Aracnidi cavernicoli Liguri. *Ann. Mus. Civ. Stor. Nat. Giacomo Doria* **1950**, *64*, 101–110.
55. Bologna, M.A.; Taglianti, A.V. *Fauna Cavernicola Delle Alpi Liguri*; Museo Civico di Storia Naturale "G. Doria": Genova, Italy, 1985; pp. 1–389.
56. Guéorguiev, V.; Beron, P. Essai sur la faune cavernicole de Bulgarie. *Ann. Spéléol.* **1962**, *17*, 285–441.
57. Fet, V.; Graham, M.R.; Webber, M.M.; Blagoev, G. Two new species of *Euscorpius* (Scorpiones: Euscorpiidae) from Bulgaria, Serbia, and Greece. *Zootaxa* **2014**, *3894*, 83–105. [CrossRef] [PubMed]
58. di Caporiacco, L. *Le Specie e Sottospecie del Genre "Euscorpius" Viventi in Italia ed in Alcune Zone Confinanti*; Accademia Nazionale dei Lincei: Rome, Italy, 1950; Volume 2, pp. 159–230.
59. Stahnke, H.L. UV light, a useful field tool. *BioScience* **1972**, *22*, 604–607. [CrossRef]
60. Stahnke, H.L. Scorpion nomenclature and mensuration. *Entomol. News* **1970**, *81*, 297–316.
61. Sissom, W.D.; Polis, G.A.; Watt, D.D. Field and laboratory methods. In *The Biology of Scorpions*; Polis, G.A., Ed.; Stanford University Press: Stanford, CA, USA, 1990; pp. 215–221.
62. Prendini, L.; Crowe, T.M.; Wheeler, W.C. Systematics and biogeography of the family Scorpionidae (Chelicerata: Scorpiones), with a discussion on phylogenetic methods. *Invertebr. Syst.* **2003**, *17*, 185–259. [CrossRef]
63. Loria, S.F.; Prendini, L. Homology of the lateral eyes of Scorpiones: A six-ocellus model. *PLoS ONE* **2014**, *9*, e112913. [CrossRef] [PubMed]
64. Vachon, M. *Études sur Les Scorpions*; Institut Pasteur d'Algérie: Algiers, Algeria, 1952; 482p.
65. Prendini, L. Phylogeny and classification of the superfamily Scorpionoidea Latreille 1802 (Chelicerata, Scorpiones): An exemplar approach. *Cladistics* **2000**, *16*, 1–78. [CrossRef]
66. Vachon, M. De l'utilité, en systématique, d'une nomenclature des dents des chélicères chez les scorpions. *Bull. Mus. Natl. Hist. Nat.* **1963**, *35*, 161–166.
67. Prendini, L.; Loria, S.F. Systematic revision of the Asian forest scorpions (Heterometrinae Simon, 1879), revised suprageneric classification of Scorpionidae Latreille, 1802, and revalidation of Rugodentidae Bastawade et al., 2005. *Bull. Am. Mus. Nat. Hist.* **2020**, *442*, 1–480. [CrossRef]
68. Vachon, M. Remarques sur la classification sous-spécifique des espèces appartenant au genre *Euscorpius* Thorell, 1876 (Scorpionida, Chactidae). *Atti Soc. Tosc. Sci. Nat. Mem. Ser. B* **1981**, *88*, 193–203.
69. Stockwell, S.A. Revision of the Phylogeny and Higher Classification of Scorpions (Chelicerata). Ph.D. Dissertation, University of California, Berkeley, CA, USA, 1989; 413p.
70. Prendini, L. The systematics of southern African *Parabuthus* Pocock (Scorpiones: Buthidae): Revisions to the taxonomy and key to the species. *J. Arachnol.* **2004**, *32*, 109–186. [CrossRef]
71. Dollo, L. Les lois de l'evolution. *Bull. Soc. Belge Géol. Paléontol. Hydrol.* **1893**, *7*, 164–166.
72. Cope, E.D. *The Primary Factors of Organic Evolution*; Open Court Publishing: Chicago, IL, USA, 1896; 555p.

73. Conway Morris, S. Ecology in deep time. *Trends Ecol. Evol.* **1995**, *10*, 290–294. [CrossRef]
74. Humphreys, W.F. Background and glossary. In *Ecosystems of the World: Subterranean Ecosystems*; Wilkens, H., Culver, D.C., Humphreys, W.F., Eds.; Elsevier: Amsterdam, The Netherlands, 2000; Volume 30, pp. 3–14.
75. Ashcroft, M.B.; Gollan, J.R. Moisture, thermal inertia, and the spatial distributions of near-surface soil and air temperatures: Understanding factors that promote microrefugia. *Agric. For. Meteorol.* **2013**, *176*, 77–89. [CrossRef]
76. Keppel, G.; Anderson, S.; Williams, C.; Kleindorfer, S.; O'Connell, C. Microhabitats and canopy cover moderate high summer temperatures in a fragmented Mediterranean landscape. *PLoS ONE* **2017**, *12*, e0183106. [CrossRef]
77. Hewitt, G.M. Some genetic consequences of ice ages, and their role in divergence and speciation. *Biol. J. Linn. Soc.* **1996**, *58*, 247–276. [CrossRef]
78. Hewitt, G.M. Post-glacial re-colonization of European biota. *Biol. J. Linn. Soc.* **1999**, *68*, 87–112. [CrossRef]
79. Schmitt, T. Molecular biogeography of Europe: Pleistocene cycles and postglacial trends. *Front. Zool.* **2007**, *4*, 11. [CrossRef] [PubMed]
80. Darwin, C. *On the Origin of Species by Means of Natural Selection, or the Preservation of Favoured Races in the Struggle of Life*; John Murray: London, UK, 1859; 502p.
81. Toscano-Gadea, C.A. *Euscorpius flavicaudis* (DeGeer, 1778) in Uruguay: First record from the New World. *Newsl. Br. Arachnol. Soc.* **1998**, *81*, 6.
82. Fet, V.; Kovařík, F. First record of *Euscorpius (Polytrichobothrius) italicus* (Scorpiones: Euscorpiidae) from Iraq. *Acta Soc. Zool. Bohem.* **2003**, *67*, 179–181.
83. Fet, V.; Gantenbein, B.; Karataş, A.; Karataş, A. An extremely low genetic divergence across the range of *Euscorpius italicus* (Scorpiones, Euscorpiidae). *J. Arachnol.* **2006**, *34*, 248–253. [CrossRef]
84. Hampe, A.; Jump, A.S. Climate relicts: Past, present, future. *Annu. Rev. Ecol. Evol. Syst.* **2011**, *42*, 313–333. [CrossRef]
85. Loria, S.F.; Ehrenthal, V.L.; Nguyen, A.D.; Prendini, L. Climate relicts: Asian scorpion family Pseudochactidae survived Miocene aridification in caves of the Annamite Mountains. *Insect Syst. Divers.* **2022**, *6*, 3. [CrossRef]
86. Murienne, J.; Karaman, I.; Giribet, G. Explosive evolution of an ancient group of Cyphophthalmi (Arachnida: Opiliones) in the Balkan Peninsula. *J. Biogeogr.* **2010**, *37*, 90–102. [CrossRef]
87. Njunjić, I.; Perreau, M.; Hendriks, K.; Schilthuizen, M.; Deharveng, L. The cave beetle genus *Anthroherpon* is polyphyletic: Molecular phylogenetics and description of *Graciliella* n. gen. (Leiodidae, Leptodirini). *Contrib. Zool.* **2016**, *85*, 337–359. [CrossRef]
88. Bedek, J.; Taiti, S.; Bilandžija, H.; Ristori, E.; Baratti, M. Molecular and taxonomic analyses in troglobiotic *Alpioniscus* (*Illyrionethes*) species from the Dinaric Karst (Isopoda: Trichoniscidae). *Zool. J. Linn. Soc.* **2019**, *187*, 539–584. [CrossRef]
89. Ribera, C.; Dimitrov, D. A molecular phylogeny of the European nesticid spiders (Nesticidae, Araneae): Implications for their systematics and biogeography. *Mol. Phylogenet. Evol.* **2023**, *180*, 107685. [CrossRef]
90. Nanni, V.; Piano, E.; Cardoso, P.; Isaia, M.; Mammola, S. An expert-based global assessment of threats and conservation measures for subterranean ecosystems. *Biol. Conserv.* **2023**, *283*, 110136. [CrossRef]
91. Mammola, S.; Cardoso, P.; Culver, D.C.; Deharveng, L.; Ferreira, R.L.; Fišer, C.; Zagmajster, M. Scientists' warning on the conservation of subterranean ecosystems. *BioScience* **2019**, *69*, 641–650. [CrossRef]
92. Mammola, S.; Piano, E.; Cardoso, P.; Vernon, P.; Domínguez-Villar, D.; Culver, D.C.; Isaia, M. Climate change going deep: The effects of global climatic alterations on cave ecosystems. *Anthr. Rev.* **2019**, *6*, 98–116. [CrossRef]

diversity

MDPI

Article

Exploring Biodiversity and Food Webs in Sulfur Cave in the Vromoner Canyon on the Greek–Albanian Border

Serban M. Sarbu [1,2,3], Traian Brad [3,*], Raluca I. Băncilă [1,3,4,*] and Andrei Ştefan [3,5]

1 "Emil Racoviţă" Institute of Speleology, Calea 13 Septembrie, Nr. 13, 050711 Bucharest, Romania; serban.sarbu@yahoo.com
2 Department of Biological Sciences, California State University, Chico, CA 95929, USA
3 "Emil Racoviţă" Institute of Speleology, Str. Clinicilor, Nr. 5-7, 400006 Cluj-Napoca, Romania; andrei.stefan@antipa.ro
4 Faculty of Natural and Agricultural Sciences, Ovidius University Constanța, Aleea Universităţii, Nr. 1, 900470 Constanța, Romania
5 "Grigore Antipa" National Museum of Natural History, Şos. Kiseleff, Nr. 1, 011341 Bucharest, Romania
* Correspondence: traian.brad@acad-cj.ro (T.B.); bancila_ralucaioana@yahoo.com (R.I.B.)

Abstract: Sulfidic caves support diverse and abundant subterranean communities, including numerous endemic species and complex food webs, though the full extent of species diversity and resource utilization in these ecosystems remains largely unexplored. This paper presents the results of biological surveys conducted from 2023 to 2024 in Sulfur Cave, located in the Vromoner Canyon on the Greek–Albanian border, focusing on microbial, vertebrate, and invertebrate communities and investigating the structure of the subterranean food web. The microbial communities from the different biofilms are dominated by chemosynthetic sulfur-oxidizing microorganisms, specifically filamentous bacteria such as *Thiotrix* and *Beggiatoa*. Two species of fish, an eel (*Anguilla* sp.) and a Cyprinid (*Alburnoides* sp.), and six bat species from three families (Rhinolophidae, Miniopteridae, and Vespertilionidae) were documented. The invertebrate fauna includes five aquatic species, 25 terrestrial species, and four amphibiotic species. Among these, eight species are endemic, and seven species exhibit troglomorphic traits. Stable isotope analysis showed light carbon and nitrogen values for the terrestrial and aquatic invertebrates, suggesting that subterranean communities rely on food produced in situ by chemoautotrophic microorganisms. Our results identified cave areas of significant biological relevance and provided reference data to inform conservation actions aimed at preserving the biodiversity of this sulfidic cave.

Keywords: sulfidic subterranean ecosystems; thermo-mineral cave; chemoautotrophy-based food web; hotspot of subterranean diversity

1. Introduction

Life without sunlight was not considered possible prior to the discovery of ecosystems solely based on chemoautotrophy, such as deep-sea hydrothermal vents, in 1977 [1]. The subsequent discovery of a chemoautotrophic sulfidic groundwater ecosystem in Movile Cave, Romania [2], and later in other sulfidic caves, showed that large amounts of organic material produced in situ by autotrophic sulfur- and methane-oxidizing microorganisms can often support abundant and diverse subterranean invertebrate communities including numerous endemic species [3]. All the sulfidic underground ecosystems explored to date emerged as hotspots of subterranean biodiversity [4–6] and proved able to support complex subterranean food webs. Chemolithoautotrophs, often deemed less efficient than phototrophs, were traditionally not considered significant primary producers, but they play a crucial role in sustaining many ecosystem-level processes in the absence of light and photosynthesis in subsurface environments [5]. Although they are very rare, ecosystems based on chemoautotrophy are of particular interest to researchers as they may represent analogs for Earth's earliest biological communities or for possible extraterrestrial life [7]. The enrichment of methane- and sulfur-oxidizing bacteria to produce microbial protein enables the production of alternative proteins with a reduced environmental impact compared to plant- or animal-based sources [8]. Despite the importance of chemolithoautotrophy, our understanding of cave chemosynthetic systems remains scarce.

Several sulfidic caves have recently been discovered and explored in Albania and northwestern Greece (Figure 1A) by teams of Italian [9] and Czech speleologists [10,11]. Thermo-mineral sulfidic water emerges from springs in the Vromoner and in the Pixaria canyons, which were carved by the Sarandaporo River in limestone outcrops near the border between Greece and Albania [12]. One of the largest springs is located in the deep recesses of Sulfur Cave, which straddles the border between the two countries in the Vromoner Canyon.

Figure 1. Location of sulfidic caves in Vromoner and Pixaria Canyons along the Sarandaporo River in NW Greece and SE Albania (**A**) and the location of the four sulfidic cave areas in Albania (**B**), i.e., (1) Zalles Cave, (2) Holtas Canyon, (3) Langarica Canyon, and (4) Sarandaporo River area on the border with Greece.

Subterranean sulfidic streams and large lakes have also been documented in Turtle Cave and Atmos Cave, located in the same limestone outcrop (Figure 1A). Due to the scarcity of studies, the full extent of species diversity or resource utilization of these chemoautotrophy-based ecosystems remains limited. In Sulfur Cave, for example, early biological observations were performed by members of the Czech Speleological Society in 2021, who reported an unusual abundance of fauna including a dense population of small flies and a large section of the cave walls covered by an extensive spider web with a multitude of spiders lurking in small funnel-shaped cavities within the web. The spiders collected during the early exploration expeditions were identified by Vlastimil Rÿžiÿka. Notably, these caves are not covered by protected areas or legally protected, and some areas have already been affected by anthropogenic activity. Efforts are needed to engage local stakeholders to foster conservation actions for the preservation of these unique cave ecosystems and the invaluable biological communities they host.

The results of the observations and biological surveys conducted in Sulfur Cave between 2023 and 2024 are presented here. The microbial, vertebrate, and invertebrate communities were surveyed, and the structure of the subterranean food web was investigated. One of the goals of this research is to identify areas of biological relevance and to provide important reference data to inform conservation actions aimed at preserving the biodiversity of this subterranean sulfidic ecosystem.

2. Materials and Methods

2.1. Sulfur Cave

Sulfur Cave is a 520 m-long hypogenic cave (Figure 2) located in the Vromoner Canyon, on the border between Greece and Albania (40.0961 N, 20.6789 E). It has a small entrance that opens into the Sarandaporo River, a tributary of the Aaos/Vjosa River. The main gallery increases in size, culminating in a large passage called Vesmír Dom (Universe Hall) and narrowing again in the most remote cave passage. The cave is traversed by a sulfidic subterranean stream that forms a lake near the entrance. The water emerges from springs located in the deepest recesses of the cave and flows all the way to the cave's entrance, where the water is discharged into the Sarandaporo River. The subterranean stream bifurcates in the Vesmír Dom and then rejoins to create a loop and flows through a small diverticulum of the main gallery, forming a chapel-like area known as Sulfur Chapel. In the Vesmír Dom, the largest room of the cave, the sulfidic stream reaches a width of 2–3 m, and the water is slightly milky due to the particles of elemental sulfur resulting from the oxidation of hydrogen sulfide (H_2S). Walls are often dry in this section of the cave because of the large deposits of hydrophilic anhydrite ($CaSO_4$) resulting from sulfuric acid speleogenesis (SAS). In the most remote part of the cave, there is a pool called Blue Eye, which represents one of the strongest sulfidic springs in Sulfur Cave.

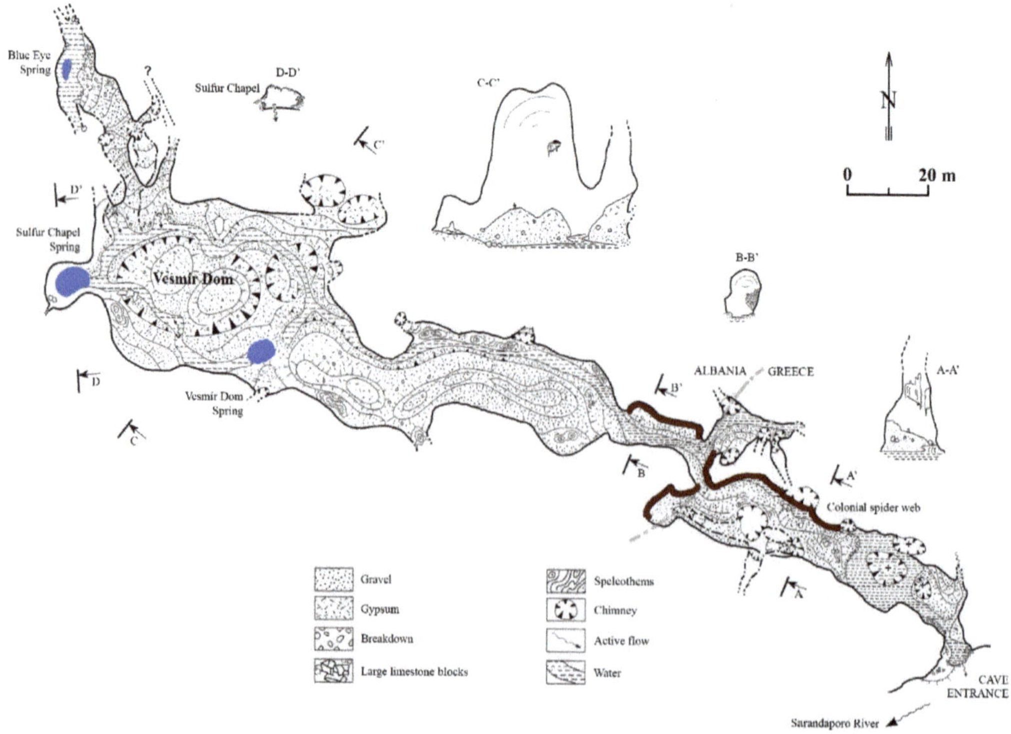

Figure 2. Plan of Sulfur Cave, with depiction of the main sulfidic springs (blue areas), and the large spider web in the vicinity of the cave entrance (brown areas) (modified from [10]).

In a few locations, small amounts of freshwater seep in from the epikarst, and the cave walls are moist and devoid of anhydrite deposits. The cave floor is primarily composed of fluvial gravel and gypsum deposits, with large limestone boulders present in some areas. Accumulations of guano on the cave floor are rare. The concentration of H_2S in the springs in Sulfur Cave can be as high as 65 mg/L; the water temperature is constant at 27 °C, with no significant annual variations (unpublished); and the pH of the water is neutral due to the strong buffering capacity of the carbonate bedrock. The cave atmosphere is highly enriched in H_2S, reaching levels up to 14 ppm in the vicinity of the strong emissions of sulfidic water within the cave, due to degassing that takes place at the surface of the springs. Air temperatures of up to 29 °C have been recorded in the upper cave passages. High water levels of the Sarandaporo River can flood the cave at a distance of up to 160 m from the cave entrance, as occurred in November 2023, depositing surface alluvial materials [11].

2.2. Sampling

2.2.1. Microbial Communities

To describe the microbial communities, biofilms were collected from above and below the water table. Thick white filamentous biofilms were harvested from sediments at the bottom of the sulfidic stream. In addition, samples of biofilms from the mud banks of small pools located in Sulfur Chapel, which are covered by thick brown biofilms, were also collected along with samples of biofilms from the nearby cave walls, which were covered by slightly blue biofilms. Sterile plastic tubes and pipettes were used for harvesting biofilms, and the samples were placed in sterile plastic tubes and kept frozen until analysis.

2.2.2. Vertebrate and Invertebrate Communities

For vertebrates, visual searches of bats and fish were performed for their presence or indirect traces (i.e., skulls), and the sounds produced by the bats were recorded. Invertebrates were sampled visually and manually using tweezers, pipettes, small paintbrushes dipped in ethanol, and occasionally a small plankton net. A UV light was used to search for scorpions. Samples were preserved in 70% ethanol for taxonomic identification and 96% ethanol for genetic and stable isotope analyses, and they were stored at −20 °C prior to processing. To minimize the negative impact on the cave populations, a limited number of specimens was collected for each invertebrate species encountered. All specimens were identified to the lowest possible taxonomic level.

The density of the larvae of *Chironomus* sp. (Diptera, Chironomidae) and *Contacyphon palustris* (C. G. Thomson, 1855) (Coleoptera, Scirtidae), found on the rocky sediments, was estimated by counting individuals within a 15 × 15 cm quadrant placed at ten randomly selected locations for each species (Figure 3A). Two high-resolution pictures of each quadrat were taken, and the number of individuals per quadrat was determined from the pictures.

2.2.3. Carbon and Nitrogen Stable Isotope Analysis

Organic samples were collected manually using fine tweezers and were dried at 60 °C for 24 h. The stable isotope ratios ($\delta^{13}C$ and $\delta^{15}N$) were determined at the Stable Isotope Lab, University of New Mexico, Albuquerque, NM, USA.

2.2.4. Molecular Identification

Morphological assignment of taxa was complemented by molecular identification. This was particularly useful for sub-adult invertebrates. Whole specimens were placed in 96% ethanol and kept at ambient temperature prior to processing in the lab. DNA extraction was carried out overnight using the QIAamp DNA Mini Kit (Qiagen, Hilden, Germany), and a fragment of the mitochondrial COI gene (subunit I of cytochrome c oxidase) was amplified using the universal primer pair LCO1490 (GGTCAACAAATCATAAAGATATTGG)/HCO2198 (TAAACTTCAGGGTGACCAAAAAATCA) [13] or the denatured pair ACOIAF (CWAATCAYAAAGATATTGGAAC)/ACOIAR (AATATAWACTTCWGGGTGACC) [14]. A PCR was performed in a 40 μL mixture containing 1X AccuStart II PCR ToughMix® (Quantabio, MA, USA), 1X GelTrack Loading Dye (Quantabio, MA, USA), 0.15 μM of each primer, and up to 20 ng/μL of DNA. The cycling conditions consisted of

5 cycles of denaturation at 94 °C for 30′, primer annealing at 45 °C for 30′, and extension at 72 °C for 50′ followed by 35 cycles of denaturation at 94 °C for 30′, primer annealing at 51 °C for 30′, and extension at 72 °C for 50′. Amplification was confirmed by electrophoresis on 1.5% (*w*/*v*) agarose gel, and PCR products were purified and Sanger-sequenced (Macrogen, Amsterdam, The Netherlands). The resulting chromatograms were visually inspected in Chromas v.2.6.6 (Technelysium Ltd., South Brisbane, Australia), low-quality ends trimmed, and assembled in CodonCode Aligner v.3.7.1 (CodonCode Corporation, MA, USA). The manually curated DNA sequences were checked against publicly available databases such as BOLD [15] and GenBank [16].

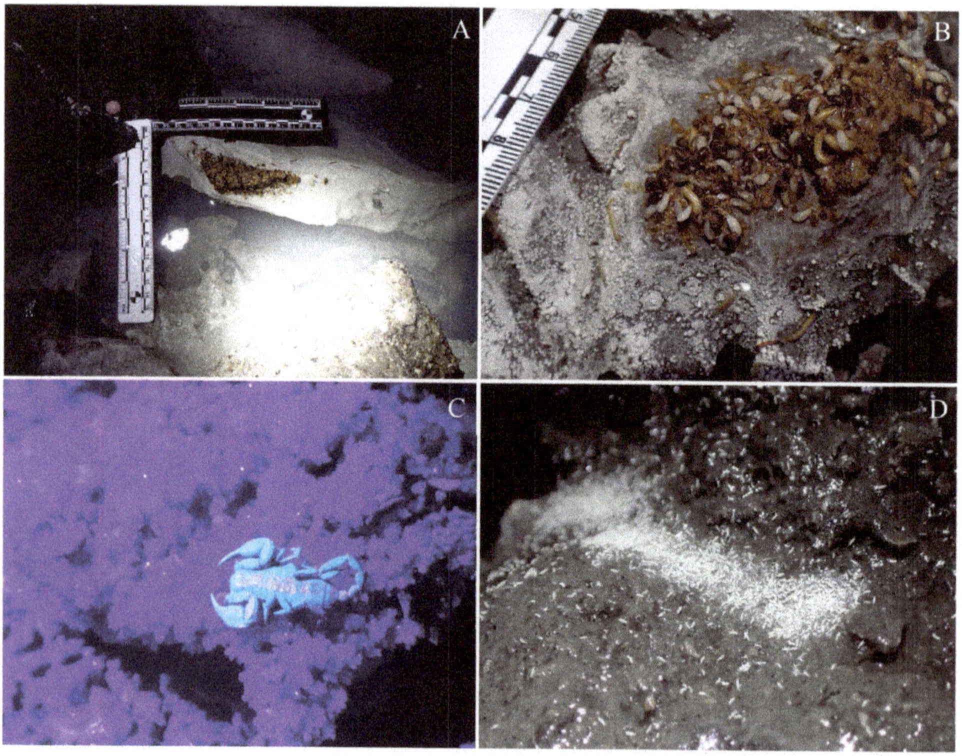

Figure 3. Particularities of methodology and cave fauna: (**A**) the method used to estimate the density of the larvae of *Chironomus* sp. (Chironomidae) and *Contacyphon palustris* (Scirtidae), i.e., counting individuals within a 15 × 15 cm quadrat; (**B**) dense agglomerations of yellow larvae, white apodous pupae, and brown adults of the *C. palustris* beetle observed on the rocky shores of the sulfidic stream; (**C**) adult specimen of *Euscorpius sulfur* (Euscorpiidae) found in the deep recesses of the cave, visualized using UV light; (**D**) dense agglomeration of collembola identified as an undescribed species of *Acheroxenylla* (Hypogastruridae) observed on thick brown biofilms covering the banks of small stagnant pools located in Sulfur Chapel.

3. Results

3.1. Microbial Communities

The white biofilms covering the bottom sediments in the cave stream were dominated by filamentous bacteria (e.g., *Thiotrix* and *Beggiatoa*) known as chemoautotrophic microorganisms that synthesize organic matter in situ using H_2S as an electron donor and dioxygen from the cave atmosphere as an electron acceptor. Preliminary metagenomic investigations have shown that the brown biofilms covering the mud banks of the small pools and the blue biofilms on walls above the pools were dominated by chemosynthetic sulfur-oxidizing microorganisms. The slightly blue biofilms on moist and anhydrite-free walls, where freshwater seeps in from the epikarst, were also predominantly composed of sulfur-oxidizing bacteria, particularly on the flowstone formations.

3.2. Vertebrate and Invertebrate Community

Two species of fish were observed in the sulfidic stream close to the entrance of the cave, i.e., an eel (*Anguilla* sp.) and numerous Cyprinid specimens (*Alburnoides* sp.). Seven species of bats belonging to three families (Rhinolophidae, Miniopteridae, and Vespertillionidae) (Table 1) were observed throughout the entire cave, including the most remote sections. The invertebrate fauna comprised five aquatic, 25 terrestrial, and four amphibiotic species (i.e., species having terrestrial adults and aquatic larvae). Of these, eight species were endemic and seven species exhibited troglomorphic traits (hereby considered as troglobites or stygobites) (Table 1).

Table 1. List of species found in Sulfur Cave, their habitat, identification method, and endemic status. Troglobites and stygobites are in bold.

Higher Rank Taxon	Family	Species	Habitat	Identification	Endemic
Vertebrates					
Anguilliformes	Anguillidae	*Anguilla* sp.	aquatic	taxonomic	no
Cypriniformes	Cyprinidae	*Alburnoides* sp.	aquatic	taxonomic	no
Chiroptera	Rhinolophidae	*Rhinolophus ferrumequinum* (Screber, 1774)	terrestrial	taxonomic	no
Chiroptera	Rhinolophidae	*Rhiolophus hipposideros* (André, 1797)	terrestrial	taxonomic	no
Chiroptera	Rhinolophidae	*Rhinolophus blasii* (Peters, 1866)	terrestrial	taxonomic	no
Chiroptera	Rhinolophidae	*Rhinolophus euryale* (Blasius, 1853)	terrestrial	taxonomic	no
Chiroptera	Miniopteridae	*Miniopterus schreibersii* (Natterer in Kuhl, 1817)	terrestrial	taxonomic	no
Chiroptera	Vespertillionidae	*Myotis emarginatus* (É. Geoffroy Saint-Hilaire, 1806)	terrestrial	taxonomic	no
Chiroptera	Vespertillionidae	*Myotis bechsteinii* (Kuhl, 1817)	terrestrial	taxonomic	no
Invertebrates					
Oligochaeta	Naididae	*Tubifex tubifex* (O. F. Müller, 1774)	aquatic	taxonomic	no
Gastropoda	Lymnaeidae	*Radix labiata* (Rossmässler, 1835)	aquatic	molecular	no
Gastropoda	Hydrobiidae	*Grossuana euxina* (Wagner, 1928)	aquatic	molecular	no
Amphipoda	**Niphargidae**	***Niphargus lourensis* Fišer, Trontelj & Sket, 2006**	**aquatic**	**molecular**	**no**
Isopoda	**Trichoniscidae**	**gen. sp.**	**terrestrial**	**taxonomic**	**yes**
Isopoda	**Trichoniscidae**	***Alpioniscus* sp.**	**terrestrial**	**taxonomic**	**no**
Pseudoscorpiones	Neobisiidae	*Neobisium (Ommatoblothrus)* sp.	terrestrial	taxonomic, molecular	yes
Pseudoscorpiones	**Chthoniidae**	***Chthonius* sp.**	**terrestrial**	**taxonomic, molecular**	**yes**
Scorpiones	Euscorpiidae	*Euscorpius sulfur* Kovařík et al., 2023	terrestrial	taxonomic	yes

Table 1. *Cont.*

Higher Rank Taxon	Family	Species	Habitat	Identification	Endemic
Acarina	Astigmata	gen. sp.	terrestrial	taxonomic	no
Acarina	Labidostommatidae	*Eunicolina nova* Sellnick, 1931	terrestrial	taxonomic	no
Araneae	**Leptonetidae**	***Cataleptoneta* sp.**	**terrestrial**	**molecular**	**yes**
Araneae	Tetragnathidae	*Metellina merianae* (Scopoli, 1763)	terrestrial	taxonomic	no
Araneae	Agelenidae	*Tegenaria domestica* (Clerck, 1757)	terrestrial	molecular	no
Araneae	Nesticidae	*Kryptonesticus eremita* (Simon, 1880)	terrestrial	molecular	no
Araneae	Linyphiidae	*Prinerigone vagans* (Andouin, 1826)	terrestrial	taxonomic	no
Araneae	Linyphiidae	*Lepthyphantes magnesiae* Brignoli, 1979	terrestrial	taxonomic	no
Chilopoda	Lithobiidae	*Lithobius viriatus* Sseliwanoff, 1880	terrestrial	taxonomic, molecular	no
Chilopoda	Cryptopidae	*Scolopocryptops* sp.	terrestrial	taxonomic, molecular	yes
Chilopoda	Cryptopidae	*Cryptops hortensis* (Donovan, 1810)	terrestrial	taxonomic	no
Collembola	**Hypogastruridae**	***Acheroxenylla* sp.**	**terrestrial**	**taxonomic**	**yes**
Collembola	Hypogastruridae	*Ceratophysella denticulata (Bagnall, 1941)*	terrestrial	taxonomic	no
Collembola	Onychiuridae	*Deuteraphorura* cf. *frasassii (Fanciulli, 1999)*	terrestrial	taxonomic	yes
Collembola	Entomobryidae	*Pseudosinella sexoculata* Schött, 1902	terrestrial	taxonomic	no
Collembola	Entomobryidae	*Heteromurus nitidus* (Tempelton, 1836)	terrestrial	taxonomic	no
Collembola	Paronellidae	*Troglopedetes* sp.	terrestrial	taxonomic	no
Collembola	Sminthuridae	*Disparrhopalites patrizii* (Cassagnau & Delamare, 1953)	terrestrial	taxonomic	no
Collembola	Neelidae	*Neelus* sp.	terrestrial	taxonomic	yes
Diptera	Chironomidae	*Tanytarsus triangularis* Goetghebuer, 1928	amphibiotic	taxonomic	no
Diptera	Chironomidae	*Chironomus* sp.	amphibiotic	taxonomic	no
Coleoptera	Scirtidae	*Contacyphon palustris* (C. G. Thomson, 1855)	amphibiotic	molecular	no
Coleoptera	Dytiscidae	*Hydroglyphus geminus* (Fabricius, 1792)	aquatic	taxonomic	no
Coleoptera	Hydrophilidae	*Coelostoma hispanicum* (Küster, 1848)	amphibiotic	taxonomic	no
Coleoptera	**Staphylinidae**	***Tychobythinus* sp.**	**terrestrial**	**taxonomic**	**yes**

Chironomus sp. (mean ± SD: 978 ± 1009; min–max: 45–2667) and larvae of *Contacyphon palustris* (1751 ± 2738; 45–9112) exhibited high densities in certain sections of the cave. The larval population of the scirtid beetle occurred on the rocky sediments in the swiftly flowing sulfidic stream near the springs. They share this habitat with numerous larvae of

Chironomus sp. (Chironomidae), the less frequent larvae of *Coelostoma hispanicum* (Küster, 1848) (Coleoptera, Hydrophilidae), and the adult aquatic beetle *Hydroglyphus geminus* (Fabricius 1792) (Coleoptera, Dytiscidae). When the *Chironomus* sp. larvae complete their development, they emerge from the aquatic environment and turn into adult flies. At this stage, they cease to feed, search for mates, reproduce, and die, often serving as food for various terrestrial predators inhabiting Sulfur Cave. Dense agglomerations of yellow larvae, white apodous pupae, and brown adults of the *C. palustris* beetle were observed on the rocky shores of the sulfidic stream (Figure 3B) and are likely preyed upon by numerous juvenile and adult specimens of *Euscorpius sulfur* Kovařík et al., 2023 (Euscorpiidae), which are often found in the deep recesses on the cave (Figure 3C).

Based on our observations, we approximated that more than 50,000 specimens of *Tegenaria domestica* (Clerck, 1757) (Agelenidae) inhabit the colony found on the cave wall near the cave entrance, along with several thousand specimens of *Prinerigone vagans* (Andouin, 1826), Linyphiidae. At the edges of this extended spider web, numerous individual spider webs host juvenile and adult specimens of *Metellina merianae* (Scopoli, 1763), Tetragnathidae, while an abundant population of *Lithobius viriatus* Sseliwanoff, 1880, Lithobiidae is found on the cave floor. Upon entering this cave section, a very dense population of chironomid flies, *Virgatanytarsus triangularis* (Goetghebuer, 1928), fills the air, and may represent a significant food resource for spiders and centipedes. Additionally, a very dense population of *T. triangularis* larvae is distributed across the rocky submerged sediments and may feed on the filamentous biofilms. The recently described endemic scorpion *Euscorpius sulfur*, occasionally occurs in Sulfur Cave and Turtle Cave [17].

The sulfidic cave stream in the deep sections of the cave is populated by eight aquatic species, three of which would leave the water upon completing their larval development. *Tubifex tubifex* (O. F. Müller, 1774) (Naididae) has a patchy distribution and was observed in areas with sandy sediments. The aquatic chironomid and beetle larvae may feed on the rich sulfur-oxidizing biofilms covering the stream sedimens, while *H. geminus* is likely to prey upon some of the young aquatic larvae. Two snail species were observed in the shallow flowing streams, scraping the bacterial biofilms: *Grossuana euxina* (Wagner, 1928) and *Radix labiata* (Rossmässler, 1835), both exhibiting sulfur deposits on the shells. The sole invertebrate species observed on the anhydrite deposits was the surface web-building spider *M. merianae* that likely fed on the chironomid flies. Instead, numerous Collembola belonging to the genus *Neelus* and small trichoniscid isopods were found on the moist surfaces of the flowstone formations created by the freshwater seeps in the cave. These organisms likely graze on microbial biofilms. In turn, they might be preyed upon by the eight predator species including the mite species *Eunicolina nova* Sellnick, 1931 (Labidostommatidae); two pseudoscorpion species, *Neobisium* sp. (Neobisiidae) and *Chthonius* sp. (Chthoniidae); two cryptopid centipede species, *Scolopocryptops* sp. and *Cryptops hortensis*, (Donovan, 1810); a pselaphin beetle, *Tychobytinus* sp.; and two web-building spider species, *Kryptonesticus eremita* (Simon, 1880) (Nesticidae) and the new troglobite species of *Cataleptoneta* sp. (Leptonetidae). Accumulations of guano on the cave floor attract terrestrial trichoniscid isopods belonging to the genus *Alpioniscus*, which may be preyed upon by *M. merianae*. A dense agglomeration of Collembola (Figure 3D) identified as an undescribed species of *Acheroxenylla* (Hypogastruridae) was observed on the thick brown biofilms covering the banks of small stagnant pools located in Sulfur Chapel.

3.3. Carbon and Nitrogen Stable Isotope Analysis

Preliminary stable isotope analysis showed light carbon and nitrogen stable isotope values for the terrestrial and aquatic invertebrates collected in Sulfur Cave. The organic carbon $\delta^{13}C$ in these samples ranges between values of −27 and −32‰ compared with the standard PDB for carbon, and the organic nitrogen $\delta^{15}N$ ranges between values of −3 and −10‰ compared with the standard of atmospheric air for nitrogen (Table 2).

Table 2. Nitrogen and carbon stable isotope values of selected taxa in Sulfur Cave.

Sample ID	Taxon	Life Stage	δ^{15}N Org ‰	δ^{13}C Org. ‰
01-2023	*Lithobius viriatus*	adult	−3.37	−31.47
02-2023	*Lithobius viriatus*	adult	−3.97	−32.00
03-2023	*Lithobius viriatus*	adult	−4.07	−32.08
04-2023	*Contacyphon palustris*	larvae	−9.58	−27.24
05-2023	*Contacyphon palustris*	pupae	−9.72	−27.66
06-2023	*Contacyphon palustris*	adults	−9.75	−28.12

4. Discussions

Similarly to other sulfidic subterranean ecosystems [6], Sulfur Cave is host to numerous different species that form stable subterranean populations. Thirty invertebrate species have been identified to date in this cave, of which eight are endemic: an amphipod, a scorpion, two pseudoscorpions, a spider, a springtail, a centipede, and a beetle, making Sulfur Cave a hotspot for subterranean diversity [18]. The presence of eight endemic species in Sulfur Cave (Table 1) suggests in situ evolution of the fauna due to secluded subsurface conditions. However, the number of endemic species is smaller than in Movile Cave [4] and Ayyalon Cave [19,20], and similar to Frasassi [21] and Cueva de Villa Luz caves [22,23]. The latter caves, having a closer connection with the surface fauna, thus sustain a smaller number of endemic species.

Surprisingly, however, most of the invertebrate populations encountered in this cave belong to surface species that do not display any troglomorphic traits such as reduction or loss of eyes and pigment, elongation of appendages [24]. This may be a consequence of the relatively small size of the caves in this region and their open access to the surface (they do not exceed a few hundred meters of subterranean passages), as well as the relative young age of the caves (hypogenic caves take a rather short time to develop; [25]), but we hypothesize the surface fauna is drawn to these caves by the presence of, and easy access to, significant amounts of food resources available year-round in the subterranean environment. The most surprising biological feature encountered in Sulfur Cave is an immense colonial spider web [10] spun by a very large population of the agelenid spider *T. domestica*. The species was previously documented in the entrance of dry caves [26] but has never been reported to build colonial nets.

The surprising abundance of the subterranean fauna encountered throughout Sulfur Cave and the presence of sulfidic water and ubiquitous microbial biofilms consisting of sulfur-oxidizing bacteria [27] suggests that this chemolithoautotrophic ecosystem is self-sustaining (i.e., the subterranean community relies on food produced in situ by chemoautotrophic microorganisms) [6,28]. Detailed genetic analysis of the cave microbiome will be the subject of future investigations conducted in Sulfur Cave and other sulfidic caves in this region. The results of the preliminary stable isotope analysis strongly support this hypothesis showing that negligible amounts of trophic resources of photosynthetic origin contribute to the base of the underground food web [20,29].

Sulfur Cave offers a unique opportunity to study the transfer of trophic resources from the aquatic to the terrestrial cave environment. We hypothesize that organic molecules generated chemoautotrophically by sulfur-oxidizing microorganisms forming biofilms that cover the bottom sediments in the cave stream are consumed by the aquatic larvae of two chironomids (*V. triangularis* and *Chironomus* sp.) and two beetle species (*C. palustris* and *H. geminus*). Upon completing their larval development, they emerge from the water and become aerial and terrestrial insects, respectively [30]. They cease to feed and most likely become the food base for the unusually abundant and diverse terrestrial cave community that consists almost exclusively of predators: 14 species of pseudoscorpions, scorpions, mites, spiders, centipedes, and beetles. Aquatic-to-terrestrial transfer of trophic resources is not uncommon in surface ecosystems [31], but the limited number of species involved in this process in Sulfur Cave makes it easy to use this case study as a model in understanding this food transfer in more complex food webs that involve a multitude of species.

Biological studies currently performed in Sulfur Cave and other sulfidic caves in the Vromoner and Pixaria canyons on the Sarandaporo River and in the Langarica Canyon on a different tributary of the Aaos/Viosa River focus on subterranean biodiversity, ecology, and geomicrobiology. Ongoing research involves the continuation of the survey of the fauna inhabiting these caves and study of the spatial distribution of the different species in the subterranean environment. The results of the preliminary stable isotope studies led to the initiation of an ambitious isotopic survey of all the invertebrates reported from these caves, which is expected to shed light on their position within the subterranean food web. Morphological and molecular descriptions of new species are currently underway. Metagenomic methods are employed to analyze the species composition of the cave biofilms, determine their role within the cave microbiome, and attempt to decipher their contribution to limestone dissolution in the process of sulfuric acid speleogenesis.

5. Conclusions

The karst region at the border of Albania and Greece is unique due to its caves, which host aquatic and terrestrial-cave-restricted species associated with sulfidic subterranean aquifers and several cases of endemism. Given that this area is already impacted by human-driven changes, there is a pressing need for effective conservation strategies. Despite the establishment of national parks, this karstic area remains outside protected zones. Recognized by conservation initiatives such as "Save the Balkan Rivers" and "The Blue Heart of Europe", our results highlight the value of its subterranean biodiversity, which enhances its ecological significance and biological uniqueness, reinforcing its priority for conservation actions.

Author Contributions: Conceptualization, S.M.S.; validation, S.M.S. and T.B.; investigation, S.M.S., A.Ş. and R.I.B.; writing—original draft preparation, S.M.S., R.I.B., A.Ş. and T.B.; writing—review and editing, S.M.S., R.I.B., A.Ş. and T.B. All authors have read and agreed to the published version of the manuscript.

Funding: This research was funded by Biodiversa+, the European Biodiversity Partnership under the 2021-2022 BiodivProtect joint call for research proposals, co-funded by the European Commission (GA N°101052342) and with the funding organisations Ministry of Universities and Research (Italy), Agencia Estatal de Investigación—Fundación Biodiversidad (Spain), Fundo Regional para a Ciência e Tecnologia (Portugal), Suomen Akatemia—Ministry of the Environment (Finland), Belgian Science Policy Office (Belgium), Agence Nationale de la Recherche (France), Deutsche Forschungsgemeinschaft e.V. (Germany), Schweizerischer Nationalfonds (Grant N° 31BD30_209583, Switzerland), Fonds zur Förderung der Wissenschaftlichen Forschung (Austria), Ministry of Higher Education, Science and Innovation (Slovenia), and the Executive Agency for Higher Education, Research, Development and Innovation Funding (Romania) and partially supported by a grant of the Ministry of Research, Innovation and Digitization, CNCS—UEFISCDI, project number PN-III-P1-1.1-TE-2021-1047, within PNCDI III.

Institutional Review Board Statement: Not applicable.

Data Availability Statement: The original contributions presented in the study are included in the article. Further inquiries can be directed to the corresponding authors.

Acknowledgments: The authors would like to thank R. Nitescu, M. Hristescu, J-F. Flot, S. Flot, M. Kenesz, M. Fotiadi, M. Vaxevanopoulos, S. Galdenzi, A. Frumkin, N. Rosene, M. Audy, R. Bouda, C. Chauveau, M. Tawfeeq, I. Urák, A. Zsigmond, and A.D. Crînguș for their help during the field expeditions and V. Atudorei from the University of New Mexico in Albuquerque for his help with the preliminary stable isotope analysis. The identification of the specimens observed and collected in Sulfur Cave was performed by I. Urák (spiders), G. Popovici (centipedes and pseudoscorpions), M. Bertrand and P. Luptacik (mites), L. Kováč and V. Papáč (springtails), B. Klausnitzer and M. Toledo (beetles), E. Stur and T. Ekrem (chironomids), M. Patrick (oligochaetes), O. Popa (snails), F. Kovařík (scorpions), F. Stoch (amphipods), and G. Gentile (isopods). Special thanks to the Hellenic

Federation of Speleology, the Hellenic Institute of Speleological Research, and the Directorate General for the Protection and Development of Forests and the Rural Environment, Directorate of Forest Management of the Greek Ministry of Environment and Energy, for issuing research and sampling permits in caves in the Sarandaporo Valley, in compliance with the Nagoya Protocol on Access to Genetic Resources and the Fair and Equitable Sharing of Benefits Arising from their Utilization to the Convention on Biological Diversity.

Conflicts of Interest: The authors declare no conflicts of interest.

References

1. Ballard, R.D. Notes on a major oceanographic find. *Oceanus* **1997**, *20*, 34–44.
2. Sarbu, S.M. *Movile Cave: A Chemoautotrophically Based Groundwater Ecosystem in Subterranean Ecosystems*; Wilkens, H., Culver, D.C., Humphreys, W.F., Eds.; Elsevier: Amsterdam, The Netherlands, 2000; pp. 319–343.
3. Hutchins, B.T.; Engel, A.S.; Nowlin, W.H.; Schwartz, B.F. Chemolithoautotrophy supports macroinvertebrate food webs and affects diversity and stability in groundwater communities. *Ecology* **2016**, *97*, 1530–1542. [CrossRef] [PubMed]
4. Brad, T.; Iepure, S.; Sarbu, S.M. The chemoautotrophically based Movile Cave groundwater ecosystem, a hotspot of subterranean biodiversity. *Diversity* **2021**, *13*, 128. [CrossRef]
5. Culver, D.C.; Pipan, T. *The Biology of Caves and Other Subterranean Habitats*; Oxford University Press: Oxford, UK; New York, NY, USA, 2019.
6. Engel, A.S. Observations on the biodiversity of sulfidic karst habitats. *J. Cave Karst Stud.* **2007**, *69*, 187–206.
7. Nealson, K.H.; Inagaki, F.; Takai, K. Hydrogen-driven subsurface lithoautotrophic microbial ecosystems (SLiMEs): Do they exist and why should we care? *Trends Microbiol.* **2005**, *13*, 405–410. [CrossRef] [PubMed]
8. Areniello, M.; Matassa, S.; Esposito, G.; Lens, P.N. Biowaste upcycling into second-generation microbial protein through mixed-culture fermentation. *Trends Biotechnol.* **2023**, *41*, 197–213. [CrossRef] [PubMed]
9. Benassi, A. Lengarices 2023. Inseguendo il soffio del Drago (Albania). *Speleologia SSI* **2024**, *89*, 4–6.
10. Audy, M.; Bouda, R.; Bruthans, J.; Rŷžiŷka, V. Albanian hypogene caves in the area of Vromoner thermal springs on the Sarandaporo River. *Speleoforum* **2022**, *41*, 42–49.
11. Audy, M.; Bruthans, J.; Mareš, J.; Sarbu, S.; Galdenzi, S.; Bouda, R. Sulfur 2023, hypogenni jeskyne Albanie a Recka Kavasila, Vromoner, Langarica. *Speleoforum* **2024**, *43*, 104–111.
12. Eftimi, R.; Frashëri, A. Regional hydrogeological characteristics of thermal waters of Albania. *Acta Geogr. Silesiana* **2018**, *12*, 29.
13. Folmer, O.; Black, M.; Hoeh, W.; Lutz, R.; Vrijenhoek, R. DNA primers for amplification of mitochondrial cytochrome c oxidase subunit I from diverse metazoan invertebrates. *Mol. Mar. Biol. Biotechnol.* **1994**, *3*, 294–299. [PubMed]
14. Colgan, D.J.; Hutchings, P.A.; Brown, S. Phylogenetic relationships within the Terebellomorpha. *J. Mar. Biol. Assoc. United Kingd.* **2001**, *81*, 765–773. [CrossRef]
15. Ratnasingham, S.; Wei, C.; Chan, D.; Agda, J.; Agda, J.; Ballesteros-Mejia, L.; Ait Boutou, H.; El Bastami, Z.M.; Ma, E.; Manjunath, R.; et al. BOLD v4: A Centralized Bioinformatics Platform for DNA-based Biodiversity Data. *arXiv* **2024**, arXiv:2404.05696.
16. Sayers, E.W.; Bolton, E.E.; Brister, J.R.; Canese, K.; Chan, J.; Comeau, D.C.; Connor, R.; Funk, K.; Kelly, C.; Kim, S.; et al. Database resources of the national center for biotechnology information. *Nucleic Acids Res.* **2022**, *50*, D20–D26. [CrossRef]
17. Kovařík, F.; Audy, M.; Sarbu, S.M.; Fet, V. *Euscorpius sulfur* sp. n.(Scorpiones: Euscorpiidae), a new cave scorpion from Albania and northwestern Greece. *Euscorpius* **2023**, *376*, 1–14.
18. Culver, D.C.; Sket, B. Hotspots of Subterranean Biodiversity in Caves and Wells. *J. Cave Karst Stud.* **2000**, *62*, 11–17.
19. Por, F.; Dimentman, C.; Frumkin, A.; Naaman, I. Animal life in the chemoautotrophic ecosystem of the hypogenic groundwater cave of Ayyalon (Israel): A summing up. *Nat. Sci.* **2013**, *5*, 7–13. [CrossRef]
20. Frumkin, A.; Chipman, A.D.; Naaman, I. An isolated chemolithoautotrophic ecosystem deduced from environmental isotopes: Ayyalon cave (Israel). *Front. Ecol. Evol.* **2023**, *10*, 1040385. [CrossRef]
21. Sarbu, S.; Galdenzi, S.; Menichetti, M.; Gentile, G. Geology and biology of the Frasassi caves in Central Italy: An ecological multi-disciplinary study of a Hypogenic underground ecosystem. In *Subterranean Ecosystems*; Wilkens, H.C.D., Ed.; Elsevier: New York, NY, USA, 2000; pp. 359–378.
22. Hose, L.D.; Pisarowicz, J.A. Cueva de villa luz, Tabasco, Mexico: Reconnaissance study of an active sulfur spring cave and ecosystem. *J. Cave Karst Stud.* **1999**, *61*, 13–21.
23. Palmer, A.N. Sulfuric acid caves: Morphology evolution. In *Treatise on Geomorphology*; Shroder, J., Frumkin, A., Eds.; Academic Press: San Diego, CA, USA, 2013; pp. 241–257.
24. White, W.B.; Culver, D.C. (Eds.) *Encyclopedia of Caves*; Academic Press: Cambridge, MA, USA, 2012.
25. Klimchouk, A.B. *Hypogene Speleogenesis: Hydrogeological and Morphogenetic Perspective NCKRI-Special Paper 1*; National Cave and Karst Research Institute: Carlsbad, CA, USA, 2007.
26. Bristowe, W.S. *World of Spiders*; Collins New Naturalist: London, UK, 1958.
27. Fenchel, T.; Blackburn, H.; King, G.M. *Bacterial Biogeochemistry: The Ecophysiology of Mineral Cycling*; Academic Press: Cambridge, MA, USA, 2012.

28. Frumkin, A.; Chipman, A.D. A Subsurface Stepping Stone Hypothesis for the Conquest of Land by Arthropods. *Diversity* **2023**, *16*, 6. [CrossRef]
29. Sarbu, S.M.; Kane, T.C.; Kinkle, B.K. A chemoautotrophically based cave ecosystem. *Science* **1996**, *272*, 1953–1955. [CrossRef]
30. Karima, Z. Chironomidae: Biology, ecology and systematics. In *The Wonders of Diptera: Characteristics, Diversity, and Significance for the World's Ecosystems*; IntechOpen: Rijeka, Croatia, 2021; pp. 1–25.
31. Bartrons, M.; Sardans, J.; Hoekman, D.; Peñuelas, J. Trophic transfer from aquatic to terrestrial ecosystems: A test of the biogeochemical niche hypothesis. *Ecosphere* **2018**, *9*, e02338. [CrossRef]

diversity

MDPI

Article

On the Diversity of Semiochemicals of the Pygidial Gland Secretions of Subterranean Ground Beetles (Coleoptera: Carabidae)

Sofija Vranić [1], Ljubodrag Vujisić [2,*], Nikola Vesović [1], Marina Todosijević [2], Miloš Pavićević [3], Dejan Radović [4] and Srećko Ćurčić [1]

1 Faculty of Biology, Institute of Zoology, University of Belgrade, Studentski Trg 16, 11000 Belgrade, Serbia
2 Faculty of Chemistry, University of Belgrade, Studentski Trg 12–16, 11000 Belgrade, Serbia
3 Biospeleological Society of Montenegro, Cara Lazara 22, 81000 Podgorica, Montenegro
4 Faculty of Security Studies, University of Belgrade, Gospodara Vučića 50, 11118 Belgrade, Serbia
* Correspondence: ljubaw@chem.bg.ac.rs; Tel.: +381-11-3336-899

Abstract: Pygidial glands are of great importance to ground beetles for defense against predators, especially for the species that live in subterranean habitats. The purpose of our study is to better understand the chemistry of the pygidial gland secretions of subterranean ground beetles, as well as the function and structure of the glands. We studied both the chemical composition of the pygidial gland secretion and morphology of the glands in adults of the troglophilic ground beetle species *Laemostenus* (*Antisphodrus*) *cavicola* (Schaum, 1858). The chemical composition of its defensive secretion was revealed using gas chromatography-mass spectrometry (GC-MS), while pygidial gland morphology of the beetle was investigated using bright-field microcopy. In total, seven chemical compounds were detected in the secretion mixture. Formic acid was the most dominant compound, followed by dodecyl acetate and undecane. Other chemicals were present in minor amounts. The morphological structure of the pygidial glands of *L.* (*A.*) *cavicola* was compared with the structure of the glands of the related congeneric troglophilic species *Laemostenus* (*Pristonychus*) *punctatus* (Dejean, 1828). Summary data on the semiochemicals that have been recorded so far in subterranean ground beetle species are presented, and the differences in the chemical composition of the secretions between and among troglobitic and troglophilic species are discussed. So far, forty-four compounds have been detected in four subterranean ground beetle species (two troglobites belonging to the tribe Trechini and two troglophiles belonging to the tribe Sphodrini). The results of this study indicate the great diversity of chemicals in the pygidial gland secretions of subterranean ground beetles.

Keywords: carabid beetles; Platyninae; Trechinae; defensive glands; secretion mixtures; gas chromatography-mass spectrometry

1. Introduction

Chemically-mediated communication in insects is well documented [1,2]. Small organic compounds, also known as semiochemicals, act as chemical cues. These are often divided into three groups: pheromones, allomones and kairomones. While the chemicals of the former group are recognized by conspecifics, the chemical compounds of the remaining two groups mediate interspecific interactions [2,3].

Ground beetles (Carabidae) release a wide variety of chemicals from a pair of abdominal glands, called pygidial glands. Each pygidial gland comprises secretory lobes (in which the production of defensive secretions takes place), a reservoir (for the storage of defensive secretions) and transportation elements. The latter include radial collecting canals that carry glandular secretions from individual secretory lobes and merge into a main collecting canal that leads secretion from the secretory lobes to the reservoir, as well as an efferent duct via which secretion is released into the environment [4,5].

The pygidial gland secretions of ground beetles are primarily regarded as allomones as they exhibit deterrent, toxic and irritant properties serving in the defense against predators, yet additional functions have also been taken into consideration [1,2,6]. Some chemical products of pygidial glands have been proven to have antimicrobial properties [7–9]. Furthermore, it was hypothesized that certain chemical compounds play a role in sexual communication or serve as alarm pheromones in conspecifics [2,10]. To date, pygidial gland secretions have been researched in about 500 species of ground beetles from all over the world [11,12].

Species of the tribe Sphodrini were rarely chemoecologically studied in the past. Schildknecht et al. [13] first analyzed the defensive secretions of two European species of the genus *Calathus* Bonelli, 1810: *Calathus* (*Calathus*) *fuscipes* (Goeze, 1777) and *C.* (*Neocalathus*) *melanocephalus* (Linnaeus, 1758). Following that, Will et al. [14] carried out a study on the defensive secretion of the Nearctic *Calathus* (*Neocalathus*) *ruficollis* Dejean, 1828. Finally, Vesović et al. [15] analyzed the chemical composition of the defensive secretion of the troglophilic *Laemostenus* (*Pristonychus*) *punctatus* (Dejean, 1828). In the same paper [15], the secretions of two troglobitic ground beetle taxa of the tribe Trechini, *Duvalius* (*Paraduvalius*) *milutini* S. Ćurčić, Vrbica, Antić & B. Ćurčić, 2014 and *Pheggomisetes globiceps ninae* S. Ćurčić, Schönmann, Brajković, B. Ćurčić & Tomić, 2004 (Serbian stenoendemics), were chemoecologically investigated. That was the first and, so far, the only study to provide data on the semiochemicals of the pygidial gland secretion mixtures from both cave-dwelling ground beetles (in total, three taxa: one troglophile and two troglobites) and representatives of the tribe Trechini [15].

Caves and pits are subterranean habitats characterized by stable ecological conditions, which significantly differ from those in the surrounding surface habitats [16]. Such conditions led to the development of unique adaptations in the organisms inhabiting those habitats. Some of the adaptations are directed towards the reduction of morphological structures that are not of use in subterranean habitats. For instance, eye and hind wing reduction is documented in many cave-dwelling insects [16]. Considering the aforementioned trend of reduction, as well as decreasing predation pressure, Vesović et al. [15] suggested that the defensive secretions of subterranean ground beetles would be simplified in adapted (troglobitic) species compared to less adapted (troglophilic) ones [15]. However, the results of this study were contrary to that hypothesis. No further studies on subterranean ground beetles' defensive secretions have been conducted in the meantime.

The pygidial glands of representatives of the tribe Sphodrini have not been the focus of many investigations in the past. Gland morphology of *Calathus* (*Neocalathus*) *ambiguus* (Paykull, 1790) and *Laemostenus* (*Pristonychus*) *terricola* (Herbst, 1784) was briefly mentioned by Forsyth [4], with no detailed description or measurements provided. The first in-depth morphological study on the pygidial glands of Sphodrini was carried out by Nenadić et al. [8] for the species *L.* (*P.*) *punctatus*. Later, scanning electron microscopy and nonlinear microscopy were introduced as methods that enabled investigations of the ultrastructure of the pygidial glands of the mentioned ground beetle species [17].

In the present study, we chose to chemoecologically investigate the defensive secretion mixture and study pygidial gland morphology of the ground beetle species *Laemostenus* (*Antisphodrus*) *cavicola* (Schaum, 1858), which belongs to the tribe Sphodrini (Figure 1). The mentioned species is a troglophile, it is adapted to complete its life cycle in caves, but can also be found outside of caves. Contrary to troglophiles, troglobites are strictly bound to subterranean habitats, such as caves and pits. The abovementioned troglophilic species is distributed in southern and southeastern Europe, mainly on the Balkan Peninsula [18]. This species was not previously investigated in terms of chemical ecology or pygidial gland morphology.

Figure 1. Habitus (dorsal view) of the adult specimen of *L.* (*A.*) *cavicola*. Photo N. Vesović.

Our aims were to: (i) identify chemical compounds in the defensive pygidial gland secretion of *L.* (*A.*) *cavicola*; (ii) examine pygidial gland morphology of the same species; (iii) compare the chemical composition of the secretion and the structure of the glands with related subterranean species; (iii) summarize data on all semiochemicals present so far in subterranean ground beetles; and (iv) discuss the differences in the chemical composition of the secretions between troglophilic and troglobitic representatives of ground beetles.

2. Materials and Methods

2.1. Sample Collection

Eight adult specimens (four males and four females) of *L.* (*A.*) *cavicola* were collected in the Grbočica Cave, village of Trnovo, area of Crmnica, close to the settlement of Virpazar, southern Montenegro. Ground beetle individuals were manually collected by S. Ćurčić and M. Pavićević on 30 September 2018. These were then placed in a portable chamber along with moist substrate from the collecting site. Temperature was kept at a constant level (10 °C). The sufficient level of humidity was maintained by occasional spraying of water. The ground beetles were fed on earthworms.

2.2. Chemical Analyses of Pygidial Gland Secretion

Sample preparation for gas chromatography-mass spectrometry (GC-MS) was conducted at room temperature in the laboratory of the University of Belgrade - Faculty of Chemistry (Belgrade, Serbia). Each individual beetle of the same sex was stimulated to discharge its defensive pygidial gland secretion by squeezing the tip of its abdomen and by pinching the legs with a tweezers into a single 12-mL glass vial with dichloromethane (0.5 mL) as a solvent (Merck, Darmstadt, Germany). The samples were subjected to GC-MS analyses immediately after their preparation. Samples were analyzed on the GC-MS system (Agilent 7890A–5975C, Agilent Technologies, Santa Clara, CA, USA) in splitless mode (with 1 μL injection volume) on a polar HP-INNOWax capillary column (30 m × 0.32 mm × 0.25 μm). Oven temperature was linearly programed in the range of 40–240 °C at a rate of 10 °C min^{-1}, with a final 10-min hold. The electron ionization (EI) (70 eV) mass spectral range was 40–550 m/z. Compounds were identified by comparison with commercially available NIST 17 and Willey 07 mass spectral libraries containing more than half a million spectra. In addition, all compounds were characterized by retention indices (RIs) obtained from the corresponding series of n-alkanes analyzed under the same chromatographic conditions immediately after the sample run. RIs obtained on standard polar capillary columns were compared with the available literature data from NIST Chemistry WebBook and PubChem (Table 1). The relative mass percentages of the identified chemicals were calculated from the corresponding areas of the GC-MS peaks. This is particularly important if analyzed compounds have different polarities and/or concentrations that cause peak broadening, because in that case measuring peak heights instead of areas would be misleading (Figure 2, Table 1).

Table 1. The chemical composition of the pygidial gland secretion of *L.* (*A.*) *cavicola* analyzed by GC-MS on a polar column (HP-INNOWax 30 m × 0.32 mm × 0.25 μm).

Peak	Rt (min)	Compound	RI	RI_{lit}	Relative Percentage (%)
1	4.16	Undecane	1100	1100	27.0
2	8.79	Acetic acid	1465	1400–1498	0.5
3	9.38	Formic acid	1508	1470–1544	37.4
4	13.81	Dodecyl formate	1858	-	0.8
5	14.31	Dodecyl acetate	1901	1876–1900	33.5
6	16.54	1-Tetradecyl acetate	2104	2062–2106	0.1
7	26.16	Palmitic acid	2935	2871–2954	0.7

RI—retention indices calculated from the GC-MS retention times of a series of n-alkanes obtained under the same chromatographic conditions; RI_{lit}—retention indices from summarized NIST Chemistry WebBook and PubChem literature data obtained on standard polar capillary columns; Rt—retention time.

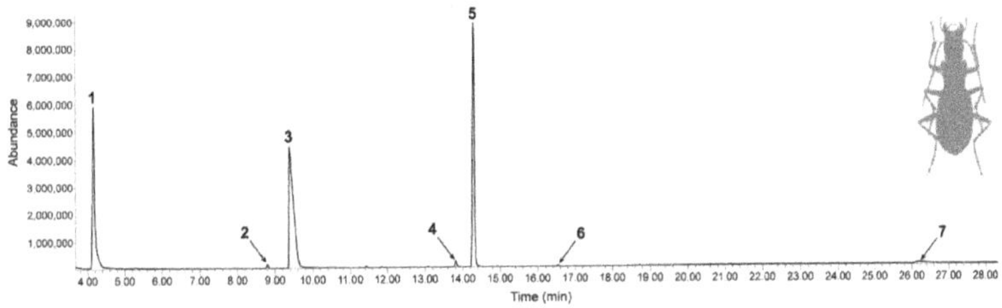

Figure 2. Gas chromatography-mass spectrometry (GC-MS) total ion chromatogram of the dichloromethane pygidial gland secretion extracts from the adults of *L.* (*A.*) *cavicola*. Ordinal numbers above peaks correspond to numbers in Table 1.

2.3. Morphological Analyses

A pair of pygidial glands were carefully extracted in 70% ethanol. All morphological features of the glands were observed and precisely measured. Gland structures were photographed with a Nikon SMZ800N stereomicroscope equipped with a Nikon DS-Fi2 digital camera (Nikon Corp., Tokyo, Japan). The measuring of different parts of the pygidial glands was conducted using a Nikon DS-L3 control unit (Nikon Corp., Tokyo, Japan).

The analyzed ground beetle specimens were deposited in the entomological collection of the Institute of Zoology, University of Belgrade-Faculty of Biology (Belgrade, Serbia).

3. Results

3.1. Chemical Composition of Pygidial Gland Secretion

In total, seven chemical compounds were detected in the pygidial gland secretion mixture of *L. (A.) cavicola* (Table 1). The secretion extract of the mentioned species contained one hydrocarbon (alkane), three carboxylic acids (two short-chain and one long-chain) and three esters. The most abundant compound in the mixture was formic acid (37.4%), followed by dodecyl acetate (33.5%) and undecane (27.0%). The remaining compounds (dodecyl formate, palmitic acid, acetic acid and 1-tetradecyl acetate) were found in minor amounts (each with less than 1%) (Figure 2, Table 1). No qualitative and quantitative differences in the chemical composition of the secretion were observed between the sexes of *L. (A.) cavicola*.

In comparison to the only previously analyzed congener, *L. (P.) punctatus* (at the same time, this is the only chemoecologically studied troglophilic species of ground beetles to date), *L. (A.) cavicola* had a simpler secretion mixture, with only seven compounds detected [vs. thirteen compounds found in *L. (P.) punctatus*] [15]. However, certain chemoecological features of the defensive secretions appeared to be similar among these related species. The presence of formic acid, acetic acid, alkanes, esters and fatty acids was recorded in both species. Formic acid, dodecyl acetate and undecane were major compounds in both species, even though *L. (P.) punctatus* had a somewhat lower percentage of the former compound [19.4% vs. 37.4% in *L. (A.) cavicola*] [15]. Acetic acid and fatty acids [caproic, palmitic, stearic and oleic acids in *L. (P.) punctatus* vs. only palmitic acid in *L. (A.) cavicola*] were minor constituents of the secretion mixtures in both species. While the secretion of *L. (P.) punctatus* possessed only long-chain acetates (decyl acetate, undecyl acetate and dodecyl acetate) in its pygidial gland secretion, the one of *L. (A.) cavicola* was distinguished by the presence of dodecyl formate. The defensive secretion of *L. (P.) punctatus* was characterized by the greater diversity of chemicals in terms of the recorded number of alkanes [3 vs. 1 in *L. (A.) cavicola*] and fatty acids [4 vs. 1 in *L. (A.) cavicola*]. Finally, the secretion of *L. (A.) cavicola* lacked an alcohol (1-dodecanol) and caproic acid, while these chemicals were present in the secretion of *L. (P.) punctatus* [15].

Comparing the semiochemical content of the secretions of troglophilic Sphodrini with compounds in the secretion samples of surface-dwelling Sphodrini, many similarities could be noted. The mixture of formic acid, various alkanes and long-chain esters, as well as the presence of acetic acid and different fatty acids, is typical for the species of both groups [13–15]. Most troglophilic species of Sphodrini can often be found in surface habitats, so it is not surprising that the chemical composition of their secretions is similar to the content recorded in species that live in outdoor habitats. However, it seems that the production of alcohols and caproic acid is restricted only to cave-dwelling Sphodrini [15].

Two chemoecologically investigated troglobitic ground beetle species significantly differed in regards to the content of their defensive secretions. It was proven that *Pheggomisetes globiceps* Buresch, 1925 is capable of producing an aldehyde (benzaldehyde), a phenol (*p*-cresol), hydrocarbons (alkanes and alkenes) and carboxylic acids in its pygidial glands. In contrast to that, the secretion of *D. (P.) milutini* was characterized by the presence of carboxylic acids alone, including those of variable chain length (four medium-chain and four long-chain) and an aromatic one (benzoic acid). Nevertheless, the defensive secretions of these two species shared some features. Benzoic, caproic and all four long-chain fatty acids, present in *D. (P.) milutini*, were also found in *P. globiceps*. Three medium-chain fatty acids (pelargonic, capric and lauric) were only present in *D. (P.) milutini* [15].

Comparison between troglophilic and troglobitic ground beetles in terms of their defensive compounds meets certain difficulties, as the analyzed species of both groups belong to different subfamilies (troglophiles to Platyninae, and troglobites to Trechinae). However, it is worthwhile to compare certain chemoecological features of both groups' semiochemicals. While representatives of both groups possess hydrocarbons, long-chain fatty acids and various aliphatic low-molecular-weight carboxylic acids in their defensive secretions, some additional compounds (e.g., aldehydes, *p*-cresol and aromatic carboxylic acids) can be found only in troglobitic taxa [15]. The overall diversity of hydrocarbons, low-molecular-weight carboxylic acids and fatty acids was greater in troglobites. In total, seventeen hydrocarbons, sixteen carboxylic acids and six fatty acids were isolated from troglobitic taxa. In the two analyzed troglophilic species, each of these three groups of compounds contained a total of three chemicals [15]. On the other hand, formic acid was not detected in the troglobites analyzed so far [15]. Troglobitic ground beetle taxa also lacked esters and alcohols. According to Vesović et al. [15], the observed differences between troglobites and troglophiles may be intergeneric, as ground beetles are characterized by a high level of conservation, so their defensive secretions should not be greatly affected by the existing selective pressures. Out of the analyzed troglobitic ground beetles, *P. globiceps* possessed the largest number of compounds (32). On the other hand, *D.* (*P.*) *milutini* contained only nine carboxylic acids in its pygidial gland secretion. Two troglophilic *Laemostenus* species were more alike regarding the number of detected compounds [*L.* (*A.*) *cavicola* and *L.* (*P.*) *punctatus* had seven and thirteen compounds, respectively]. Even though the pygidial gland secretion of *L.* (*A.*) *cavicola* had the smallest number of compounds, it possessed all classes of organic compounds characteristic for representatives of the tribe Sphodrini (short-chain carboxylic acids, alkanes, esters and fatty acids). Considering all analyzed cave-dwelling ground beetle species (two troglophiles and two troglobites), forty-four compounds were isolated from their pygidial gland secretion extracts. Hydrocarbons and carboxylic acids (nineteen and seventeen, respectively) were the most numerous compounds. Esters, which were limited to troglophilic species, were less abundant (five). Aldehydes, phenols and alcohols were represented by a single compound each [15]. A list of all chemical compounds detected in the pygidial gland secretions of subterranean ground beetle species studied to date is shown in Table 2.

3.2. Pygidial Gland Morphology

The paired pygidial glands of *L.* (*A.*) *cavicola* are composed of clustered secretory lobes, which are white and spherical (Figure 3A,C). The diameter of secretory lobes varies between 210 and 300 μm. The number of secretory units (lobes) per cluster ranges between 25 and 30. The main collecting canal is 2 cm long and 50–80 μm wide (Figure 3A). The diameter of its lumen is 20–30 μm. The main collecting canal enters the basal part of the reservoir near the beginning of the efferent duct (Figure 3B). The reservoir is elongated, ellipsoidal in shape and medially flattened at its inner margin. The lumen of the reservoir cannot be seen due to the thickness of its well-developed muscle wall (Figure 3B). The length of the reservoir is 1.97 mm and its width is 0.88 mm. It sharply and distally narrows, forming a 0.18-mm wide and 1.3-mm long efferent duct through which secretion is released to the outside (Figure 3B).

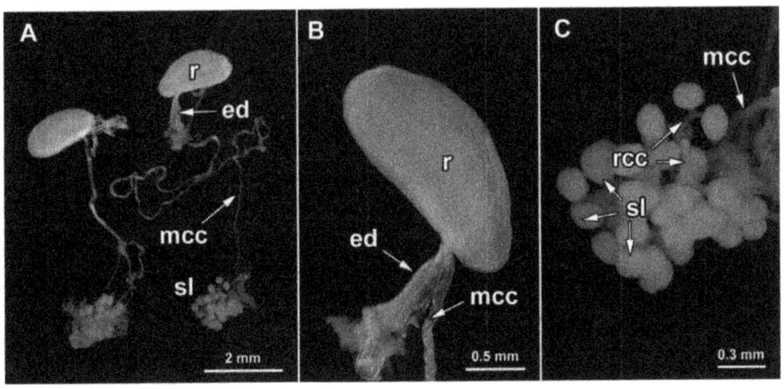

Figure 3. Pygidial gland morphology of *L.* (*A.*) *cavicola*: (**A**) a pair of the glands; (**B**) a reservoir with the distal part of a main collecting canal and an efferent duct; (**C**) a cluster of secretory lobes. Abbreviations: ed—efferent duct; mcc—main collecting canal; r—reservoir; rcc—radial collecting canal; sl—secretory lobe. Photo N. Vesović.

Table 2. The chemical composition of the pygidial gland secretions of subterranean ground beetle species (the first two are troglophilic, while the latter two are troglobitic) analyzed to date after [15].

Compound	*L. (A.) cavicola*	*L. (P.) punctatus*	*D. (P.) milutini*	*P. globiceps*
Undecane	*	*		*
Tridecane				*
Acetic acid	*	*		*
Formic acid	*	*		
Benzaldehyde				*
Propionic acid				*
Isobutyric acid				*
Butyric acid				*
Isovaleric acid				*
Decyl acetate		*		
Undecyl acetate		*		
Dodecyl formate	*			
Dodecyl acetate	*	*		
1-Tetradecyl acetate	*			
Isocaproic acid				*
1-Dodecanol		*		
Caproic acid		*	*	*
p-Cresol				*
Pelargonic acid			*	
Capric acid			*	
7-Hexyldocosane		*		
9-Methyltetracosane		*		
Benzoic acid			*	*
Lauric acid			*	
Pentacosane				*
Pentacosene				*
3-Ethyltetracosane				*
Hexacosane				*
Myristic acid			*	*
11-Methylheptacosane				*
Heptacosene				*
Heptacosadiene isomer 1				*
Heptacosadiene isomer 2				*
Octacosane				*
Palmitic acid	*	*	*	*
Nonacosene				*
Nonacosadiene isomer 1				*
Nonacosadiene isomer 2				*
Nonacosadiene isomer 3				*
Nonacosatetraene				*
Nonacosapentaene				*
Stearic acid		*	*	*
Oleic acid		*	*	*
Linoleic acid				*

*—The presence of compounds in the species.

Each pygidial gland of L. (*A.*) *cavicola* comprises 30–35 secretory lobes, which is significantly fewer than the number of lobes reported by other authors (60–70 lobes per cluster) for L. (*P.*) *terricola* and L. (*P.*) *punctatus* [4,8]. The number of lobes in L. (*A.*) *cavicola* is also greater than the number of the same structures (13) reported in C. (*N.*) *ambiguus*, which belongs to the same tribe (Sphodrini) [4]. Forsyth [4] reported the presence of only 12 secretory lobes in representatives of the tribe Platynini that belongs to the same subfamily (Platyninae). On the other hand, 13 Japanese species of Platyninae have a great number of secretory lobes (50 or more) [19]. According to the results of the aforementioned studies, it seems that the number of lobes varies within the entire subfamily Platyninae, but also among its taxa. The spherical shape of the lobes in L. (*A.*) *cavicola* indicates the presence of carboxylic acids as dominant compounds [19], which was proven by chemical analyses. The average size of a single lobe was somewhat greater in L. (*A.*) *cavicola* (210–300 μm) compared to L. (*P.*) *punctatus* (150–200 μm) (Table 3) [8]. The presence of a long main collecting canal is a common feature for all Platyninae [4,8]. In L. (*A.*) *cavicola*, it is 20 mm long, which is longer than the same structure in L. (*P.*) *punctatus* (10–15 mm) (Table 3) [8]. Interestingly, even though it is shorter, the main collecting canal in L. (*P.*) *punctatus* has a greater diameter (120 μm) than the one in L. (*A.*) *cavicola* (50–80 μm) [17]. The muscle-coated reservoir in L. (*A.*) *cavicola* is elongated, with the top and the base almost equal, unlike those that occur in the related species L. (*P.*) *punctatus*. Furthermore, the inner side of the reservoir in L. (*A.*) *cavicola* lacks a depression, which is very conspicuous in L. (*P.*) *punctatus* [8]. The lengths of the reservoirs in L. (*A.*) *cavicola* and L. (*P.*) *punctatus* are nearly identical (1.97 and 2.00 mm, respectively). However, the reservoir of L. (*P.*) *punctatus* is almost twice as wide as the one in L. (*A.*) *cavicola* (1.50 and 0.88 mm, respectively) (Table 3) [8]. The entering points of the main collecting canal and the efferent duct in L. (*A.*) *cavicola* are close to each other, as has been reported for other Platyninae [4,8]. The efferent duct in L. (*A.*) *cavicola* is about three times shorter than the one in L. (*P.*) *punctatus* (1.30 and 4.00 mm, respectively) (Table 3) [8]. A description of the main collecting canal and the efferent duct was not given in detail in previous studies [4,8]. Furthermore, these structures are conservative parts of pygidial glands and exhibit less variability within ground beetles, and are therefore less informative [4].

Table 3. Comparative measurements of different pygidial glands structures in L. (*A.*) *cavicola* and L. (*P.*) *punctatus* after [8]. All measurements are expressed in millimeters.

Species	Gland Structure					
	Secretory Lobes	Main Collecting Canal		Reservoir		Efferent Duct
	Diameter	Length	Width	Length	Width	Length
L. (*A.*) *cavicola*	0.21–0.30	20	0.05–0.08	1.97	0.88	1.30
L. (*P.*) *punctatus*	0.15–0.20	10–15	0.12	2.00	1.50	4.00

4. Discussion

The great diversity of hydrocarbons in the analyzed troglobitic ground beetles might not be related to the colonization of cave habitats. Since a similar situation was found in some epigean Bembidiini [*Bembidion* (*Peryphanes*) *deletum* Audinet-Serville, 1821 and *B.* (*Peryphus*) *subcostatum* (Motschulsky, 1850)], it might rather be considered as a characteristic of the subfamily Trechinae [20], where chemoecologically analyzed troglobites of the tribe Trechini also belong. The pygidial gland secretions of the mentioned species of Bembidiini contained a complex secretion mixture of low-molecular-weight carboxylic acids, as was the case with troglobitic representatives of the same subfamily. Interestingly, in other chemically tested species of the subfamily Trechinae, such a diversity of compounds has not been found. There is also the possibility that some of the detected compounds do not originate from the secretions of the pygidial glands, but from the cuticle, which depends on the sampling method that was used. At this point, any comprehensive conclusions cannot be made due to the small number of chemically tested species and individuals of the subfamily Trechinae. Furthermore, the absence of formic acid in troglobites could be attributed to the lack of need for aggressive substances under conditions of low predation risk, but also to the fact that this chemical is rare among the members of the subfamily Trechinae (it has never been detected in species of the tribe Trechini to date) [20]. Other carboxylic acids are also considered aggressive irritants [1]. As mentioned before, both chemoecologically analyzed troglobites are characterized by the great diversity of chemicals belonging to that class of organic compounds. Various hydrocarbons (as in the case of *P. globiceps*) or esters (as in two troglophilic *Laemostenus* species), which are known to increase the repellent properties of the secretions, certainly represent effective predator deterrents [15]. The presence of benzaldehyde in the secretion of *P. globiceps* is fairly unusual as the distribution of the compound in the pygidial gland secretions of ground beetles was thought to be limited only to representatives of tiger beetles (subfamily Cicindelinae) [21–23]. At the same time, the question of its origin arises. Tiger beetles are assumed to utilize a cyanogenic pathway that is known to occur in other benzaldehyde-secreting species [21]. However, cyanogenesis is not widely distributed, and it would be informative to chemoecologically investigate related species in order to establish the distributional pattern of benzaldehyde in ground beetles.

Nenadić et al. [8,9] and Dimkić et al. [17] proved certain antimicrobial properties of the defensive secretion of the troglophilic species *L. (P.) punctatus*, which is a relative of *L. (A.) cavicola*. Carboxylic acids and an alcohol (1-dodecanol) from its secretion may synergistically have a negative impact on the growth of selected groups of microorganisms (bacteria and fungi). Some of the analyzed microbes cohabitate with *L. (P.) punctatus* and are potentially entomopathogenic [9,17]. It is yet to be examined whether the synergistic or individual effect of the compounds is responsible for antimicrobial features of the pygidial gland secretion [9]. Considering similar habitat preferences of *L. (A.) cavicola*, it might be possible that certain compounds of its pygidial gland secretion exhibit antimicrobial properties and serve for protection against microbial pathogens.

Pygidial glands are of great significance to ground beetles. They play an important role in protecting these insects from predators [1,11]. This especially applies to the subterranean species of ground beetles, whose survival in inhospitable cave environments is made possible thanks to the action of these glands and other defensive mechanisms [15].

Author Contributions: Conceptualization, S.Ć.; methodology, L.V., N.V. and D.R.; software, N.V.; validation, S.Ć., N.V. and L.V.; formal analysis, L.V., M.T. and N.V.; investigation, S.Ć. and M.P.; resources, S.V.; data curation, S.V.; writing—original draft preparation, S.V.; writing—review and editing, S.Ć., L.V. and N.V.; visualization, N.V.; supervision, S.Ć.; project administration, S.Ć.; funding acquisition, S.Ć. All authors have read and agreed to the published version of the manuscript.

Funding: This research was funded by the Serbian Ministry of Education, Science and Technological Development (Contracts Nos. 451-03-68/2022-14/200178, 451-03-68/2022-14/200168 and 451-03-68/2022-14/200160), as well as by the Montenegrin Academy of Sciences and Arts (Grant "Catalogue of Pseudoscorpions of Montenegro").

Institutional Review Board Statement: Not applicable.

Data Availability Statement: Not applicable.

Acknowledgments: We would like to thank Boban Rakić (Serbian Institute of Occupational Health "Dr Dragomir Karajović", Belgrade, Serbia), who helped us during field research carried out in a few subterranean sites in Montenegro.

Conflicts of Interest: The authors declare no conflict of interest.

References

1. Blum, M.S. *Biochemical Defenses of Insects*; Academic Press: New York, NY, USA, 1978. [CrossRef]
2. Blum, M.S. Semiochemical parsimony in the Arthropoda. *Annu. Rev. Entomol.* **1996**, *41*, 353–374. [CrossRef] [PubMed]
3. Brown, W.L.; Eisner, T.; Whittaker, W.H. Allomones and kairomones: Transspecific chemical messengers. *BioScience* **1970**, *20*, 21–22. [CrossRef]
4. Forsyth, D.J. The structure of pygidial defence glands of Carabidae (Coleoptera). *Trans. Zool. Soc. London* **1972**, *32*, 249–309. [CrossRef]

5. Kanehisa, K.; Shiraga, T. Morphological study of the pygidial defensive systems in carabid beetles. *Ber. Ohara Inst. Landwirtsch. Biol. Okayama Univ.* **1978**, *17*, 83–94.
6. Rork, A.M.; Renner, T. Carabidae semiochemistry: Current and future directions. *J. Chem. Ecol.* **2018**, *44*, 1069–1083. [CrossRef]
7. Nenadić, M.; Soković, M.; Glamočlija, J.; Ćirić, A.; Perić-Mataruga, V.; Ilijin, L.; Tešević, V.; Vujisić, L.; Todosijević, M.; Vesović, N.; et al. Antimicrobial activity of the pygidial gland secretion of three ground beetle species (Insecta: Coleoptera: Carabidae). *Sci. Nat.* **2016**, *103*, 34. [CrossRef]
8. Nenadić, M.; Soković, M.; Glamočlija, J.; Ćirić, A.; Perić-Mataruga, V.; Tešević, V.; Vujisić, L.; Todosijević, M.; Vesović, N.; Ćurčić, S. Antimicrobial activity of the pygidial gland secretion of the troglophilic ground beetle *Laemostenus* (*Pristonychus*) *punctatus* (Dejean, 1828) (Insecta: Coleoptera: Carabidae). *Bull. Entomol. Res.* **2016**, *106*, 474–480. [CrossRef]
9. Nenadić, M.; Ljaljević-Grbić, M.; Stupar, M.; Vukojević, J.; Ćirić, A.; Tešević, V.; Vujisić, L.; Todosijević, M.; Vesović, N.; Živković, N.; et al. Antifungal activity of the pygidial gland secretion of *Laemostenus punctatus* (Coleoptera: Carabidae) against cave-dwelling micromycetes. *Sci. Nat.* **2017**, *104*, 52. [CrossRef] [PubMed]
10. Holliday, A.E.; Mattingly, T.M.; Toro, A.A.; Donald, L.J.; Holliday, N.J. Age- and sex-related variation in defensive secretions of adult *Chlaenius cordicollis* and evidence for their role in sexual communication. *Chemoecology* **2016**, *26*, 107–119. [CrossRef]
11. Giglio, A.; Vommaro, M.L.; Brandmayr, P.; Talarico, F. Pygidial glands in Carabidae, an overview of morphology and chemical secretion. *Life* **2021**, *11*, 562. [CrossRef] [PubMed]
12. Vesović, N. Morphological Study of the Pygidial Glands and Analysis of the Chemical Composition of the Secretions of Selected Ground Beetle Species (Insecta: Coleoptera: Carabidae). Ph.D. Thesis, Faculty of Biology, University of Belgrade, Belgrade, Serbia, 12 February 2019. (In Serbian).
13. Schildknecht, H.; Maschwitz, U.; Winkler, H. Zur Evolution der Carabiden-Wehrdrüsensekrete. Über Arthropoden-Abwehrstoffe XXXII. *Naturwissenschaften* **1968**, *55*, 112–117. [CrossRef] [PubMed]
14. Will, K.W.; Gill, A.S.; Lee, H.; Attygalle, A.B. Quantification and evidence for mechanically metered release of pygidial secretions in formic acid-producing carabid beetles. *J. Insect Sci.* **2010**, *10*, 12. [CrossRef] [PubMed]
15. Vesović, N.; Ćurčić, S.; Vujisić, L.; Nenadić, M.; Krstić, G.; Perić-Mataruga, V.; Milosavljević, S.; Antić, D.; Mandić, B.; Petković, M.; et al. Molecular diversity of compounds from pygidial gland secretions of cave-dwelling ground beetles: The first evidence. *J. Chem. Ecol.* **2015**, *41*, 533–539. [CrossRef] [PubMed]
16. Assmann, T.; Casale, A.; Drees, C.; Habel, J.C.; Matern, A.; Schuldt, A. The dark side of relict species biology: Cave animals as ancient lineages. In *Relict Species: Phylogeography and Conservation Biology*, 1st ed.; Habel, J.C., Assmann, T., Eds.; Springer: Berlin/Heidelberg, Germany, 2010; pp. 91–103. [CrossRef]
17. Dimkić, I.; Stanković, S.; Kabić, J.; Stupar, M.; Nenadić, M.; Ljaljević-Grbić, M.; Žikić, V.; Vujisić, L.; Tešević, V.; Vesović, N.; et al. Bat guano-dwelling microbes and antimicrobial properties of the pygidial gland secretion of a troglophilic ground beetle against them. *Appl. Microbiol. Biotechnol.* **2020**, *104*, 4109–4126. [CrossRef] [PubMed]
18. Casale, A. Subtribe Sphodrina Laporte, 1834. In *Catalogue of Palaearctic Coleoptera. Volume 1. Archostemata–Myxophaga–Adephaga. Revised and Updated Edition*; Löbl, I., Löbl, D., Eds.; Brill: Leiden, The Netherlands, 2017; pp. 773–790.
19. Kanehisa, K.; Murase, M. Comparative study of the pygidial defensive systems of carabid beetles. *Appl. Entomol. Zool.* **1977**, *12*, 225–235. [CrossRef]
20. Vranić, S.; Vujisić, L.; Vesović, N.; Jeremić, M.; Pantelić, D.; Todosijević, M.; Pavlović, D.; Ćurčić, N.B.; Radovanović, M.; Petrović, M.D.; et al. Secretions of pygidial defensive glands in three species of the genus *Bembidion* (Carabidae), and morphology of pygidial glands in *B.* (*Peryphanes*) *dalmatinum*. *Ann. Zool. Fenn.* **2022**, *59*, 131–147. [CrossRef]
21. Blum, M.S.; Jones, T.H.; House, G.J.; Tschinkel, W.R. Defensive secretions of tiger beetles: Cyanogenetic basis. *Comp. Biochem. Physiol. B* **1981**, *69*, 903–904. [CrossRef]
22. Kelley, K.C.; Schilling, A.B. Quantitative variation in chemical defense within and among subgenera of *Cicindela*. *J. Chem. Ecol.* **1998**, *24*, 451–472. [CrossRef]
23. Pearson, D.L.; Blum, M.S.; Jones, T.H.; Fales, H.M.; Gonda, E.; Witte, B.R. Historical perspective and the interpretation of ecological patterns: Defensive compounds of tiger beetles (Coleoptera: Cicindelidae). *Am. Nat.* **1988**, *132*, 404–416. [CrossRef]

diversity

MDPI

Hypothesis

A Subsurface Stepping Stone Hypothesis for the Conquest of Land by Arthropods

Amos Frumkin [1,*] and Ariel D. Chipman [2,3]

1 Institute of Earth Sciences, The Hebrew University of Jerusalem, Jerusalem 91904, Israel
2 Department of Ecology, Evolution and Behavior, The Silberman Institute of Life Sciences, The Hebrew University of Jerusalem, Jerusalem 91904, Israel; ariel.chipman@mail.huji.ac.il
3 The National Natural History Collections, The Hebrew University of Jerusalem, Jerusalem 91904, Israel
* Correspondence: amos.frumkin@mail.huji.ac.il; Tel.: +972-548-820-489

Abstract: The conquest of land by arthropods is commonly believed to be a surface phenomenon associated with the arrival of photosynthetic plants, atmospheric oxygenation, and an ozone shield in the mid-Paleozoic Era. However, recent molecular and fossil evidence suggests terrestrial fauna may have first appeared in the Cambrian, before the proliferation of plants and ozone, which are thought to be essential for survival. This raises the question—how could arthropods survive on land without established plants and an ozone shield? We propose a hypothesis that chemolithoautotrophic cave ecosystems, independent of photosynthesis, may have served as a subsurface stepping stone, providing a possible explanation for the land invasion enigma. Chemolithoautrophic caves have offered abundant food and radiation protection, enabling ancient arthropods to evolve strategies to adapt to new frontiers through gradual dispersion from the sea to shielded cave waters, then to cave hygropetric margins of cave waters, and, finally, to the surface.

Keywords: subterranean habitat; cave fauna; early Palaeozoic; terrestrial colonization; arthropods; shielded caves; chemolithoautotrophy

1. Introduction

1.1. The Terrestrialization Problem

The terrestrialization of plants and arthropods represents a major milestone in the diversity, evolution, and adaptations of Earth's ecosystems. Understanding how species colonized the land surface would shed new light on the links between the evolution of life and the environment, as well as how biology affects the physical and chemical environment and vice versa.

In order for animals living in the sea to colonize the land surface, a number of physiological and environmental barriers need to be overcome. These include aerial respiration and gas exchange, water management and osmoregulation, digestion, temperature control, terrestrial locomotion without the benefit of buoyancy, aerial sensory perception, reproduction outside of water, including water resistance of eggs and embryos, utilization of new food sources, and protection against environmental and other stresses such as desiccation, rapid temperature fluctuations, and increased exposure to ultraviolet radiation [1–3]. Water is essential to life in several ways, such as a medium for biochemical reactions, for the transport of cell solutes, for the maintenance of cell turgor, and for inter-cellular transport and signaling. The variability of water availability on land is problematic for terrestrial life—dehydration can be as fatal as inundation by floods for land organisms.

Adaptations caused by changes in anatomy, physiology, and sensory systems allowed the colonizing species to overcome the terrestrial environmental challenges. In extant animals, degrees of terrestrial adaptation can be determined from their ecology, behavior, anatomy, and physiology. Comparing the anatomical and physiological features of aquatic and terrestrial organisms has helped scientists understand the underlying mechanisms responsible for the transition from sea to land. Understanding the evolutionary changes in body structure, respiratory system, and locomotion of aquatic versus terrestrial animals sheds light on the adaptations for living on land and provides insights into the development of these traits in different lineages [1–4]. The fossil record helps track some of the evolutionary changes that occurred during the transition to land, but this record is partial and unsatisfactory in most terrestrial environments.

By comparatively examining the ways in which animals respond to the challenges and opportunities presented by the land, we can gain a better understanding of ecological dynamics and how ecosystems have evolved over time. On the other hand, understanding paleohydrology, paleoclimate, and ancient geomorphology can shed light on the possible routes from sea to land that allowed transition and adaptation to living on the land.

1.2. The Earliest Terrestrial Ecosystems

Precambrian Metazoa are known only from the marine environment, and there is no evidence of higher plants in the fossil record until much later in the Palaeozoic. Nevertheless, probable prokaryotic colonizers would have been extremely important in developing soil profiles and contributing towards a terrestrial environment that was amenable to later colonization by higher plants and animals.

The fossil record holds only a few clues for the sequence and timing of events during the major phase of terrestrialization in the Palaeozoic. Four phases of plant terrestrialization were recognized by Edwards and Selden [4], based mainly on the plant fossil record. Exceptional examples of invertebrate animal fossils appear in the later phases, with the first terrestrial animals almost undoubtedly being arthropods [2]. Complex terrestrial biotas, based mainly on arthropods and plants, had developed by the Devonian period; colonizations by vertebrates and molluscs followed these early pioneers much later, into already well-established ecosystems.

In spite of the many attempts to decipher the terrestrialization process, details of the journey from water to land remain obscure, with the precise timing of the water-to-land transition being disputed for different animal taxa. Molecular clocks may reliably date evolutionary timescales only if ground-truthed by fossils, which are rare for early terrestrial organisms [5–7].

Land colonization by plants and arthropods is commonly considered together, assuming arthropod terrestrialization was associated with the arrival of photoautotrophic multicellular plants. Early plants increased terrestrial food availability and generated a more effective ozone shield. With low early atmospheric O_2 levels, stratospheric ozone was low, enabling short-wavelength ultraviolet light to enter the troposphere and trigger photochemical reactions [8]. Strong feedback exists between ozone formation, atmospheric chemistry, and biospheric oxygen production [9]. Thus, the ozone shield was likely strengthened as atmospheric oxygen rose since ~450 Ma in the Late Ordovician [10].

The oldest undisputed land plant fossils are Ordovician cryptospores [11,12] and spore-bearing plant fragments [13,14], although signs of fungus-like fossils have been known in paleo-caves since the Ediacaran [15,16]. Molecular clocks date crown group plants to the mid-Cambrian to early Ordovician [17–19]. However, molecular clocks suggest arthropods may have colonized land as early as the mid-Cambrian ~510 Ma [20,21], probably earlier than macroscopic terrestrial plants that produce significant amounts of oxygen and provide adequate shelter.

1.3. Arthropod Terrestrialization

Arthropods underwent at least three independent terrestrializations in the Paleozoic [5], but relevant body fossils from the first stages of terrestrialization are almost nonexistent. Notably, arthropod trace (trackways) fossils do not definitively indicate full land 'colonization'—brief forays into upper tidal or other paralic zones for feeding, reproduction, or molting while still remaining largely marine cannot be excluded [22–24].

Definitive proof of terrestrialization requires finding fossils of terrestrial crown-group members with terrestrial taphonomy, and these are few and far between.

Molecular phylogenies show that myriapods—an entirely terrestrial group—diverged from other mandibulates and started diversifying in the Cambrian [5], but see [2]. Some late Cambrian trace fossils may have myriapod or euthycarcinoid (a stem-group lineage of myriapods) origins, likely made by a semi-terrestrial arthropod [2,22]. The earliest undisputed crown-group myriapod body fossils are much younger, from the Silurian [2].

Hexapods (including insects), the largest modern terrestrial arthropod group, have an extremely poor early terrestrial fossil record. Molecular dating suggests that the hexapod diversification slightly postdated the diversification of myriapods, both during the Ordovician [5], with their closest relatives being remipede crustaceans [25–29]. No consensus exists as to which are the earliest fossil hexapods [27–30], but they are Devonian at the earliest.

Arachnids—the land members of Chelicerata—constitute the third arthropod terrestrialization event in the Paleozoic. While chelicerates diverged from mandibulates in the Precambrian based on molecular trees, the terrestrial members only started diversifying in the Ordovician, suggesting a later transition to land [2,5]. However, recent phylogenetic [31] and morphological-developmental [3] data raise the possibility of two independent terrestrialization events—one in pulmonate arachnids and one in non-pulmonates. Silurian trigonotarbids represent the earliest known terrestrial arachnids [32], although their precise phylogenetic position is unclear.

Although the major arthropod groups probably invaded land independently, they faced similar challenges and developed analogous, sometimes identical, adaptations once established [33]. The conundrum is whether early arthropod terrestrialization could occur before suitable plants developed, associated with habitats sheltered from UV radiation, and if it did, what did they feed on and where did they shelter? If molecular clocks are reliable, myriapod and possibly other arthropod diversification began before plants significantly populated land and while ozone levels were apparently too low to adequately shield against damaging UV radiation outside of water. The amount of damage caused by UV radiation to extant arthropods is highly variable and context-dependent [34,35], and it is difficult to assess how it would have affected Cambrian–Ordovician arthropods. However, it is likely that the UV radiation they would have sustained before the establishment of an adequate ozone layer would have reached levels that are lethal to extant arthropods.

1.4. Previous Hypotheses

Some previous hypotheses stressed the anatomical and physiological modifications that could facilitate terrestrialization [36]: The tracheal hypothesis suggests that the evolution of tracheal respiratory systems in arthropods allowed them to breathe air and invade land. Tracheal systems allow for direct gas exchange, bypassing the need for breathing through gills or across moist surfaces typical of marine arthropods. This hypothesis proposes that tracheal breathing was what enabled mandibulate arthropods like millipedes, centipedes, and primitive insects to move onto land during the Silurian and Devonian periods [37]. Note that this hypothesis does not cover the terrestrialization of pulmonate arachnids.

The general terrestrialization hypothesis focuses more broadly on adaptations that allowed arthropods to deal with the stresses of the terrestrial environment, not just respiratory adaptations [38]. Evolutionary innovations like thickened cuticles, waxy coatings, spine-like outgrowths, sensory organs, and hardened exoskeletons may have enabled the first terrestrial colonizations [38,39].

The herbivory hypothesis suggests that feeding on primitive vascular plants provided an attractive nutritional resource that may have pulled some arthropods towards terrestrial life [40]. Fossil evidence of terrestrial arthropod feeding marks on early land plants provides some support for this idea. The availability of abundant, nutritious food on land could have facilitated some of the early dispersals from marine environments.

Some combination of tracheal breathing, structural adaptations to the stresses of land, and the availability of new terrestrial food sources likely worked together to enable early arthropods like myriapods to migrate from coastal and tidal environments to inland terrestrial habitats [32]. All of these co-evolving adaptations likely played a role in the process at different points in time.

In this paper, we focus on the optimal environmental route that could facilitate terrestrialization. Previous hypotheses relating to this question included scenarios such as the marine–interstitial route [1], the freshwater-to-terrestrial route [41,42], mycotrophic plant colonization associations [43], the aerial plankton/wind transportation hypothesis, and the assembling ecosystem hypothesis [44–46].

The freshwater-to-terrestrial route hypothesis proposes that the ancestors of terrestrial animals evolved originally in freshwater habitats like rivers and lakes. Gradual adaptation occurred from aquatic to moist terrestrial environments like river banks. This is supported by some genetic similarities between freshwater and terrestrial taxa [1,46].

The marine interstitial hypothesis suggested that terrestrial animal groups evolved from ancestors adapted to marine intertidal zones, especially moist spaces between sand/soil particles. This habitat buffered climate change while allowing adaptation to terrestrial life. Some modern intertidal invertebrates display relevant adaptations [47]. A similar hypothesis suggests that terrestrialization occurred via interstitial terrestrial soil spaces [48,49]. The mycotrophic hypothesis proposes that fungal symbioses were key to the first terrestrial animals providing nutritious food sources in plant-absent environments. Modern fungal farming invertebrates showcase how fungi could facilitate terrestrial transition. The co-evolution of land fungi and animals was postulated [37,50].

The aerial plankton/wind transportation hypothesis suggested that some marine organisms, propagules, or dormant life stages evolved to become part of aerial plankton, passively carried by wind to land [44,45]. Some modern aerial plankton show that such dispersal is possible from the ocean to land.

None of the previous land invasion scenarios provided a complete explanation covering all aspects. Here we propose a subsurface stepping stone hypothesis (Figure 1), attempting to provide a full explanation, or at least a complementary one to previous hypotheses. In this model, a first step towards terrestrialization is life in a self-sufficient subterranean cave ecosystem. The term 'cave' refers here to subsurface systems of interconnected voids that can be penetrated by arthropods. Our hypothesis builds upon recent advances in the study of chemoautotrophic caves, discussed shortly below.

1.5. *Advances in the Study of Chemoautotrophic Caves*

Chemoautotrophic ecosystems were first discovered in the 1960s in an underwater cave in Israel [51]. Such caves contain chemoautotrophic bacteria that derive energy from inorganic molecules in the cave systems and fix carbon dioxide into organic compounds. This provides the basis for the complex food webs within the caves [52].

Since their discovery, over 300 large chemoautotrophic cave systems (accessible to humans) have been documented around the world, especially along continental margins. The diversity of bacteria involved has expanded greatly beyond initial discoveries, including new phyla and metabolic pathways [53].

Recent advances in techniques for studying such ecosystems have progressed from initial morphological observations to include genetic sequencing, isotopic tracing, microbial activity measurements, and modeling. This has revealed complex interdependencies between bacterial primary producers, metazoan consumers at multiple trophic levels, and abiotic factors in marine and cave environments [54,55].

Notable chemoautotrophic cave ecosystems studied in depth include Movile Cave in Romania, which contains rich sulfur-oxidizing communities with over 50 endemic species [52,56], and Ayyalon Cave, whose entire chemoautotrophic ecosystem is endemic [55,57].

There is evidence that some chemoautotrophic cave ecosystems receive a substantial nutrient supply from marine photosynthetic primary production or breakdown of aboveground organic matter. However, many ecosystems appear capable of functioning independently, representing unique extreme habitats [55,58].

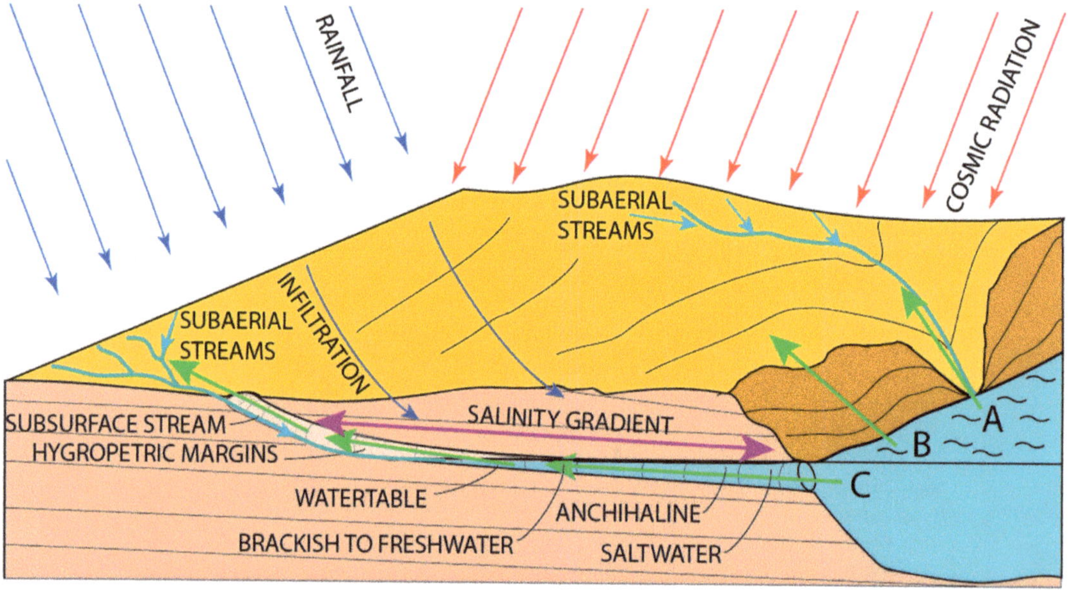

Figure 1. Diagram of three potential routes for the colonization of land by arthropods (green arrows). A. Colonization from sea to land through subaerial rivers, with exposure to physiological challenges such as ultraviolet radiation, lack of food sources, and other terrestrial stresses [1,46]. B. Direct colonization from sea to land, possibly using the soil interface [48], with exposure to physiological challenges. This route involves arthropods migrating directly from the ocean onto land, requiring adaptation to harsh conditions including UV exposure, lower humidity, a lack of food sources, and other terrestrial stresses. C. Gradual, stepwise colonization via caves containing anchialine water, hygropetric margins (where a thin film of water covers the rock surface), and subsurface streams (this paper). This route provides a more incremental pathway to land through brackish zones and freshwater inside caves. Adaptation to terrestrial life could occur gradually by first exploiting food sources in cave water sheltered from UV radiation. Eventually, arthropods could emerge from caves to colonize land after suitable habitats developed with abundant plant food sources, high oxygen levels, and protective ozone. The cave environment allows for incremental adaptation before facing the selection pressures on land.

Recent phylogenetic and biogeographic studies show that chemoautotrophic cave ecosystems, which had also acted as anchialine systems connected with the sea, have supported arthropod communities based on sulfur-oxidizing microorganisms for millions to tens of millions of years [59–61]. This demonstrates the long-term stability and robustness of such cavities and their ecosystems.

Overall, chemoautotrophic caves provide models for studying life in durable, extreme environments in the partial or complete absence of sunlight. Continued research promises to reveal more about subsurface biomes and the fundamentals of carbon and nutrient cycling. Here, we suggest that such caves could have served as a stepping stone for the transition of arthropods from sea to land prior to the appearance of terrestrial photosynthetic plants.

2. Discussion

The accepted prerequisites for the arthropod land conquest, including a well-developed land flora and an effective ozone shield against UV radiation, were irrelevant for underground self-sustaining systems. Such a subsurface system of terrestrial chemosynthetic microbial life could provide a reliable long-term trophic basis without the need for a land flora and without the hazards of UV radiation. Microbial chemolithoautotrophy is indeed ancient in Earth's history and predates photolithotrophy. The oldest known chemotrophic organisms are Archaean thermophilic prokaryotes, which inhabited sub-sea-floor volcanogenic–hydrothermal environments [62]. Anoxygenic forms of photosynthesis and chemoautotrophy played dominant roles in primary production for much of the Archaean era [63]. Biomats producing energy from sulfur compounds have been known since the Archean [64] and likely thrived in cave systems during the Paleozoic as they do today (Figure 2). Therefore, it is reasonable to hypothesize that the faunal invasion into cave systems could have taken place prior to and independently of the evolution of terrestrial ecosystems. Such an invasion would have been possible during early Paleozoic times of relatively low atmospheric oxygen concentrations and without terrestrial plants, utilizing terrestrial, highly productive sulfidic groundwater, where microbial biomats provided the needed resources for the food web. Under low-oxygen conditions, appropriate faunal adaptations would be necessary [65]. Some cave crustaceans and arachnids possess the hemocyanin copper protein, which functions as an oxygen carrier in the blood, allowing them to extract small amounts of dissolved oxygen under hypoxic conditions [66].

Species can rapidly colonize caves if sufficient variability and phenotypic plasticity are present in the population, as has been demonstrated to be possible within a few generations [67]. Species already adapted to detritivorous life below the photic zone could have switched to cave habitats with almost no special adaptations. Troglomorphic features and behaviors, such as loss of eyes and pigments or elongated sensory organs, could be caused by changes in the expression of developmental genes within the available time frame.

Caves provide gradual transitions from saltwater through anchihaline brackish to freshwater and from water-inundated voids through the hygropetric environment at the water margins to dry habitats (Figure 1). This is and has been promoted by the natural water cycle: Rainfall can infiltrate the subsurface and reach caves, and it can also be drained in subaerial catchments as runoff, collecting into streams that can form large cave systems and become subsurface rivers. Any form of this groundwater, while flowing into the sea, may create freshwater in anchihaline brackish environments. These can provide transitional routes, allowing arthropods to gradually adapt and move from salt water to freshwater. The subsurface rivers also carry food, such as fungus, from land into caves. Thus, cave arthropods could benefit from several food sources, including sea-sourced materials, chemosynthetic microorganisms, and terrestrial materials.

Metazoans with terrestrial surface locomotion, such as arthropods, could develop amphibious foraging behavior on the hygropetric borders of water bodies in caves to access biofilms stranded by changes in water levels or to hunt smaller organisms utilizing these stranded biofilms. Such foraging would be a first step towards venturing into drier subsurface voids. This hypothetical behavior is consistent with the ecology of the Ayyalon Cave arachnids—the pseudoscorpion *Ayyalonia dimentmani* and the scorpion *Akrav israchanani* [68,69], as well as the observation of an amphibious Mexican cave chactoid scorpion, *Alacran* [70].

Following improved adaptation to subaerial locomotion, feeding, and breathing, which could have all taken place within the isolated and self-sufficient cave system, some species could have ventured out of caves, initially relying on chemoautotrophic biofilms extending out through cave streams. Once there was enough photoautotrophic food on the surface, these pioneering arthropods could have stayed out of caves, taking advantage of the increasing surface resources.

Leaving the caves and surviving in exposed environments could have taken place during the Ordovician to Devonian periods, coupled with the rise of terrestrially productive photosynthetic plants and concomitantly with the increase in atmospheric oxygen and ozone levels [71–73]. This could occur many millions of years after the adaptation to cave systems. Upon emergence from the cave systems, the pioneering arthropods could lose their troglomorphic adaptations due to changes in the expression of developmental genes.

Figure 2. Chemoautotrophic microbial mat over the water of Ayyalon Cave, Israel, during the recent research expedition of 30 November 2023. This cave site sustains an endemic ecosystem with arthropods for millions of years [60–62]. Photo: A. Frumkin.

This scenario is consistent with modern observations, which show repeated recolonization of surface habitats by cave-dwelling arthropods. These include troglobitic arachnids [74] and groundwater arthropods [75], indicating that the adaptation to the subsurface is not a one-way evolutionary route. We do not claim that the first steps to terrestrialization had to have taken place in deep, isolated chemoautotrophic systems. There is no reason to assume that the stepping-stone caves were necessarily as isolated as some known current sulfidic karst systems [52,55]. Indeed, we would expect some light to have penetrated a few of the caves; otherwise, we would expect visual sense organs to have degenerated irreversibly, as seen in some subterranean arthropods [76].

Our hypothesis requires the existence of cave systems of karst or volcanic origin during early Paleozoic times, a known feature of this period [77,78]. Volcanic caves may have been more common than karst caves in ancient terrestrial environments, but other cave-forming processes are possible, as in other planetary systems, e.g., tectonic fracturing [79,80].

Chemolithoautotrophic systems exist in karst caves today and can be used as field models for the hypothesis presented here [52,55]. These examples are relatively independent of atmospheric oxygen levels. Recent studies show that microbial life is common at great depths [81]. Sulfuric caves connected to the sea can accommodate chemolithoautotrophic microbial mats, supporting chemosynthetic primary production [82]. Modern lava tubes that accommodate lava flows in Hawaii demonstrate terrestrial thermal sulfidic water connected with seawater via anchihaline environments. The lava-tube basaltic bedrock contains reduced sulfur, iron, and manganese, which can serve as energy sources for chemolithoautotrophic biomats [83]. Prior to limestone cave development, lava tubes could thus be the earliest caves to support a subsurface haven for life and possibly a terrestrialization route as well. As noted above, because light is not available beyond the twilight zone of caves, the dark-zone inhabitants have to adapt to alternative food sources, such as chemolithotrophic bacteria [82–84].

Finding supporting fossil evidence for our hypothesis is not easy, as Paleozoic arthropod fossils in caves are rare [85,86], and Paleozoic caves are not well preserved [77,78,87]. The lack of fossilization of cave fauna is demonstrated by the isopod suborder Phreatoicoidea, which is currently exclusive to caves, but whose fossils are commonly found in late Carboniferous marine sediments. After moving into freshwater environments during the Permian [88], their fossils are unknown after the Triassic, during or after which they became subterranean. The low preservation potential of cave fauna is consistent with the almost complete lack of terrestrial or semi-terrestrial fossils of stem-group terrestrial taxa.

Perhaps the most suggestive evidence for caves being stepping stones to terrestrialization is that the closest living sister group to terrestrial insects are the obligatory cave-dwelling remipedes. While there is no a priori reason to assume that the common ancestor of remipedes and insects resembled remipedes in its ecology and habitat, such a reconstructed ancestor would be fully consistent with our hypothesis.

Finally, the highly disjunct distribution of crustaceans adapted specifically to the poorly oxygenated water of anchihaline caves, including Spelaeogriphaceae and Remipedia, suggests a Pangean origin, possibly of Paleozoic age [89]. This should be carefully evaluated on a comparative phylogenetic level, which could potentially reveal multiple independent adaptations to cave habitats. This is again consistent with a very early shift to cave-dwelling habitats in some arthropod taxa.

3. Conclusions

The development of complex terrestrial arthropod life on Earth required overcoming several major hurdles. First, the challenges of transitioning from an aquatic environment onto dry land were immense. Arthropods needed to evolve adaptations to overcome UV radiation, prevent desiccation, move on solid substrates, obtain oxygen, and find new food sources outside of water when no vegetation was available.

Direct colonization from the ocean onto land would require rapid, radical adaptations to the harsh conditions on land. Arthropods would need to quickly evolve solutions for various problems all at once.

We propose an alternative route—gradual adaptation through cave environments. Such a subsurface stepping stone hypothesis provides an answer to the conundrum of early Paleozoic arthropod terrestrialization, which took place before suitable surface habitats formed with abundant plants, oxygen, and ozone shielding. Coastal caves could have provided a more stepwise pathway to terrestrial life. The cave habitat shielded arthropods from UV radiation, while they gradually adapted to drier conditions through brackish and freshwater (Figure 1). Food sources could be obtained from chemolithoautotrophic microbial mats, as well as other cave organisms such as fungi. Once pre-adapted through cave life, emerging onto land would be less difficult after abundant plants, oxygen, and increased ozone developed outside. Traits needed in caves could have been discarded later, as surface conditions improved.

In conclusion, the proposed terrestrialization route would have allowed one or more arthropod taxa to adapt to new frontiers via a stepwise and gradual dispersion from the sea into anchihaline cave waters, hygropetric margins of water bodies in caves, and finally to the surface. This hypothesis is consistent with modern cave observations [52,61] and phylogenetic reconstructions from DNA data [49].

As the complex problem of terrestrialization has produced several persuasive hypotheses, our hypothesis may stand alone or be complementary to former scenarios of terrestrialization. Several available routes could be used for the colonization of land by various taxa diachronically across the heterogenic coastal terrain. In any particular case, a specific hypothesis may correspond better to reality than others, and future studies can address the question of which scenario corresponds best to the available key evidence.

Our new hypothesis can potentially be supported by several possible types of additional data or analysis:

- Look for geological and fossil evidence of suitable ancient coastal/cave/paleokarst habitats existing in the Cambrian through Ordovician periods when molecular data suggests early arthropod diversification on land. The recovery of early Paleozoic fossils representing crown group or upper-stem group terrestrial arthropod taxa with identifiable adaptations to cave life or to bacterial mat feeding would help to substantiate the new hypothesis.
- Identify fossils with signs of incipient troglomorphy (eye/pigment loss, sensory elongation) that do not show full adaptation to caves. Transitional forms would be expected.
- Use modeling approaches to determine if subsurface habitats could have supported arthropod nutritional and respiratory needs before surface habitats developed. Energetic feasibility modeling could be informative.

Author Contributions: Conceptualization, A.F.; Validation, A.F. and A.D.C.; Investigation, A.F.; Writing—Original Draft Preparation, A.F.; Writing—Review and Editing, A.F. and A.D.C. All authors have read and agreed to the published version of the manuscript.

Funding: This research received no external funding.

Institutional Review Board Statement: Not applicable.

Data Availability Statement: Not applicable.

Conflicts of Interest: The authors declare no conflict of interest.

References

1. Little, C. *The Terrestrial Invasion—An Ecophysiological Approach to the Origins of Land Animals;* Cambridge University Press: Cambridge, UK, 1990.
2. Lozano-Fernandez, J.; Carton, R.; Tanner, A.R.; Puttick, M.N.; Blaxter, M.; Vinther, J.; Olesen, J.; Giribet, G.; Edgecombe, G.D.; Pisani, D. A molecular palaeobiological exploration of arthropod terrestrialization. *Philos. Trans. R. Soc. B Biol. Sci.* **2016**, *371*, 20150133. [CrossRef] [PubMed]
3. Sharma, P.P. Chelicerates and the conquest of land: A view of arachnid origins through an evo-devo spyglass. *Integr. Comp. Biol.* **2017**, *57*, 510–522. [CrossRef] [PubMed]
4. Edwards, D.; Selden, P.A. The development of early terrestrial ecosystems. *Bot. J. Scotl.* **1992**, *46*, 337–366. [CrossRef]
5. Rota-Stabelli, O.; Daley, A.C.; Pisani, D. Molecular timetrees reveal a Cambrian colonization of land and a new scenario for ecdysozoan evolution. *Curr. Biol.* **2013**, *23*, 392–398. [CrossRef]
6. Wolfe, J.M.; Daley, A.C.; Legg, D.A.; Edgecombe, G.D. Fossil calibrations for the arthropod Tree of Life. *Earth-Sci. Rev.* **2016**, *160*, 43–110. [CrossRef]

7. Howard, R.J.; Edgecombe, G.D.; Legg, D.A.; Pisani, D.; Lozano-Fernandez, J. Exploring the evolution and terrestrialization of scorpions (Arachnida: Scorpiones) with rocks and clocks. *Org. Divers. Evol.* **2019**, *19*, 71–86. [CrossRef]
8. Kendall, B. Recent advances in geochemical paleo-oxybarometers. *Annu. Rev. Earth Planet. Sci.* **2021**, *49*, 399–433. [CrossRef]
9. Gregory, B.S.; Claire, M.W.; Rugheimer, S. Photochemical modelling of atmospheric oxygen levels confirms two stable states. *Earth Planet. Sci. Lett.* **2021**, *561*, 116818. [CrossRef]
10. Krause, A.J.; Mills, B.J.; Zhang, S.; Planavsky, N.J.; Lenton, T.M.; Poulton, S.W. Stepwise oxygenation of the Paleozoic atmosphere. *Nat. Commun.* **2018**, *9*, 4081. [CrossRef]
11. Steemans, P.; Hérissé, A.L.; Melvin, J.; Miller, M.A.; Paris, F.; Verniers, J.; Wellman, C.H. Origin and radiation of the earliest vascular land plants. *Science* **2009**, *324*, 353. [CrossRef]
12. Lüttge, U. Terrestrialization: The conquest of dry land by plants. In *Progress in Botany*; Springer International Publishing: Cham, Switzerland, 2020; Volume 83, pp. 65–89.
13. Wellman, C.H.; Osterloff, P.L.; Mohiuddin, U. Fragments of the earliest land plants. *Nature* **2003**, *425*, 282–285. [CrossRef] [PubMed]
14. Buschmann, H.; Holzinger, A. Understanding the algae to land plant transition. *J. Exp. Bot.* **2020**, *71*, 3241–3246. [CrossRef] [PubMed]
15. Gan, T.; Luo, T.; Pang, K.; Zhou, C.; Zhou, G.; Wan, B.; Li, G.; Yi, Q.; Czaja, A.D.; Xiao, S. Cryptic terrestrial fungus-like fossils of the early Ediacaran Period. *Nat. Commun.* **2021**, *12*, 641. [CrossRef] [PubMed]
16. Gan, T.; Zhou, G.; Luo, T.; Pang, K.; Zhou, M.; Luo, W.; Wang, S.; Xiao, S. Earliest Ediacaran speleothems and their implications for terrestrial life after the Marinoan snowball Earth. *Precambrian Res.* **2022**, *376*, 106685. [CrossRef]
17. Clarke, J.T.; Warnock, R.C.M.; Donoghue, P.C.J. Establishing a time-scale for plant evolution. *New Phytol.* **2011**, *192*, 266–301. [CrossRef]
18. Morris, J.L.; Puttick, M.N.; Clark, J.W.; Edwards, D.; Kenrick, P.; Pressel, S.; Wellman, C.H.; Yang, Z.H.; Schneider, H.; Donoghue, P.C.J. The timescale of early land plant evolution. *Proc. Natl. Acad. Sci. USA* **2018**, *115*, E2274–E2283. [CrossRef]
19. Donoghue, P.C.J.; Harrison, C.J.; Paps, J.; Schneider, H. The evolutionary emergence of land plants. *Curr. Biol.* **2021**, *31*, R1281–R1298. [CrossRef]
20. Lozano-Fernandez, J.; Tanner, A.R.; Puttick, M.N.; Vinther, J.; Edgecombe, G.D.; Pisani, D. A Cambrian–Ordovician terrestrialization of arachnids. *Front. Genet.* **2020**, *11*, 182. [CrossRef]
21. Wheat, C.W.; Wahlberg, N. Phylogenomic insights into the Cambrian explosion, the colonization of land and the evolution of flight in Arthropoda. *Syst. Biol.* **2013**, *62*, 93–109. [CrossRef]
22. MacNaughton, R.B.; Cole, J.M.; Dalrymple, R.W.; Braddy, S.J.; Briggs, D.E.; Lukie, T.D. First steps on land: Arthropod trackways in Cambrian-Ordovician eolian sandstone, southeastern Ontario, Canada. *Geology* **2002**, *30*, 391–394. [CrossRef]
23. Briggs, D.; Wright, J.; Suthren, R. Forum Comment—Death near the shoreline, not life on land: Ordovician arthropod trackways in the Borrowdale Volcanic Group, UK. *Geology* **2019**, *47*, e464. [CrossRef]
24. Shillito, A.P.; Davies, N.S. Death near the shoreline, not life on land: Ordovician arthropod trackways in the Borrowdale Volcanic Group, UK. *Geology* **2019**, *47*, 55–58. [CrossRef]
25. Lozano-Fernandez, J.; Giacomelli, M.; Fleming, J.F.; Chen, A.; Vinther, J.; Thomsen, P.F.; Glenner, H.; Palero, F.; Legg, D.A.; Iliffe, T.M.; et al. Pancrustacean evolution illuminated by taxon-rich genomic-scale data sets with an expanded remipede sampling. *Genome Biol. Evol.* **2019**, *11*, 2055–2070. [CrossRef] [PubMed]
26. Schwentner, M.; Combosch, D.J.; Nelson, J.P.; Giribet, G. A phylogenomic solution to the origin of insects by resolving crustacean-hexapod relationships. *Curr. Biol.* **2017**, *27*, 1818–1824. [CrossRef]
27. Haas, F.; Waloszek, D.; Hartenberger, R. *Devonohexapodus bocksbergensis*, a new marine hexapod from the Lower Devonian Hunsrück Slates, and the origin of Atelocerata and Hexapoda. *Org. Divers. Evol.* **2003**, *3*, 39–54. [CrossRef]
28. Willmann, R. Reinterpretation of an alleged marine hexapod stem-group representative. *Org. Divers. Evol.* **2005**, *5*, 199–202. [CrossRef]
29. Garrouste, R.; Clément, G.; Nel, P.; Engel, M.S.; Grandcolas, P.; D'Haese, C.; Lagebro, L.; Denayer, J.; Gueriau, P.; Lafaite, P.; et al. A complete insect from the Late Devonian period. *Nature* **2012**, *488*, 82–85. [CrossRef]
30. Haug, C.; Haug, J.T. The presumed oldest flying insect: More likely a myriapod? *PeerJ* **2017**, *5*, e3402. [CrossRef]
31. Ballesteros, J.A.; Sharma, P.P. A critical appraisal of the placement of Xiphosura (Chelicerata) with account of known sources of phylogenetic error. *Syst. Biol.* **2019**, *68*, 896–917. [CrossRef]
32. Jeram, A.J.; Selden, P.A.; Edwards, D. Land animals in the Silurian: Arachnids and myriapods from Shropshire, England. *Science* **1990**, *250*, 658–661. [CrossRef]
33. Dunlop, J.A.; Scholtz, G.; Selden, P.A. Water-to-land transitions. In *Arthropod Biology and Evolution*; Springer: Berlin/Heidelberg, Germany, 2013; pp. 417–439.
34. Ben-Yakir, D. Direct and indirect effects of UV radiation. In *Optical Manipulation of Arthropod Pests and Beneficials*; Ben-Yakir, D., Ed.; CABI Digital Library: Vancouver, BC, Canada, 2020; pp. 49–59.
35. Ben-Yakir, D.; Fereres, A. The effects of UV radiation on arthropods: A review of recent publications (2010–2015). *Acta Hortic.* **2016**, *1134*, 335–342. [CrossRef]
36. Ward, P. *Out of Thin Air: Dinosaurs, Birds, and Earth's Ancient Atmosphere*; National Academies Press: Cambridge, MA, USA, 2006.
37. Selden, P.A.; Edwards, D. Colonisation of the land. *Evol. Foss. Rec.* **1989**, *6*, 122–152.

38. Little, C. *The Colonisation of Land: Origins and Adaptations of Terrestrial Animals*; Cambridge University Press: Cambridge, MA, USA, 1983.
39. Dunlop, J.A. Geological history and phylogeny of Chelicerata. *Arthropod Struct. Dev.* **2010**, *39*, 124–142. [CrossRef] [PubMed]
40. Labandeira, C.C. The origin of herbivory on land: Initial patterns of plant tissue consumption by arthropods. *Insect Sci.* **2007**, *14*, 259–275. [CrossRef]
41. De Deckker, P. Terrestrial ostracods in Australia. *Aust. Mus. Mem.* **1983**, *18*, 87–100. [CrossRef]
42. Diesel, R.; Schubart, C.D.; Schuh, M. A reconstruction of the invasion of land by Jamaican crabs (Grapsidae: Sesarminae). *J. Zool.* **2000**, *250*, 141–160. [CrossRef]
43. Retallack, G.J. Ordovician-Devonian lichen canopies before evolution of woody trees. *Gondwana Res.* **2022**, *106*, 211–223. [CrossRef]
44. Delwiche, C.F.; Graham, L.E.; Thomson, N. Lignin-like compounds and sporopollen in Coleochaete, an algal model for land plant ancestry. *Science* **1989**, *245*, 399–401. [CrossRef]
45. Graham, L.E.; Cook, M.E.; Busse, J.S. The origin of plants: Body plan changes contributing to a major evolutionary radiation. *Proc. Natl. Acad. Sci. USA* **2000**, *97*, 4535–4540. [CrossRef]
46. Martin, M.W. Early Evolution of Terrestrial Arthropods—Paleontological and Molecular Evidence. Ph.D. Thesis, University of Toronto, Toronto, ON, Canada, 1999.
47. Schram, F.R. Cladistic analysis of metazoan phyla and the placement of fossil problematica. In *The Early Evolution of Metazoa and the Significance of Problematic Taxa*; Cambridge University Press: Cambridge, UK, 1991; pp. 177–197.
48. Ghilarov, M.S. *The Peculiarities of the Soil as an Environment and Its Significance in Insect Evolution*; Academic Science Publishing House: Moscow, Russia, 1949; 279p.
49. van Straalen, N.M. Evolutionary terrestrialization scenarios for soil invertebrates. *Pedobiologia* **2021**, *87*, 150753. [CrossRef]
50. Labandeira, C. Early history of arthropod and vascular plant associations. *Annu. Rev. Earth Planet. Sci.* **1998**, *26*, 329–377. [CrossRef]
51. Tsurnamal, M.; Por, F.D. The subterranean fauna associated with the blind palaemonid prawn Typhlocaris galilea Calman. *Int. J. Speleol.* **1968**, *3*, 219–223. [CrossRef]
52. Sarbu, S.M.; Kane, T.C.; Kinkle, B.K. A chemoautotrophically based cave ecosystem. *Science* **1996**, *272*, 1953–1955. [CrossRef] [PubMed]
53. Engel, A.S. Microbial diversity of cave ecosystems. In *Geomicrobiology: Molecular and Environmental Perspective*; Barton, L.L., Mandl, M., Loy, A., Eds.; Springer: Dordrecht, The Netherlands, 2010; pp. 219–238.
54. Thurber, A.R.; Levin, L.A.; Rowden, A.A.; Sommer, S.; Linke, P.; Kröger, K. Microbes, macrofauna, and methane: A novel seep community fueled by aerobic methanotrophy. *Limnol. Oceanogr.* **2013**, *58*, 1640–1656. [CrossRef]
55. Frumkin, A.; Chipman, A.D.; Naaman, I. An Isolated Chemolithoautotrophic ecosystem deduced from environmental isotopes: Ayyalon Cave (Israel). *Front. Ecol. Evol.* **2023**, *10*, 1040385. [CrossRef]
56. Sarbu, S.M. Movile Cave: A chemoautotrophically based groundwater ecosystem. In *Subterranean Ecosystems*; Wilkens, H., Culver, D.C., Humphreys, W.F., Eds.; Elsevier: Amsterdam, The Netherlands, 2000; pp. 319–343.
57. Por, F.D.; Dimentman, C.; Frumkin, A.; Naaman, I. Animal life in the chemoautotrophic ecosystem of the hypogenic groundwater cave of Ayyalon (Israel): A summing up. *Nat. Sci.* **2013**, *5*, 7–13–13. [CrossRef]
58. Porter, M.L.; Engel, A.S.; Kane, T.C.; Kinkle, B.K. Productivity-diversity relationships from chemolithoautotrophically based sulfidic karst systems. *Int. J. Speleol.* **2009**, *38*, 27–40. [CrossRef]
59. Guy-Haim, T.; Simon-Blecher, N.; Frumkin, A.; Naaman, I.; Achituv, Y. Multiple transgressions and slow evolution shape the phylogeographic pattern of the blind cave-dwelling shrimp Typhlocaris. *PeerJ* **2018**, *11*, e16690. [CrossRef]
60. Guy-Haim, T.; Kolodny, O.; Frumkin, A.; Achituv, Y.; Velasquez, X.; Morov, A.R. Shedding light on the Ophel biome: The Trans-Tethyan phylogeography of the sulfide shrimp Tethysbaena (Peracarida: Thermosbaenacea) in the Levant. *PeerJ* 2023, *in press*. [CrossRef]
61. Frumkin, A.; Dimentman, C.; Naaman, I. Biogeography of living fossils as a key for geological reconstruction of the East Mediterranean: Ayyalon-Nesher Ramla system, Israel. *Quat. Int.* **2022**, *624*, 168–180. [CrossRef]
62. Rasmussen, B. Filamentous microfossils in a 3235-million-year-old volcanogenic massive sulphide deposit. *Nature* **2000**, *405*, 676–679. [CrossRef] [PubMed]
63. Galvez, M.E.; Fischer, W.W.; Jaccard, S.L.; Eglinton, T.I. Materials and pathways of the organic carbon cycle through time. *Nat. Geosci.* **2020**, *13*, 535–546. [CrossRef]
64. Lepot, K. Signatures of early microbial life from the Archean (4 to 2.5 Ga) eon. *Earth-Sci. Rev.* **2020**, *209*, 103296. [CrossRef]
65. Hourdez, S.; Lallier, F.H. Adaptations to hypoxia in hydrothermal-vent and cold-seep invertebrates. *Rev. Environ. Sci. Biotechnol.* **2007**, *6*, 143–159. [CrossRef]
66. Burmester, T. Origin and evolution of arthropod hemocyanins and related proteins. *J. Comp. Physiol. B* **2002**, *172*, 95–107. [PubMed]
67. Bilandžija, H.; Hollifield, B.; Steck, M.; Meng, G.; Ng, M.; Koch, A.D.; Gračan, R.; Ćetković, H.; Porter, M.L.; Renner, K.J.; et al. Phenotypic plasticity as a mechanism of cave colonization and adaptation. *eLife* **2020**, *9*, e51830. [CrossRef] [PubMed]
68. Ćurčić, B.P.M. *Ayyalonia dimentmani* ng, n. sp. (Ayyaloniini n. trib., Chthoniidae, Pseudoscorpiones) from a cave in Israel. *Arch. Biol. Sci.* **2008**, *60*, 331–339. [CrossRef]

69. Fet, V.; Soleglad, M.E.; Zonstein, S.L. The genus Akrav Levy, 2007 (Scorpiones: Akravidae) revisited. *Euscorpius* **2011**, *134*, 1–49. [CrossRef]
70. Santibáñez-López, C.E.; Francke, O.F.; Prendini, L. Shining a light into the world's deepest caves: Phylogenetic systematics of the troglobiotic scorpion genus Alacran Francke, 1982 (Typhlochactidae: Alacraninae). *Invertebr. Syst.* **2014**, *28*, 643–664. [CrossRef]
71. Harfoot, M.B.; Beerling, D.J.; Lomax, B.H.; Pyle, J.A. A two-dimensional atmospheric chemistry modeling investigation of Earth's Phanerozoic O_3 and near-surface ultraviolet radiation history. *J. Geophys. Res. Atmos.* **2007**, *112*, D07308. [CrossRef]
72. Lenton, T.M.; Dahl, T.W.; Daines, S.J.; Mills, B.J.; Ozaki, K.; Saltzman, M.R.; Porada, P. Earliest land plants created modern levels of atmospheric oxygen. *Proc. Natl. Acad. Sci. USA* **2016**, *113*, 9704–9709. [CrossRef] [PubMed]
73. Sønderholm, F.; Bjerrum, C.J. Minimum levels of atmospheric oxygen from fossil tree roots imply new plant–oxygen feedback. *Geobiology* **2021**, *19*, 250–260. [CrossRef] [PubMed]
74. Prendini, L.; Francke, O.F.; Vignoli, V. Troglomorphism, trichobothriotaxy and typhlochactid phylogeny (Scorpiones, Chactoidea): More evidence that troglobitism is not an evolutionary dead-end. *Cladistics* **2010**, *26*, 117–142. [CrossRef] [PubMed]
75. Copilaş-Ciocianu, D.; Fišer, C.; Borza, P.; Petrusek, A. Is subterranean lifestyle reversible? Independent and recent large-scale dispersal into surface waters by two species of the groundwater amphipod genus Niphargus. *Mol. Phylogenetics Evol.* **2018**, *119*, 37–49.
76. Chipman, A.D.; Ferrier, D.E.; Brena, C.; Qu, J.; Hughes, D.S.; Schröder, R.; Torres-Oliva, M.; Znassi, N.; Jiang, H.; Almeida, F.C.; et al. The first myriapod genome sequence reveals conservative arthropod gene content and genome organisation in the centipede Strigamia maritima. *PLoS Biol.* **2014**, *12*, e1002005. [CrossRef] [PubMed]
77. James, N.P.; Choquette, P.W. (Eds.) *Paleokarst*; Springer Science & Business Media: Berlin/Heidelberg, Germany, 1988.
78. Bosák, P.; Ford, D.C.; Glazek, J.; Horácek, I. (Eds.) *Paleokarst: A Systematic and Regional Review*; Elsevier: Amsterdam, The Netherlands, 1989.
79. Sauro, F.; Pozzobon, R.; Massironi, M.; De Berardinis, P.; Santagata, T.; De Waele, J. Lava tubes on Earth, Moon and Mars: A review on their size and morphology revealed by comparative planetology. *Earth-Sci. Rev.* **2020**, *209*, 103288. [CrossRef]
80. Wynne, J.J.; Titus, T.N.; Agha-Mohammadi, A.A.; Azua-Bustos, A.; Boston, P.J.; de León, P.; Demirel-Floyd, C.; De Waele, J.; Jones, H.; Malaska, M.J.; et al. Fundamental science and engineering questions in planetary cave esxploration. *J. Geophys. Res. Planets* **2021**, *127*, e2022JE007194. [CrossRef]
81. Dai, X.; Wang, Y.; Luo, L.; Pfiffner, S.M.; Li, G.; Dong, Z.; Xu, Z.; Dong, H.; Huang, L. Detection of the deep biosphere in metamorphic rocks from the Chinese continental scientific drilling. *Geobiology* **2021**, *19*, 278–291. [CrossRef]
82. D'Angeli, I.M.; Ghezzi, D.; Leuko, S.; Firrincieli, A.; Parise, M.; Fiorucci, A.; Vigna, B.; Addesso, R.; Baldantoni, D.; Carbone, C.; et al. Geomicrobiology of a seawater-influenced active sulfuric acid cave. *PLoS ONE* **2019**, *14*, e0220706. [CrossRef]
83. Northup, D.E.; Melim, L.A.; Spilde, M.N.; Hathaway, J.J.M.; Garcia, M.G.; Moya, M.; Stone, F.D.; Boston, P.J.; Dapkevicius, M.L.N.E.; Riquelme, C.; et al. Lava cave microbial communities within mats and secondary mineral deposits: Implications for life detection on other planets. *Astrobiology* **2011**, *11*, 601–618. [CrossRef]
84. Hathaway, J.J.M.; Garcia, M.G.; Balasch, M.M.; Spilde, M.N.; Stone, F.D.; Dapkevicius, M.D.L.N.; Amorim, I.R.; Gabriel, R.; Borges, P.A.; Northup, D.E. Comparison of bacterial diversity in Azorean and Hawai'ian lava cave microbial mats. *Geomicrobiol. J.* **2014**, *31*, 205–220. [CrossRef] [PubMed]
85. Retallack, G.J. Ordovician land plants and fungi from Douglas Dam, Tennessee. *J. Palaeosci.* **2019**, *68*, 173–205. [CrossRef]
86. Sendi, H.; Vršanský, P.; Podstrelena, L.; Hinkelman, J.; Kudelova, T.; Kúdela, M.; Vidlička, Ľ.; Ren, X.; Quicke, D.L. Nocticolid cockroaches are the only known dinosaur age cave survivors. *Gondwana Res.* **2020**, *82*, 288–298. [CrossRef]
87. Osborne, A.R. The world's oldest Caves: How did they survive and what can they tell us? *Acta Carsologica* **2007**, *36*, 133–142. [CrossRef]
88. Schram, F.R. *Crustacea*; Oxford University Press: New York, NY, USA, 1986.
89. Culver, D.C.; Pipan, T. *The Biology of Caves and Other Subterranean Habitats*; Oxford University Press: New York, NY, USA, 2019.

MDPI AG
Grosspeteranlage 5
4052 Basel
Switzerland
Tel.: +41 61 683 77 34

Diversity Editorial Office
E-mail: diversity@mdpi.com
www.mdpi.com/journal/diversity